Ho Math Chess 何数棋谜　妈！我会棋谜式加法啦！
Mom! I Learn Addition Using Math-Chess-Puzzles Connection
Contents include both traditional and Math-Chess-Puzzles combined methods. Extra strength
©2008 − 2018 Frank Ho, Amanda Ho　All rights reserved. www.homathchess.com
Student Name_____ Date_____

Mom! I Learn Addition Using Math-Chess-Puzzles Connection

棋谜式加法

Frank Ho   Amanda Ho

何数棋谜 培训

Ho Math Chess Learning Centre

Ho Math Chess 何数棋谜　妈！我会棋谜式加法啦！
Mom! I Learn Addition Using Math-Chess-Puzzles Connection
Contents include both traditional and Math-Chess-Puzzles combined methods. Extra strength
©2008 – 2018 Frank Ho, Amanda Ho　　All rights reserved. www.homathchess.com
Student Name _____ Date _____

Table of Contents

HOW TO USE THIS WORKBOOK? 7
- 2017 PREFACE 前言 12
- HO MATH CHESS WORKSHEETS AND TRADITIONAL WORKSHEETS 15
- BENEFITS OF USING HO MATH CHESS WORKBOOK 18
- CHESS SYMBOLS AND THEIR MATHEMATICAL VALUES 22

PREREQUISITES BEFORE WORKING ON ADDITIONS 23
- COUNTING NUMBERS AND WHOLE NUMBERS 24
- WRITING D + D NUMBER SENTENCE (COMMUTATIVE PROPERTY) 交换率 25
- STRATEGIES FOR D + D ADDITION 加法的策略 27
- ENGLISH NAMES FOR NUMBERS (FOR STUDENTS WHO CAN READ AND WRITE ENGLISH.) ... 29
- NUMBER ORDER AND RANK 32
- ORDER THE FOLLOWING NUMBERS FROM SMALLEST TO LARGEST 33

PART 1: ADDING 123, DOUBLING, ADDING TEN, AND 9 + D STRATEGIES 35
- MIXED COUNTING 1, 2, 3 FORWARD D + D (START WITH A BIGGER NUMBER AND COUNT ON DOTS) ... 36
- MIXED COUNTING 1, 2, 3 FORWARD AND BACKWARD OF D + D OR D – D 38
- COUNTING FORWARD BY 2'S (EVEN NUMBERS OR ODD NUMBERS) 45
- COUNTING BACKWARDS BY 2'S (EVEN NUMBERS OR ODD NUMBERS) 46
- COUNTING FORWARD BY 5'S (FOR WORKING ON CLOCK PROBLEMS) 47
- COUNTING BACKWARDS BY 5'S (FOR WORKING ON CLOCK PROBLEMS) 48

ADDITION STRATEGY: REVERSE CALCULATION OF SUMS 倒算和 49
- ADDING TO 4 凑 4 50
- ADDING TO 5 凑 5 51
- ADDING TO 6 凑 6 52
- ADDING TO 7 凑 7 53
- ADDING TO 8 凑 8 54
- ADDING TO 9 凑 9 55

PARTITIONING NUMBERS 56
- PARTITION 3 拆 3 56
- PARTITION 4 拆 4 57
- PARTITION 5 拆 5 58
- PARTITION 6 拆 6 59
- PARTITION 7 拆 7 60
- PARTITION 8 拆 8 61
- PARTITION 9 拆 9 62
- PARTITION 10 拆 10 63

NUMBER PUZZLES 64

ADDITION STRATEGY: ADDING 1, 2, 3 加 1, 2, 3 65
- ADDING D + 1 (THE LARGER NUMBER IN COUNTING) 65
- ADDING D + 1 AND 1 + D 66
- 7 DAYS MASTERING COMPUTATION SKILLS 70

Ho Math Chess 何数棋谜   妈！我会棋谜式加法啦！
Mom! I Learn Addition Using Math-Chess-Puzzles Connection
Contents include both traditional and Math-Chess-Puzzles combined methods. Extra strength
©2008 – 2018 Frank Ho, Amanda Ho   All rights reserved. www.homathchess.com

Student Name _____   Date _____

DAY 1 OF DOUBLING 1, 2, OR 3	71
DAY 2 OF 2 ADDING 1, 2, OR 3, ADDING TO 10, DOUBLING 2	72
DAY 3 OF 3 ADDING 1, 2, OR 3, ADDING TO 10, DOUBLING 3	74
DAY 4 OF 4 ADDING 1, 2, OR 3, ADDING TO 10, DOUBLING 4	76
DAY 5 OF 5 ADDING 1, 2, OR 3, ADDING TO 10, DOUBLING 5	77
DAY 6 OF 6 ADDING 1, 2, OR 3, ADDING TO 10, DOUBLING 6	79
DAY 6 OF 7 ADDING 1, 2, OR 3, ADDING TO 10, DOUBLING 7	83
DAY 7 OF 8 ADDING 1, 2, OR 3, ADDING TO 10, DOUBLING 8	85

9 PLUS STRATEGY 89

PART 2: MORE PRACTICES OF ADDITION STRATEGY : ADDING 1, 2, 3 加 1, 2, 3 92

ADDING 1, 2, OR 3 BY COUNTING	97
COUNTING 0, 1 FORWARD D + D	97
COUNTING FORWARD 2 FOR D + D WITH NO CARRYING	99
ADDING UP TO 10 BY ADDING 1	101

ADDITION STRATEGY: DOUBLING = 2 TIMES (EVEN NUMBERS) 双数 110

GETTING EVEN NUMBERS BY DOUBLING A NUMBER	111
DOUBLING = 2 TIMES (EVEN NUMBERS)	112

WORKING BACKWARDS ON DOUBLING 119

DOUBLING BY RELATING NUMBERS	124
DOUBLING, UP, AND DOWN	142

ADDING STRATEGY: ADDING TO 10 凑十 148

ADDITION STRATEGY: 9 PLUS 1 (9 + D = 9 + 1 + (D−1))   (9 加 1, 9 + D= 9 + 1 + (D−1)) 162

ADDING 9 BY RELATING	166
USING 9 PLUS TECHNIQUE	170

USING DOUBLES, 9 + D TO ADD 183

ADDING 5 BY DOUBLING AND 10'S COMPLEMENT	190
ADDING 6 BY DOUBLING AND 10'S COMPLEMENT	192
ADDING 7 BY DOUBLING AND 10'S COMPLEMENT	194
ADDING 8 BY DOUBLING AND 10'S COMPLEMENT	196

MAKING 10 OR ADDING AROUND 10 OR DOUBLING 198

MIXED DOUBLING, ADDING TO 10, ADDING 1, 2, 3 200

MIXED DOUBLING AND ADDING TO 10	201

MIXED DOUBLING, ADDING TO 10, ADDING 1, 2, 3 203

MULTI-DIRECTION ADDITION OF DOUBLING AND ADDING TO 10	204
IF-THEN FOR PLUS (DO THE LEFT + FIRST THEN THE RIGHT +) REVIEW	206

ASSESSMENT OF 9 + D, DOUBLING, ADDING TO 10 214

PART 3: MORE PRACTICES OF CHESS AND MATH INTEGRATED PROBLEMS 215

ADDITIONS TO 10	215
ADDING UP TO 10	216
MAKING 10 OR ADDING AROUND 10	239

Ho Math Chess 何数棋谜　妈！我会棋谜式加法啦！
Mom! I Learn Addition Using Math-Chess-Puzzles Connection
Contents include both traditional and Math-Chess-Puzzles combined methods. Extra strength
©2008 – 2018 Frank Ho, Amanda Ho　　All rights reserved. www.homathchess.com

Student Name _____　Date _____

- KNIGHT MOVES TO MAKE 10 251
- DOUBLING, ADDING 10, OR 9 + D 257
- MAKING 10 OR ADDING AROUND 10 OR DOUBLING 258
- CHESS AND MATH INTEGRATED PROBLEMS 275
- 8 PLUS 279
- 7 PLUS 281
- 6 PLUS 282
- 5 PLUS 283
- 4 PLUS 284
- ADDITIONS TO 11 285
- ADDITIONS TO 12 286
- ADDITIONS TO 13 287
- ADDITIONS TO 14 288
- ADDITIONS TO 15 289
- ADDITIONS TO 16 290
- USING 8 TO ADD 292
- USING 9 TO ADD 293
- NUMBER RELATIONSHIPS 294

PART 4: ADDITION BY CARRYING 317

- D + DD 317
- CARRYING OR REGROUPING FOR D + DD 318
- DD + D 323
- DD + D HIGHER DECADES IN PATTERN 326
- DD + D HIGHER DECADES IN PATTERN 327
- MAKING MULTIPLES OF 100 336
- DD + D 337
- DD + DD 339
- CARRYING OR REGROUPING FOR DD 340
- DD + DDD ADDING 10 341
- DD + DDD ADDING DOUBLES 342
- DD + DDD ADDING 11 343
- DD + DDD ADDING 12 344
- DD + DDD ADDING 13 345
- DD + DDD ADDING 14 346
- DD + DDD ADDING 15 347
- DD + DDD ADDING 16, 17, 18 348
- DD + DDD 349
- DDD + DDD 353
- MENTAL ADDITION 358
- D + D + D 361

NUMBER PATTERN 365

- HIGH PERFORMANCE MULTI-DIGIT ADDITION 368
- COMPUTING THROUGH FUNCTION CONCEPT 374

FROM FINGERING MATH TO COMPUTING THROUGH FUNCTION CONCEPT 375

- COMPUTING ADDITIONS THROUGH MATH AND CHESS INTEGRATED PUZZLE 381

100 TABLE AND PATTERN 389

USING DD + D OR D + DD ADDING UP TO 10 CONCEPT USING CHESS MOVES 393

- ADDITION USING CHESS MOVES 413

Ho Math Chess 何数棋谜 妈！我会棋谜式加法啦！	
Mom! I Learn Addition Using Math-Chess-Puzzles Connection	
Contents include both traditional and Math-Chess-Puzzles combined methods. Extra strength	
©2008 – 2018 Frank Ho, Amanda Ho All rights reserved. www.homathchess.com	
Student Name _____ Date _____	

ADDITION AND SUBTRACTION CONCEPT BY LINK	477
ADDITION AND SUBTRACTION BY LINK USING DOTS	**493**
D + D OF MATH AND CHESS INTEGRATED PROBLEMS	**501**
ADDING (BE PATIENT WITH YOUR WRITING.)	**509**
PART 5: COGNITIVE ABILITY 思唯意識	**512**
PUZZLE # 1	512
PUZZLE # 2	513
PUZZLE # 3	514
PUZZLE # 4	515
PUZZLE # 5	516
PUZZLE # 6	517
PUZZLE # 7	518
PUZZLE # 8	519
PUZZLE # 9	520
PUZZLE # 10	521
PUZZLE # 11	521
ONE-DRAWABLE FIGURE	522
MATRIX REASONING	524
PART 6: PATTERN, GEOMETRY, TIME, CALENDAR, A FRACTION	**537**
LOOKING FOR PATTERNS	538
PART 7: GEOMETRY	**541**
LINES	541
ANGLES	544
SHAPES	545
QUADRILATERALS	546
SYMMETRY LINE	549
PART 8: TIME AND CLOCK	**550**
TIME	550
TELLING TIME	551
TELLING TIME	**552**
DRAWING THE HANDS ON THE CLOCK TO SHOW THE TIME	553
DIGITAL TIME	556
CALENDAR	557
PART 9: FRACTION	**558**
REPRESENTING FRACTIONS	561
PART 10: WORD PROBLEM SOLVING	**563**
PART 11: TESTS	**564**
TEST 1	564
TEST 2	570
TEST 3	573
TEST 4	577
TEST 5	579

Ho Math Chess 何数棋谜　妈！我会棋谜式加法啦！
Mom! I Learn Addition Using Math-Chess-Puzzles Connection
Contents include both traditional and Math-Chess-Puzzles combined methods. Extra strength
©2008 – 2018 Frank Ho, Amanda Ho   All rights reserved. www.homathchess.com
Student Name _____ Date _____

PART 12: A FREE PAPER SET OF HO MATH CHESS 583

介紹何数棋谜 587

INTRODUCING HO MATH CHESS™ 588

Ho Math Chess 何数棋谜   妈！我会棋谜式加法啦！
Mom! I Learn Addition Using Math-Chess-Puzzles Connection

Contents include both traditional and Math-Chess-Puzzles combined methods. Extra strength
©2008 – 2018 Frank Ho, Amanda Ho   All rights reserved. www.homathchess.com

Student Name _____ Date _____

How to use this workbook?

- It is important that teachers, tutors, or parents read this "how to" before teaching. Follow the following 3 steps to making tutoring or teaching successfully.

  Step 1: Buy this workbook which costs only equivalently to a one-hour tutoring fee.
  Step 2: Thoroughly read the information in this workbook before Part 1 before tutoring or teaching.
  Step 3: Read the following information and follow them too. If you have any comments or problems while tutoring, please contact us through fho1928@gmail.com.

- The teaching order of addition strategies is important. The teaching order is as follows:
  1. Teach adding by 1, 2, or 3 first.
  2. Teach doubling strategy (2 times table) because the speed of doubling is faster than adding 10.
  3. Teach adding to 10 for 2 of 1 digit. This technique can also be used in subtraction.
  4. Teach 9 + 1 digit such as 2, 3, 4, 5, 6, 7, 8.
     9 + 2 is to give 1 to 9 or split 2 to be 1 + 1, then the answer is 11.

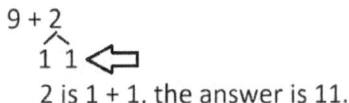

2 is 1 + 1, the answer is 11.

- Must thoroughly understand the strategies used in this workbook, this workbook does not stress rote teaching or use fingers to count forward or backwards, instead, I emphasize to teach students to use some basic strategies or number senses to get answers.
- Students are required to remember some very basic addition facts such as doubling (double 2 = 2 + 2 = 4, double 3 = 3 + 3 = 6, double 4 = 4 + 4 = 8, double 5 = 5 + 5 = 10, double 6 = 6 + 6 = 12, double 7 = 7 + 7 = 14, double 8 = 8 + 8 = 16, double 9 = 9 + 9 = 18), adding to 10 by 2 of 1-digit(1 + 9 = 10, 9 + 1 = 10, 2 + 8 = 10, 8 + 2 = 10, 3 + 7 = 10, 7 + 3 = 10, 4 + 6 = 10, 6 + 4 = 10, 5 + 5 = 10). It is important to conduct oral tests after student finished each strategy. Do not test each problem in sequential order, but in random order. So, do not ask students what is 1 + 9, 2 + 8, 3 + 7 etc. in sequentially increasing or decreasing order. Ask any problem in random order such as 4 + 6 = ?, 8 + 2 = ?. I have parents told me that they tested their children just last week but when I tested their children, they were not fluent. One of the reasons is the parent gave problem in sequential order.
- Do not let the students drag the study of strategy too long, normally students should show improving results after 8 – 10 hours of study in class and 4 – 6 hours of work at home. If parents do not encourage students to do homework (5 or more pages each week) at home as assigned by the tutor, then the results will be difficult to get.
- Few students only need to do the Part 1, then they could grasp all strategies of addition and move on to more advanced additions such as Part 3 – d + dd etc. Many students need to do Part 2 to reinforce the concepts and do more practices to have a good grasp of the strategies.

Ho Math Chess 何数棋谜　妈！我会棋谜式加法啦！
Mom! I Learn Addition Using Math-Chess-Puzzles Connection

Contents include both traditional and Math-Chess-Puzzles combined methods. Extra strength

©2008 – 2018 Frank Ho, Amanda Ho　All rights reserved. www.homathchess.com

Student Name _____ Date _____

- Emphasize the idea of using "reasoning" to get answers after the student understands the strategies such as 6 + 7 = 13 or 7 + 6 = 13 by using 6 + 6 = 12. This is the reason that we have many doubling related problems to teach students how to use the doubling strategies to get other answers. The "reasoning" strategies equally apply to add to 10.
- No chess knowledge is required to work on worksheets, students just follow the dark error of a chess symbol to get numbers or values. The teacher or parent can skip all chess and math integrated materials although it is not recommended. The math, chess, and puzzles integrated worksheets will get students to work more on the brain than the traditional worksheets.
- Not all students in the same grade are at the same math ability level, so do not teach this workbook page after each page. Not all pages need to be finished. I feel very sad if a tutor does not bother to study the teaching ideas of this workbook but just "blindly" ask students to work on these worksheets page after page. Do not go on to a new strategy until the student has mastered each old strategy. If the student still can not master the strategy after finishing all worksheets of this workbook, then most likely, according to my experience, the student and parents must review the student's study ability, methods, and desire.
- What are the reasons that students may not master the addition skills after learning these strategies?.There is a small percentage of students who do not seem to be able to master the addition skills after finishing the Part 1 or Part 2 of its workbook. I can analyze the problems from the three parties which are student, tutor, or guardian.

(1) The student did not follow the tutor's instructions when working on worksheets. For example, the student is supposed to try to "memorize" the adding to 10 and doubling when working on problems, that was one reason I created many questions so that students will not feel bored easily, but some students did not bother to "memorize" or naturally "instil" into their brain, but just copied answers from the previous answer.

(2) The student still tried to use fingers especially when the student put hands under the table and gave answers slowly or tried to repeat a tutor's question to buy some time.

(3) The student has no desire to study arithmetic.

(4) The student seems to have a very short memory and could understand while being taught, but one or two weeks later, the student would act as almost never being taught at all.

(5) The tutor continues to use manipulatives other than fingers to teach instead of using reasoning such as 7 + 6 can be logically be answered by using doubling 6 + 6 = 12. Take what you learned and use it on something else is what I want to stress to do additions by using these strategies.

(6) The tutor used symbols such as drawing circles to do additions, instead of training a student's brain to think logically and abstractly. 4 + 2 can be taught by telling a student to go up 2 floors from the 4th floor.

(7) The guardian just told the child to do work at home and hardly paid attention o how the work was completed by the student.

Ho Math Chess 何数棋谜　妈！我会棋谜式加法啦！
Mom! I Learn Addition Using Math-Chess-Puzzles Connection
Contents include both traditional and Math-Chess-Puzzles combined methods. Extra strength
©2008 — 2018 Frank Ho, Amanda Ho　All rights reserved. www.homathchess.com

Student Name _____ Date _____

- What we like to achieve after using this workbook is to get some students' feedback, hopefully, as follows:

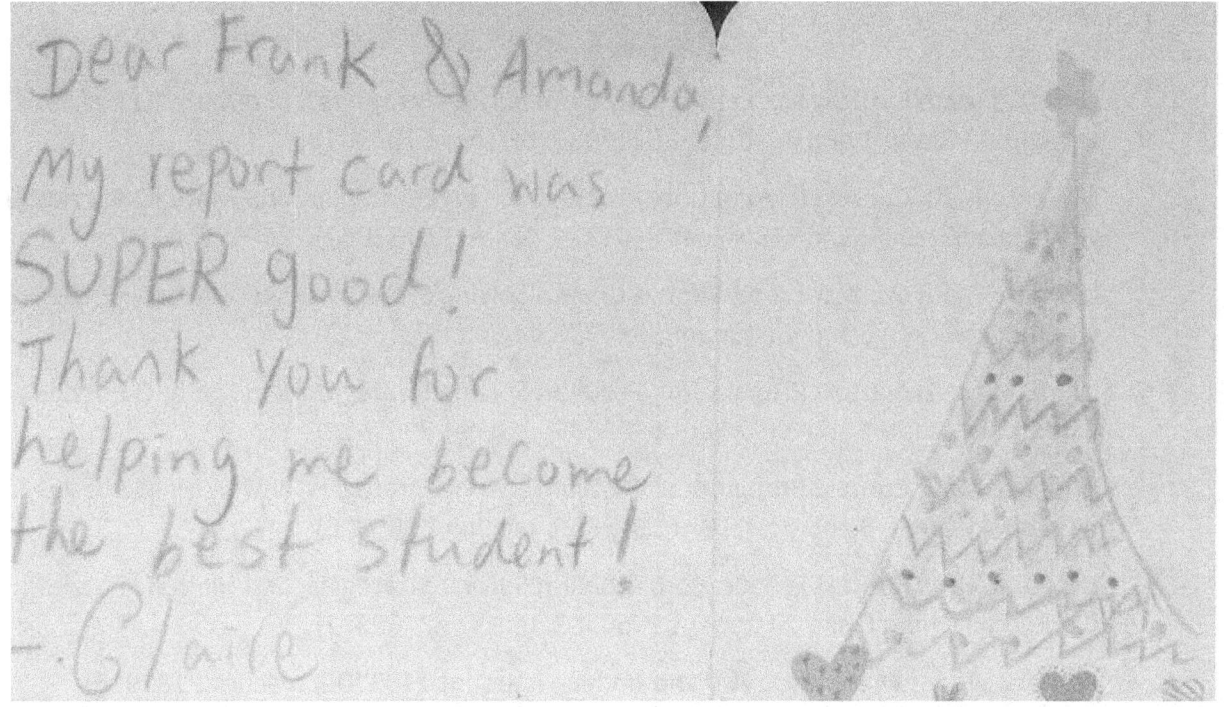

Ho Math Chess 何数棋谜　　妈! 我会棋谜式加法啦!
Mom! I Learn Addition Using Math-Chess-Puzzles Connection
Contents include both traditional and Math-Chess-Puzzles combined methods. Extra strength
©2008 – 2018 Frank Ho, Amanda Ho　　All rights reserved. www.homathchess.com
Student Name_____ Date_____

- I also have written many other workbooks which are available on www.amazon.com, so there is a continuity of finishing the elementary mathematics by using these workbooks.

Other Ho Math Chess Publications

    Ultimate Math Contest Preparation, Problem Solving Strategies, and Math IQ Puzzles (3 in 1 workbook) Grade 1 and 2 (student 6717980, teacher 6720849)

    Ultimate Math Contest Preparation, Problem Solving Strategies, and Math IQ Puzzles (3 in 1 workbook) Grade 2 and 3 (student 6745934, teacher 6783062)

    Ultimate Math Contest Preparation, Problem Solving Strategies, and Math IQ Puzzles (3 in 1 workbook) Grade 3 and 4 (student 8990731, teacher 6813413)

    Ultimate Math Contest Preparation, Problem Solving Strategies, and Math IQ Puzzles (3 in 1 workbook) Grade 4 and 5 (student 6906686, teacher 6907410)

    Ultimate Math Contest Preparation, Problem Solving Strategies, and Math IQ Puzzles (3 in 1 workbook) Grade 5 and 6 (student 6979530, teacher 6985906)

    Ultimate Math Contest Preparation, Problem Solving Strategies, and Math IQ Puzzles (3 in 1 workbook) Grade 6 and 7 (student 7060502, teacher 7060972)

    Primary Grades Math (Grades 4 and under) (student 7182306, teacher 7182916)

    Elementary Grades Math (Grades 5 and Up)

    Math Chess and Puzzles for Primary Grades (student 9195575, teacher 9195575Junior Kindergarten Math (for Pre-K and junior kindergarten) (student 5738681, teacher 6432094)

    Learning Chess to Improve math

    Frankho ChessDoku 3 by 3 (student 4233466, teacher 4276722)

    Frankho ChessDoku 4 by 4

    Frankho ChessDoku 5 by 5

    Mom! I Learn Addition Using Math-Chess-Puzzles Connection (student 5919284, teacher 5669603)

    Mom! I Learn Subtraction Using Math-Chess-Puzzles Connection (student 658604, teacher 568614)

Ho Math Chess 何数棋谜　　妈！我会棋谜式加法啦！
Mom! I Learn Addition Using Math-Chess-Puzzles Connection
Contents include both traditional and Math-Chess-Puzzles combined methods. Extra strength
©2008 – 2018 Frank Ho, Amanda Ho　　All rights reserved. www.homathchess.com
Student Name _____ Date _____

Mom! I Learn Multiplication Using Math-Chess-Puzzles Connection student 5891187, teacher 5891177)

Mom! I Learn Division Using Math-Chess-Puzzles Connection (student 698270, teacher 5697799)

2018 Christmas

Frank Ho

Ho Math Chess 何数棋谜　妈！我会棋谜式加法啦！
Mom! I Learn Addition Using Math-Chess-Puzzles Connection
Contents include both traditional and Math-Chess-Puzzles combined methods. Extra strength
©2008 – 2018 Frank Ho, Amanda Ho　　All rights reserved. www.homathchess.com
Student Name_____Date_____

2017 Preface 前言

Is it insane to have over 500 pages of an addition workbook? The answer is simple, it is not. For a minority of students, they can master the addition strategies taught in this workbook in a short time. Most of the students need to finish a certain quantity to reach fluency. I found out, some students with dyscalculia may need to do multiple times of math worksheets which are not needed for an average student. This is one reason this workbook gets larger than the first version. It is designed for students with different abilities. The other reason why this workbook is so huge is that this revised Addition workbook integrates the following previously published three workbooks:

Mom! I Learn Addition Using Math-Chess-Puzzles Connection
Learning Calculation without Counting Fingers
Math Entrance Examination for Private School

The new and revised Addition workbook particularly stresses the importance of teaching the Addition strategies in the following order:

Strategy 1: Adding 1, 2, or 3
Strategy 2: Doubling
Strategy 3: Adding 2 numbers to 10
Strategy 4: 9 Plus 1 after the number split into 1 and (number -1)

The reason for integrating all 3 books into one workbook is so that students can achieve many learning purposes without the need to buy 3 workbooks. The other reason is to train students to do calculations without counting fingers.

The following is a preface which I wrote for the workbook *Learning Calculation without Counting Fingers*.

The design of Ho Math Chess worksheets encourages students to do calculations without using their fingers to count numbers. Almost none of our A students and students participating in math competitions use their fingers to do basic calculations and their ability of mental math is strong than those who use fingers to calculate. The main reason is, without using fingers; students will develop a strong number sense and are able to see steps ahead when working on math problems.

The following gives a list of potential problems for some elementary students who continue to use fingers to do calculations might encounter when going to higher grades:

Ho Math Chess 何数棋谜　　妈！我会棋谜式加法啦！
Mom! I Learn Addition Using Math-Chess-Puzzles Connection
Contents include both traditional and Math-Chess-Puzzles combined methods. Extra strength
©2008 – 2018 Frank Ho, Amanda Ho　　All rights reserved. www.homathchess.com
Student Name _____ Date _____

- Difficult and slow to find answers which are related to problems requiring work backwards such as subtractions or finding quotients. For example, it takes finger calculating students much longer to find what is 12 - 9? When doing simple divisions such as 24 ÷ 2, 36 ÷ 3 finger
- calculating, students cannot do them mentally without using pencil and paper.
- Have hard time to find product of primes of a number; slow to find an answer and have hard time to see if a number is divisible by a certain number using divisibility rules; difficult to find GCF or LCM and later LCD; difficult and slow to reduce a fraction; have hard time to do the conversions between fractions, decimals, and % etc.
- Unable to see what is the trick to add certain type of questions such as 99 + 7 + 8 + 4 + 6 + 1 + 3 + 2 etc. Students who do not use fingers can scan this problem and they know the answer almost immediately and automatically.
- Because of the slow speed of doing basic calculations and the lack of mental math ability, the student has less chance to get in the enrichment math program due to competitiveness. By the same token, the student has less chance to do reasonably well in any math contests which stress calculations speed and the calculator is not allowed. Students will do poorly when oral math problems are presented in any math contests.
- Students will continue to experience difficulties when going to high schools with those problems requiring mental math ability to quickly recall four basic computation facts. Problems such as finding the square roots; adding like terms; factoring trinomials using cross-multiplication method; 30-60-90 special degree; graphing $2x + 3y = 6$ etc.

When we talk about calculation without counting fingers, we are not saying the only reason is so that students can recall arithmetic facts quickly. From the above observations, we can see that mental math or calculate without counting fingers is not limited to the scope of being a human calculator of doing some calculations in fast speed. To acquire the mental math skills, so it truly benefits students, the students must learn the skill of reversing calculating like how to get factors of a number; how to quickly find prime factors of a number; how to find a number if it has a perfect number as a factor; how not only do calculation in linear fashion; but also know how to do calculations in cross or vertical way quickly and be able to see patterns. Mental math also helps the student do trinomial factoring when they are in higher grades such as 9 or 10.

When in high school, students with math deficiencies often find it is difficult for them to go back to work on addition, subtraction, subtraction, or multiplication again since it is embarrassing for them to work on grade 3 math while they are already in grade 10.

Mom! I Learn Addition Using Math-Chess-Puzzles Connection
Contents include both traditional and Math-Chess-Puzzles combined methods. Extra strength
©2008 – 2018 Frank Ho, Amanda Ho   All rights reserved. www.homathchess.com

Student Name _____ Date _____

With the above in mind, one purpose of this workbook is to create an environment where students are encouraged to think more even when they are doing calculations. This workbook encourages students to do calculations without counting their fingers, so the calculations are totally done in their brains without relying on any manipulative. The requirement for students to write the numbers to arrive answers will help them reinforce of memorization of number facts.

These mental calculations worksheets will help students develop strong mental math ability. As consequence, this helps students do well later in their SSAT, SAT and increase the chance of getting into private schools if they wish to. The chance of getting into an enrichment math program in high schools is also higher.

April 2017

Mom! I Learn Addition Using Math-Chess-Puzzles Connection

Ho Math Chess 何数棋谜　　妈！我会棋谜式加法啦！
Mom! I Learn Addition Using Math-Chess-Puzzles Connection
Contents include both traditional and Math-Chess-Puzzles combined methods. Extra strength
©2008 – 2018 Frank Ho, Amanda Ho　　All rights reserved. www.homathchess.com
Student Name _____ Date _____

Ho Math Chess worksheets and traditional worksheets

Frank Ho, Amanda Ho

Ho Math Chess Tutoring Centre
Vancouver, Canada

What's wrong with the traditional drill computation? From tutoring point of view, there is nothing wrong to give them to children for practice on fluency and grasp of basics. However, there is something wrong from children's point of view that is they are boring, dull and not fun. Why do children feel that way? Well, the time has changed but the format of traditional computation worksheets has not caught up with the pace of the society. The future belongs to a generation who understands how to process information and the information might include digits, bytes, numbers, graphics, images, languages, symbols, equations etc. Children today might chat with others on the internet while downloading or uploading files and viewing movie clips at the same time. Multitasking and the multi-way of processing information seem to have come as second nature to children, but is our computation format reflecting the way that children are living today? Certainly not, this is one reason why children feel so bored and lost interest to continue to work on the same" old" style of computation worksheets.

These simple monotonic basic number facts computation worksheets are no longer reflecting the real world which the young generation is facing today or will be living in the future. The computing world children are facing today is much like a rich tapestry, where diversified fabric and colours are integrated. Children today are absorbing not just numbers but an array of information like image, sound, music, symbols, spatial information, or even abstract ideas all bundled together and delivered through many types of media. Children today are not happy just working on pure number drill without any other stimulus or motivator. After realizing the importance of having fun while learning, Ho Math Chess has been embarked on an important teaching philosophy that is to integrate chess and puzzles into math worksheets so that children can learn math while having fun.

Ho Math Chess 何数棋谜   妈！我会棋谜式加法啦！
Mom! I Learn Addition Using Math-Chess-Puzzles Connection
Contents include both traditional and Math-Chess-Puzzles combined methods. Extra strength
©2008 — 2018 Frank Ho, Amanda Ho   All rights reserved. www.homathchess.com
Student Name_____ Date_____

The values of chess pieces used in this workbook are listed as follows:

	Pawn	Knight	Bishop	Rook	Queen	king
Ho Math Chess	↓	✜	⤢	⬌	✴	✱
Traditional chess	♟	♞	♝	♜	♛	♚
Value	1	3	3	5	9	0

The following offers a brief comparison between Ho Math Chess worksheets and traditional worksheets

	Traditional worksheets	**Ho Math Chess worksheets**
Computing Direction	Mostly left-right or top-down	Multi-direction math and chess integrated. Chess queen has 8 directions. The reason for training children to calculate in multi-way is as follows: Math computation is not just one way of computing from left to right or top to down. Many fractions require reducing in a diagonal way such as $\frac{4}{3} \times \frac{9}{8}$.
Concept	Mostly one concept	Multi-concept. For example, many work backwards problems are presented in addition worksheets to train students to get answers by using calculating backwards strategy.
Abstract symbols	Very little abstract symbols are used in arithmetic computations.	Chess abstract symbols are used starting from pre-k.
Subject	Only math	Multi-subject: math, chess, puzzles, and brainpower
Entertainment	No entertainment value	Offering entertainment value through puzzle-like problems. Many times, no problems are present obviously, thus Ho Math and Chess worksheets train student's thinking skills while working on pure computation worksheets.
Computing	Only number crunching is offered	Students need to think a lot more to find out what is the question when working on a simple calculation problem.
Manipulative	Short term value	Chess has life-long value.
💡idea	Worksheets have no new ideas.	Innovative worksheets train students to have creative ideas.
Contents	Many similar worksheets	World first, unique and the only one.

Ho Math Chess 何数棋谜　妈！我会棋谜式加法啦！
Mom! I Learn Addition Using Math-Chess-Puzzles Connection
Contents include both traditional and Math-Chess-Puzzles combined methods. Extra strength
©2008 – 2018 Frank Ho, Amanda Ho   All rights reserved. www.homathchess.com
Student Name _____ Date _____

Ho Math Chess workbook is a multi-function workbook, it trains children not only their basic computing ability but also trains them to be an astute data warehouse manager or an excellent data miner by developing their problem-solving ability and critical thinking skills.

Ho Math Chess workbook provides education and entertainment value to get young children involved in the future world they will be facing.

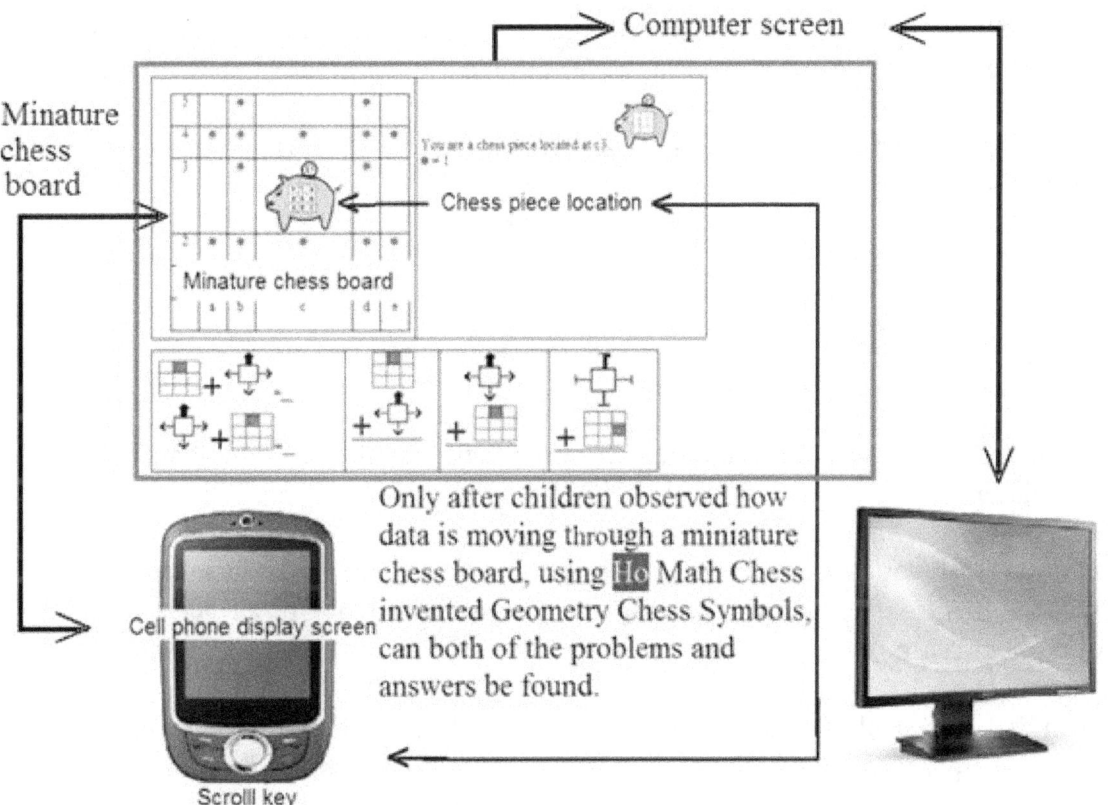

Ho Math Chess 何数棋谜　妈！我会棋谜式加法啦！
Mom! I Learn Addition Using Math-Chess-Puzzles Connection
Contents include both traditional and Math-Chess-Puzzles combined methods. Extra strength
©2008 – 2018 Frank Ho, Amanda Ho　All rights reserved. www.homathchess.com
Student Name _____ Date _____

Benefits of using Ho Math Chess workbook

We have found that the traditional worksheets seem to present difficulties for young children to learn math especially 4-year-old, many of them end up using fingers as a mechanism to count. This workbook is created very differently from the traditional workbook. It will eliminate the tendency for children to use fingers and learn the skills of addition with a shorter learning curve. The other most important difference is this workbook is much more fun than working on the traditional worksheets and Ho **Math Chess** requires students to "think" more than traditional worksheets. The other major differences are as follows:

1. For the traditional math workbooks, children only need to write the answers. For example, for additions, children only need to write the answer of additions but no need to write what numbers have been added together such as addends unless the problems are word problems solving type. In this workbook, most of the problems require children to write addends so that children have opportunities to have the effect of "reciting" the problems and thus reinforcing their memory. An example is given as follows:

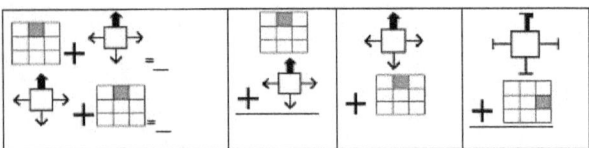

2. Children will feel more fun if working on this workbook since most of our problems having "puzzles" built in.

3. Many additions techniques such adding up to ten, doubling etc. are gradually introduced to children so that they learn addition in an easy and intuitive way and naturally increase the ability of number sense.

4. One reason that many children have math problems is that they cannot do the reverse thinking, connecting and organizing information, or analyzing problems. This workbook is created to train students to work on computation worksheets, yet students will also get the same benefits as they were working on word problems. This is because our worksheets are created such that students will have to

  - Analyze to find out how to solve the problem by following abstract symbols.
  - Visually connect pieces of information together by following symbols.
  - Organize information by table lookup.
  - Put them together by the following the logic.
  - Follow direction (in chess moves).
  - Find ways around to look for information (convergent and divergent).

| Ho Math Chess 何数棋谜　妈！我会棋谜式加法啦！ |
| Mom! I Learn Addition Using Math-Chess-Puzzles Connection |
| Contents include both traditional and Math-Chess-Puzzles combined methods. Extra strength |
| ©2008 − 2018 Frank Ho, Amanda Ho　All rights reserved. www.homathchess.com |
| Student Name _____ Date _____ |

The process of operating instructions provides fun and challenging for children.

This workbook improves children's math computing and problem-solving abilities and at the same time, increases their brain power.

Frank Ho
Amanda Ho

January 2012

Ho Math Chess 何数棋谜　　妈！我会棋谜式加法啦！
Mom! I Learn Addition Using Math-Chess-Puzzles Connection
Contents include both traditional and Math-Chess-Puzzles combined methods. Extra strength
©2008 – 2018 Frank Ho, Amanda Ho　　All rights reserved. www.homathchess.com

Student Name_____ Date_____

使用何数棋谜教材的好处

加拿大何数棋谜培训中心

創辦人何数棋（Frank Ho），何数谜（Amanda Ho）

www.mathandchess.com

只见棋谜不见题　　劝君迷路不哭涕
数学象棋加谜题　　健脑思维真神奇

今天儿童面对的世界是学习如何处理数字,图形,资料搜寻,音影上下载,资讯比较,分类等资讯.这些活动实际已成為儿童生活的一部份.所以如果说学数学就是计算数字就错了.学数学的另一个目的就是学习如何利用数字资讯去解决问题及培养创造力.但是**传统式数学的计算练习题却完全没跟上科研已经改变了儿童面对的世界.**

| Ho Math Chess 何数棋谜　妈！我会棋谜式加法啦！ |
| Mom! I Learn Addition Using Math-Chess-Puzzles Connection |
| Contents include both traditional and Math-Chess-Puzzles combined methods. Extra strength |
| ©2008－2018 Frank Ho, Amanda Ho　All rights reserved. www.homathchess.com |
| Student Name _____ Date _____ |

何数棋谜

棋谜式健脑思维趣味数学结合高科技操作图解

只见棋谜不见题　劝君迷路不哭涕
数学象棋加谜题　健脑思维真神奇

Ho Math and Chess worksheet layout simulates a computer screen and a cell phone screen.

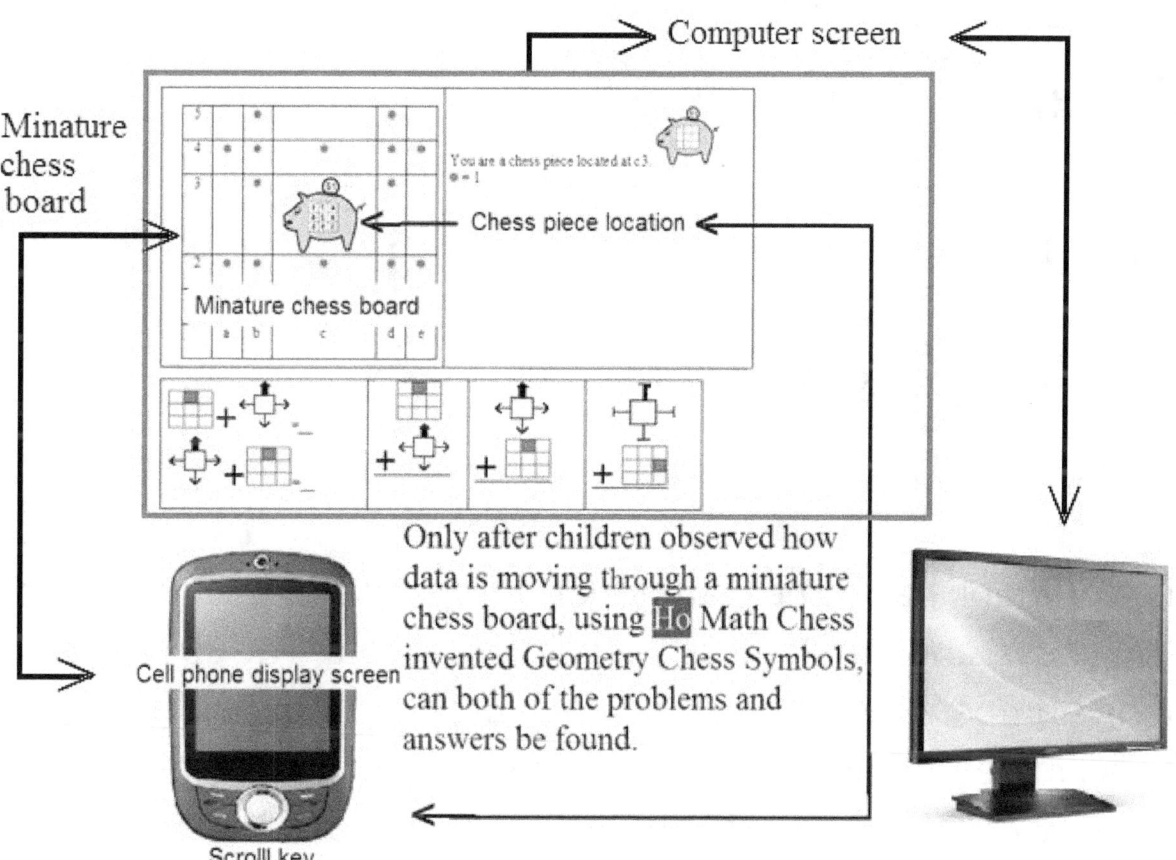

Only after children observed how data is moving through a miniature chess board, using Ho Math Chess invented Geometry Chess Symbols, can both of the problems and answers be found.

Ho Math Chess 何数棋谜 妈！我会棋谜式加法啦！
Mom! I Learn Addition Using Math-Chess-Puzzles Connection
Contents include both traditional and Math-Chess-Puzzles combined methods. Extra strength
©2008 – 2018 Frank Ho, Amanda Ho   All rights reserved. www.homathchess.com

Student Name _____ Date_____

a
Chess symbols and their mathematical values

Symbols of chess pieces	Names of chess pieces	Mathematical values
	Queen	9
	Rook	5
	Bishop	3
	Knight	3
	Pawn	1
	King	0

This workbook has some traditional style worksheets and our unique math, chess, and puzzles integrated worksheets. The strategies of adding numbers are offered and each problem's difficulty increases in small increment and in a very organized pattern to get students familiar with the relationship between numbers and develop their number sense.

Ho Math Chess 何数棋谜　妈！我会棋谜式加法啦！
Mom! I Learn Addition Using Math-Chess-Puzzles Connection
Contents include both traditional and Math-Chess-Puzzles combined methods. Extra strength
©2008 – 2018 Frank Ho, Amanda Ho　All rights reserved. www.homathchess.com
Student Name _____ Date _____

Prerequisites before working on additions

Before a student could work on "real" addition, there is some arithmetic knowledge that a student must possess such as the following:

- To be able to recite from 0 to 100 forward and also backwards fluently.
- To be able to write digits from 0 to 9 in correct writing strokes.
- To know the rank of numbers such as 3 is larger than 2 by 1 and 4 is one less than 5 etc.
- To know the names of the shapes.

If a student has some difficulties of achieving the level of the above arithmetic knowledge, we have two workbooks *Junior Kindergarten Math* and *Math Chess and Puzzles for Primary Grades* for students for them to work on.

The workbook *Math Chess and Puzzles for Primary Grades* can also be used along with this workbook, so the student will not feel too bored when just working on computations all the time.

Ho Math Chess 何数棋谜　妈！我会棋谜式加法啦！
Mom! I Learn Addition Using Math-Chess-Puzzles Connection
Contents include both traditional and Math-Chess-Puzzles combined methods. Extra strength
©2008 – 2018 Frank Ho, Amanda Ho　All rights reserved. www.homathchess.com

Student Name_____ Date_____

Counting numbers and whole numbers

Numbers starting with 1 and then counting forward by adding 1 such as 1, 2, 3, 4, 5, 6, 7, … and so on are called counting numbers (natural numbers). Counting numbers are also called whole numbers. The difference between counting numbers and whole numbers is that the whole numbers include 0 but the counting numbers do not include 0. In this book, whenever the word "number" is mentioned, it means the whole number.

Fill in the following □ by a number.

1, 3, 5, 7, □, 11, 13, □　9, 15

Replace the following "?" by a counting number.

The numbers starting at 2 and added by 2 each time are called _____ numbers. 0 can be considered as an even number for a computational purpose like checking if a number is divisible by 2.

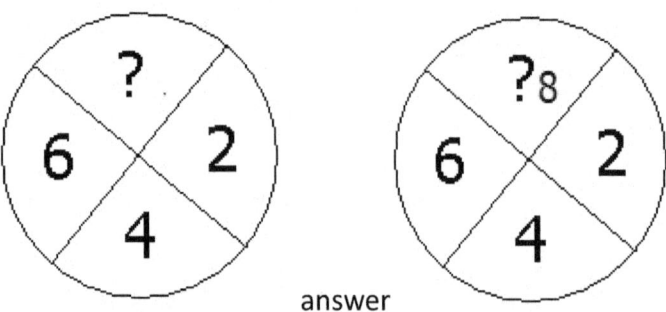

answer

Replace the following ? with a counting number.
The numbers starting at 1 and added by 2 each time are called _____ numbers.

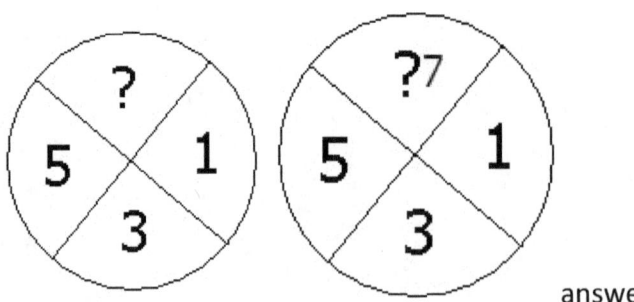

answer

Ho Math Chess 何数棋谜 妈！我会棋谜式加法啦！
Mom! I Learn Addition Using Math-Chess-Puzzles Connection

Contents include both traditional and Math-Chess-Puzzles combined methods. Extra strength

©2008 − 2018 Frank Ho, Amanda Ho   All rights reserved. www.homathchess.com

Student Name _____ Date _____

Writing d + d number sentence (Commutative property) 交换率

Fill in each □ with a number.

Math and Chess sentence	Number sentence.	Comments
♟♟♟♟♟ + ♟♟	$5 + 2 = 7$	**5 plus 2 is 7.** 5 and 2 are (equals) 7. The sum of 5 and 2 is (equals) 7. 5 added to 2 is (equals) 7. The sum of 5 and 2 is equal to 7. Addend + Addend = Sum **Think to add from left to right (Think 5 then add 2).**
♟♟ + ♟♟♟♟♟	2+5=7	Think to add 2 to 5 (from right to left).
♟♟♟♟♟♟♟ + ♟♟	7+2=9	Think to add from left to right (Think 7 then add 2).
♟♟ + ♟♟♟♟♟♟♟	2+7=9	Think to add from right to left (Think 7 then add 2).
♟♟♟♟♟♟♟♟♟ + ♟♟	9+2=11	Think to add from left to right (Think 9 then add 2).
♟♟ + ♟♟♟♟♟♟♟♟♟	2+9=11	Think to add from right to left (Think 9 then add 2).
♟♟♟♟♟♟♟♟ + ♟♟♟	8+3=11	
♟♟♟ + ♟♟♟♟♟♟♟♟	3+8=11	
♟♟♟ + ♟♟♟♟♟♟♟♟♟	3+9=12	
♟♟♟♟♟♟♟♟♟ + ♟♟♟	9+3=12	

Ho Math Chess 何数棋谜　妈！我会棋谜式加法啦！
Mom! I Learn Addition Using Math-Chess-Puzzles Connection
Contents include both traditional and Math-Chess-Puzzles combined methods. Extra strength
©2008 − 2018 Frank Ho, Amanda Ho　　All rights reserved. www.homathchess.com

Student Name _____ Date _____

Commutative law (a + b = b + a) of adding up to 5 交换律
The teacher can use a pile of pencils as a total to split the total.

Fill the following □ with a number.

0	+	□5	=	♖	=	♖	+	□0
♙	+	□4	=	♖	=	4	+	□1
2	+	□3	=	♖	=	♘	+	□2
♘	+	□2	=	♖	=	2	+	□3
4	+	□1	=	♖	=	1	+	□4
♖	+	□0	=	♖	=	♔	+	□5

Ho Math Chess 何数棋谜　妈！我会棋谜式加法啦！
Mom! I Learn Addition Using Math-Chess-Puzzles Connection
Contents include both traditional and Math-Chess-Puzzles combined methods. Extra strength
©2008 － 2018 Frank Ho, Amanda Ho　All rights reserved. www.homathchess.com

Student Name _____ Date _____

Strategies for d + d addition 加法的策略

The following addition strategies reveal how additions could be done using a few strategies without using fingers or boring repetitions. With the above in mind, the teacher or parent must be very careful not to teach children with a wrong strategy, otherwise, children could get confused.

For example, 8 + 7 = double 7 plus 1usind doubling strategy, not to teach children to use 8 + 7 = 8 + 2 + 5 = 10 + 5 = 15. It is alright to teach addition strategy to children, but children still need to master one technique and continue to use the same strategy until they can calculate without any hesitation.
Teach children so they know 8 + 7 = 7 + 8 = double 7 + 1 = 15.

	1	2	3	4	5	6	7	8	9
1	double 1 (Add the number itself.)								
2	+ 1	double 2							
3	+ 1	+ 2	double 3						
4	+ 1	+ 2	+ 3	double 4					
5	+ 1	+ 2	+ 3	1 away from double 5	double 5				
6	+ 1	+ 2	+ 3 (later remember 3 x 3 =9)	Adding to 10	double 5 +1	double 6			
7	+ 1	+ 2	Adding to 10	Adding to 10 +1	double 5 + 2	double 6 + 1	double 7		

Ho Math Chess 何数棋谜 妈！我会棋谜式加法啦！
Mom! I Learn Addition Using Math-Chess-Puzzles Connection

Contents include both traditional and Math-Chess-Puzzles combined methods. Extra strength

©2008 – 2018 Frank Ho, Amanda Ho   All rights reserved. www.homathchess.com

Student Name _____ Date _____

	1	2	3	4	5	6	7	8	9
8	+ 1	Adding to 10	Adding 10 + 1	Adding 10 + 2 (later remember 3 of 4 =12)	Double 5 + 3	double 6 + 2	double 7 +1	Double 8	
9	Adding to 10	9 + 1+1 (note that 2 = 1+1)	9 + 1+ 2 (note that 3 = 1+2)	9 + 1+ 3 (note that 4 = 1+3)	9 + 1+ 4 (note that 5= 1+4)	9 + 1+ 5 (note that 6= 1+5)	9 + 1+ 6 (note that 7= 1+6)	9 + 1+ 7 (note that 8 = 1+7)	double 9

Ho Math Chess 何数棋谜　妈！我会棋谜式加法啦！
Mom! I Learn Addition Using Math-Chess-Puzzles Connection

Contents include both traditional and Math-Chess-Puzzles combined methods. Extra strength
©2008 – 2018 Frank Ho, Amanda Ho   All rights reserved. www.homathchess.com

Student Name _____ Date _____

English names for numbers (For students who can read and write English.)

Large numbers are for advanced students only. These worksheets depend on a student's English ability, so do not enforce it.

Whole number	English words	Comments
0	Zero	
1	One	
2	Two	
3	Three	
4	Four	
5	Five	
6	Six	
7	Seven	
8	Eight	
9	Nine	
10	Ten	
11	Eleven	
12	Twelve	

13	Thirteen	
14	Fourteen	
15	Fifteen	
16	Sixteen	
17	Seventeen	
18	Eighteen	
19	nineteen	

English names for numbers

Whole number	English words	Comments
20	twenty	
30	thirty	
40	Forty	
50	Fifty	
60	Sixty	
70	Seventy	
80	Eighty	
90	Ninety	
21	twenty-one	
25	twenty-five	If a number is in the range of 21 to 99 and the second digit is not zero, a hyphen is used to separate the two words.
36	Thirty-six	
47	Forty-seven	
86	Eighty-six	

99	ninety-nine	
100	One hundred	No "s" needs to be added to "hundred".
200	two hundred	
900	Nine hundred	

English names for numbers

2,000	Two thousand	
9,000	Nine thousand	
19,000	Nineteen thousand	No "s" needs to be added to "thousand".
20,000	Twenty thousand	
24,000	Twenty-four thousand	
79,000	Seventy-nine thousand	
100,000	One hundred thousand	
1,000,000	One million	

Number order and rank

Insert >, <, or =.

Students must know the order and ranking of numbers and be able to recite fluently forward and backwards up to 20 before using counting skills in the brain (not fingers) and addition. Students should work on less than 20 numbers first.

1. 5 > 3
2. 4 > 3
3. 6 < 7
4. 7 > 5
5. 5 > 3
6. 8 > 6
7. 7 > 5
8. 2 < 3
9. 9 > 8
10. 9 < 10
11. 11 > 10
12. 16 = 16
13. 13 < 17
14. 15 < 18
15. 8 > 7
16. 17 > 14
17. 27 = 27
18. 13 < 15
19. 27 < 29
20. 23 > 21
21. 33 > 30
22. 26 < 28
23. 43 > 34
24. 31 > 29
25. 57 < 67
26. 34 < 37
27. 51 < 69
28. 75 < 77
29. 39 < 41
30. 32 < 36
31. 34 < 86
32. 67 < 76
33. 58 > 54
34. 38 < 83
35. 80 < 82
36. 32 > 27

Order the following numbers from smallest to largest.

5	8	3	6	3 5 6 8
6	2	7	1	1 2 6 7
4	9	5	3	3 4 5 9
3	5	1	7	1 3 5 7
11	15	13	18	11 13 15 18
34	44	6	54	6 34 44 54
51	61	31	21	1 2 31 51 61
36	42	65	41	36 41 42 65
46	24	26	43	24 26 43 46
25	34	64	52	25 32 52 64
57	53	50	56	50 53 56 57
35	38	32	37	32 35 37 38
42	22	52	32	22 32 42 52
21	41	51	31	21 31 41 51
42	56	53	27	27 42 53 56
50	64	36	19	19 36 50 64
57	36	71	50	36 50 57 71
47	38	92	65	38 47 65 92
36	64	38	60	36 38 60 64

Mom! I Learn Addition Using Math-Chess-Puzzles Connection

Order the following numbers from greatest to smallest.

5	8	3	7	8 7 5 3
2	5	6	3	6 5 3 2
7	9	4	5	9 7 5 4
2	1	7	6	7 6 2 1
3	6	1	8	8 6 3 1
11	16	13	15	16 15 13 11
47	49	41	46	49 47 46 41
32	62	72	42	72 62 42 32
23	93	73	53	93 73 53 23
45	84	85	63	85 84 63 45
52	54	74	37	74 54 52 37
45	75	63	28	75 63 45 28
47	37	34	86	86 47 37 34
26	53	75	70	75 70 53 26
19	76	85	46	85 76 46 19
45	86	36	58	86 58 45 36
73	27	48	43	73 48 43 27
78	46	27	58	78 58 46 27
57	38	96	34	96 57 38 34

Ho Math Chess 何数棋谜　妈！我会棋谜式加法啦！
Mom! I Learn Addition Using Math-Chess-Puzzles Connection
Contents include both traditional and Math-Chess-Puzzles combined methods. Extra strength
©2008－2018 Frank Ho, Amanda Ho　All rights reserved. www.homathchess.com
Student Name _____ Date _____

Part 1: Adding 123, doubling, adding ten, and 9 + d strategies

Most educators would think that learning computation is just "drill to kill' but the truth is because most workbooks do not provide any "thought-provoking" type of questions. The computation type of workbooks can be written in a way to train students' problem-solving skills, much like puzzles can be used to train a student's problem-solving skills. Ho Math Chess computation workbooks are written to increase a student's computation ability, but also to increase a student's problem-solving ability.

Students need to thoroughly understand the strategy of "123DT" that is to **add 1, 2, or 3** to a number, **doubling (D), adding ten (T)** and be able to recall results like recalling students' own phone numbers. Once they master these three basic skills then they will have little problems to add multi-digit numbers. This type of problems also trains the student's ability to work backwards.

Many children are not able to make progress or fall behind when using the "regular" math materials which are used at day schools or materials bought from stores. We have produced this special student math materials for those who need extra help.

We have a workbook called Frankho ChessDoku 3 by 3 which integrates math, chess, and Sudoku all in one workbook, Students love to work on this workbook to train their logical thinking skills and also have fun. This workbook can be used to complement the training of training a student's 1, 2, 3 addition skills. This workbook can be purchased from www.amazon.com.

The remaining part of this section contains a 7-day of practice worksheets of adding 1, 2, 3, doubling, adding to ten and 9 plus strategies. Some students could become proficient in additions, but the majority of students need more practices.

Ho Math Chess 何数棋谜   妈！我会棋谜式加法啦！
Mom! I Learn Addition Using Math-Chess-Puzzles Connection
Contents include both traditional and Math-Chess-Puzzles combined methods. Extra strength
©2008 − 2018 Frank Ho, Amanda Ho   All rights reserved. www.homathchess.com

Student Name _____ Date _____

Mixed counting 1, 2, 3 forward d + d (Start with a bigger number and count on dots).
(♔ = 0, ♙ = 1, ♗ = 3, ♘ = 3, ♖ = 5, ♕ = 9)

5 + .♙	.♙ + 5	6 + .1	7 + ..2	8 + .♙
.1 + 8	7 + ...♘	6 + ...3	7 + ..2	5 + ..2
6 + ..2	...3 + 6	.1 + 7	7 + ..2	.1 + 8
8 + .1	5 + ..2	6 + ..2	..2 + 7	7 + .♙
.♙ + 8	7 + ..2	6 + ..2	.♙ + 6	8 + ..2

Mom! I Learn Addition Using Math-Chess-Puzzles Connection

Contents include both traditional and Math-Chess-Puzzles combined methods. Extra strength

©2008 – 2018 Frank Ho, Amanda Ho   All rights reserved. www.homathchess.com

Student Name _____ Date _____

Mixed counting 1, 2, 3 forward d + d (Start with bigger number and count on dots).
(♔ = 0, ♙ = 1, ♗ = 3, ♘ = 3, ♖ = 5, ♕ = 9)

6	.1	6	7	8
+ .1	+ 7	+ .1	+ ..2	+ .♙
.♙	7	6	7	6
+ 8	+ ..2	+ ...♗	+ ..2	+ ...♘
6	...♘	...3	7	..2
+ ..2	+ 6	+ 7	+ ..2	+ 8
7	6	6	...♗	7
+ ..2	+ ..2	+ ...3	+ 7	+ .1
..2	2	7	...♘	8
+ 8	+ ..7	+ ..2	+ 6	+ .♙

37

Ho Math Chess 何数棋谜   妈！我会棋谜式加法啦！
Mom! I Learn Addition Using Math-Chess-Puzzles Connection
Contents include both traditional and Math-Chess-Puzzles combined methods. Extra strength
©2008 − 2018 Frank Ho, Amanda Ho   All rights reserved. www.homathchess.com

Student Name _____ Date _____

Mixed counting 1, 2, 3 forward and backward of d + d or d − d
(♔ = 0, ♙ = 1, ♗ = 3, ♘ = 3, ♖ = 5, ♕ = 9)

5 + ♙	2 + 5	6 − 1	7 + 2	8 + 2
1 + 8	7 + 2	6 + 2	9 − 2	8 − 2
6 + 2	3 + 6	1 + 9	7 − 2	♙ + 8
8 + ♙	5 − 2	6 + 2	2 + 7	7 + 2
1 + 8	8 − 2	6 + 2	♙ + 6	8 − 1

Ho Math Chess 何数棋谜   妈！我会棋谜式加法啦！
Mom! I Learn Addition Using Math-Chess-Puzzles Connection
Contents include both traditional and Math-Chess-Puzzles combined methods. Extra strength
©2008 — 2018 Frank Ho, Amanda Ho   All rights reserved. www.homathchess.com

Student Name _____ Date _____

Mixed counting 1, 2, and 3 forward and backwards of d + d or d − d
(♔ = 0, ♙ = 1, ♗ = 3, ♘ = 3, ♖ = 5, ♕ = 9)

6		2	♕	7	8
+ 1		+ 5	− 2	+ 2	+ 1
♙		6	6	7	♕
+ 8		+ 2	+ 3	− 2	− 2
6		3	♙	7	1
+ 2		+ 6	+ 7	− ♙	+ 8
8		♖	6	2	6
+ ♙		− 2	+ 2	+ 7	+ 1
♙		8	6	1	7
+ 8		− 2	+ 2	+ 6	− ♙

Mixed counting 1, 2, and 3 forward and backwards of d + d or d - d
(♔ = 0, ♙ = 1, ♗ = 3, ♘ = 3, ♖ = 5, ♕ = 9)

3	♗	8	7	8
+ 1	+ 5	- 2	+ ♙	+ 1
2	6	♗	8	♕
+ 7	+ ♙	+ ♗	- 2	- 1
6	2	2	7	1
+ ♗	+ 6	+ 7	- 2	+ 8
8	5	6	♙	6
+ ♙	- 3	+ ♗	+ 8	+ 3
♙	8	6	♗	7
+ 7	- ♗	+ 2	+ 6	- 2

Mom! I Learn Addition Using Math-Chess-Puzzles Connection

Mixed counting 1, 2, 3 forward and backwards of d + d or d - d
(♔ = 0, ♙ = 1, ♗ = 3, ♘ = 3, ♖ = 5, ♕ = 9)

6 + ♗	3 + 5	9 − 2	7 + 2	8 + 1
♙ + 8	♗ + 6	6 + 3	7 − 2	9 − 2
6 + 2	♘ + 6	♗ + 5	7 − 2	♙ + 8
4 + 3	6 − ♗	5 + 3	2 + 7	6 + ♘
♘ + 5	8 − 3	6 + 2	3 + 6	7 − ♗

Ho Math Chess 何数棋谜   妈！我会棋谜式加法啦！
Mom! I Learn Addition Using Math-Chess-Puzzles Connection
Contents include both traditional and Math-Chess-Puzzles combined methods. Extra strength
©2008 – 2018 Frank Ho, Amanda Ho   All rights reserved. www.homathchess.com

Student Name _____   Date _____

Mixed counting 1, 2, 3 forward and backwards of d + d or d - d
(♔ = 0, ♙ = 1, ♗ = 3, ♘ = 3, ♖ = 5, ♕ = 9)

8	3	8	6	7
+ ♙	+ ♖	− 2	+ 2	+ 1
1	7	♖	♕	8
+ 6	+ 2	+ 3	− 2	− 2
7	3	♙	6	♙
+ 2	+ 6	+ 8	− ♙	+ 7
♖	7	♖	2	6
+ 4	− 2	+ 2	+ 4	+ 1
♙	♕	6	2	8
+ 8	− 2	+ 2	+ 7	− ♙

Mixed counting 1, 2, 3 forward and backwards of d + d or d - d
(♔ = 0, ♙ = 1, ♗ = 3, ♘ = 3, ♖ = 5, ♕ = 9)

7	2	8	1	6
+ 2	+ 7	+ ♙	+ 8	+ ♘
3	♖	♘	♖	4
+ 6	+ ♘	+ 5	+ 4	+ 5
5	2	♙	♖	1
+ 2	+ ♖	+ 5	+ 1	+ 4
4	4	4	♙	4
+ ♙	+ 2	+ 2	+ 4	+ 3
♘	3	2	1	♘
+ 3	+ 2	+ ♘	+ 3	+ 1

Ho Math Chess 何数棋谜 妈！我会棋谜式加法啦！
Mom! I Learn Addition Using Math-Chess-Puzzles Connection
Contents include both traditional and Math-Chess-Puzzles combined methods. Extra strength
©2008 − 2018 Frank Ho, Amanda Ho   All rights reserved. www.homathchess.com

Student Name _____   Date _____

Mixed counting 1, 2, 3 forward and backwards of d + d or d - d
(♔ = 0, ♙ = 1, ♗ = 3, ♘ = 3, ♖ = 5, ♕ = 9)

8	4	8	6	7
+ 1	+ 5	- ♗	+ 3	+ ♙
♗	7	♖	9	8
+ 6	+ 2	+ ♗	- 3	- ♗
6	♗	♙	6	2
+ ♗	+ 6	+ 8	- ♙	+ 7
7	7	♖	♗	6
+ 1	- 2	+ 2	+ 4	+ 3
♙	9	6	♙	8
+ 8	- ♗	+ 2	+ 6	- 2

Ho Math Chess 何数棋谜　妈！我会棋谜式加法啦！
Mom! I Learn Addition Using Math-Chess-Puzzles Connection
Contents include both traditional and Math-Chess-Puzzles combined methods. Extra strength
©2008 – 2018 Frank Ho, Amanda Ho　All rights reserved. www.homathchess.com

Student Name _____ Date _____

Counting forward by 2's (even numbers or odd numbers)

2	4	6	8	10	12	14	16	18
13	15	17	19	21	23	25	27	29
54	56	58	60	62	64	66	68	70
78	80	82	84	86	88	90	92	94
33	35	37	39	41	43	45	47	49
78	80	82	84	86	88	90	92	94
62	64	66	68	70	72	74	76	78
19	21	23	25	27	29	31	33	35
27	29	31	33	35	37	39	41	43
80	82	84	86	88	90	92	94	96
26	28	30	32	34	36	38	40	42
45	47	49	51	53	55	57	59	61
53	55	57	59	61	63	65	67	69
14	16	18	20	22	24	26	28	30
24	26	28	30	32	34	36	38	40
63	65	67	69	71	73	75	77	79

Ho Math Chess 何数棋谜　妈！我会棋谜式加法啦！
Mom! I Learn Addition Using Math-Chess-Puzzles Connection

Contents include both traditional and Math-Chess-Puzzles combined methods. Extra strength

©2008 – 2018 Frank Ho, Amanda Ho   All rights reserved. www.homathchess.com

Student Name _____ Date _____

Counting backwards by 2's (even numbers or odd numbers)

20	18	16	14	12	10	8	6	4
23	21	19	17	15	13	11	9	7
61	59	57	55	53	51	49	47	45
44	42	40	38	36	34	32	30	28
32	30	28	26	24	22	20	18	16
56	54	52	50	48	46	44	42	40
77	75	73	71	69	67	65	63	61
99	97	95	93	91	89	87	85	83
78	76	74	72	70	68	66	64	62
16	14	12	10	8	6	4	2	0
29	27	25	23	21	19	17	15	13
54	52	50	48	46	44	42	40	38
25	23	21	19	17	15	13	11	9
81	79	77	75	73	71	69	67	65
48	46	44	42	40	38	36	34	32
41	39	37	35	33	31	29	27	25

Ho Math Chess 何数棋谜   妈！我会棋谜式加法啦！
Mom! I Learn Addition Using Math-Chess-Puzzles Connection
Contents include both traditional and Math-Chess-Puzzles combined methods. Extra strength
©2008 − 2018 Frank Ho, Amanda Ho   All rights reserved. www.homathchess.com

Student Name _____   Date _____

Counting forward by 5's (for working on clock problems)

5	10	15	20	25	30	35	40	45
15	20	25	30	35	40	45	50	55
20	25	30	35	40	45	50	55	60
41	46	51	56	61	66	71	76	81
35	40	45	50	55	60	65	70	75
60	65	70	75	80	85	90	95	100
32	37	42	47	52	57	62	67	72
47	52	57	62	67	72	77	82	87
52	57	62	67	72	77	82	87	92
15	20	25	30	35	40	45	50	55
35	40	45	50	55	60	65	70	75
45	50	55	60	65	70	75	80	85
27	32	37	42	47	52	57	62	67
21	26	31	36	41	46	51	56	61
48	53	58	63	68	73	78	83	88
43	48	53	58	63	68	73	78	83

Ho Math Chess 何数棋谜 妈！我会棋谜式加法啦！
Mom! I Learn Addition Using Math-Chess-Puzzles Connection
Contents include both traditional and Math-Chess-Puzzles combined methods. Extra strength
©2008 − 2018 Frank Ho, Amanda Ho   All rights reserved. www.homathchess.com

Student Name _____ Date _____

Counting backwards by 5's (for working on clock problems)

55	50		45	40	35	30	25	20	15
70	65		60	55	50	45	40	35	30
46	41		36	31	26	21	16	11	6
94	89		84	79	74	69	64	59	54
87	82		77	72	67	62	57	52	47
40	35		30	25	20	15	10	5	0
62	57		52	47	42	37	32	27	22
57	52		47	42	37	32	27	22	17
76	71		66	61	56	51	46	41	36
75	70	65	60	55	50	45	40	35	
85	80	75	70	65	60	55	50	45	
72	67		62	57	52	47	42	37	32
40	35		30	25	20	15	10	5	0
101	96	91	86	81	76	71	66	61	
91	86		81	76	71	66	61	56	51
52	47		42	37	32	27	22	17	12

Ho Math Chess 何数棋谜 妈！我会棋谜式加法啦！
Mom! I Learn Addition Using Math-Chess-Puzzles Connection
Contents include both traditional and Math-Chess-Puzzles combined methods. Extra strength
©2008 – 2018 Frank Ho, Amanda Ho   All rights reserved. www.homathchess.com

Student Name _____ Date _____

Addition strategy: reverse calculation of sums 倒算和

Students learn to add to a number by circling the numbers, then learn to partition a number. 先凑後拆
These strategies also help students learn our Frankho ChessDoku 3 by 3 puzzles.

Adding to 3. 凑 3
Add digits of each number then write each sum beside that number.
Circle the following numbers when the sum of their digits is 3.
Think 1 + 2 = 2 + 1 = 3
If the students have difficulties in doing adding 4, then just let them skip the questions such as 5 + 4. The goal is to master adding 1, 2, and 3 first.

21	32	43	54
31	42	53	54
41	52	21	21
21	62	63	
51	72		
61			
21			
71			
81			

Answer

㉑	32	43	54
31	42	53	54
41	52	㉑	㉑
21	62	63	
51	72		
61			
㉑			
71			
81			

12	23	34	45
13	24	35	45
12	21	21	21
14	25	35	
15	26		
12	27		
16	21		
17			
18			

7 answers

Ho Math Chess 何数棋谜　妈！我会棋谜式加法啦！
Mom! I Learn Addition Using Math-Chess-Puzzles Connection
Contents include both traditional and Math-Chess-Puzzles combined methods. Extra strength
©2008 – 2018 Frank Ho, Amanda Ho　　All rights reserved. www.homathchess.com

Student Name _____ Date _____

Adding to 4 凑 4

Add digits of each number then write each sum beside that number.
Circle the following numbers when the sum of their digits is 4.

Think 2 + 2 = 4, 1 + 3 = 3 + 1 = 4

If the students have difficulties in doing adding 4, then just let them skip the questions such as 5 + 4. The goal is to master adding 1, 2, and 3 first.

21		32	43	54
31 =		42	53	31 =
41		31 =	31 =	
51		52	63	
31 =		62	22 =	
61		31		
71		72		
31=				
81				

12	23	34	22 =	
13 =	22 =	13 =	45	
14	24	35	13 =	
15	13 =	35		
13 =	25	13 =		
16	26			
17	13 =			
13 =	27			
18	13			

18 answers

50

Ho Math Chess 何数棋谜　妈！我会棋谜式加法啦！
Mom! I Learn Addition Using Math-Chess-Puzzles Connection
Contents include both traditional and Math-Chess-Puzzles combined methods. Extra strength
©2008 − 2018 Frank Ho, Amanda Ho　All rights reserved. www.homathchess.com

Student Name _____ Date _____

Adding to 5 凑 5

Add digits of each number then write each sum beside that number.
Circle the following numbers when the sum of their digits is 5.

Think 2 + 3 = 3 + 2 = 5, 1 + 4 = 4 + 1 = 5

If the students have difficulties in doing adding 4, then just let them skip the questions such as 5 + 4. The goal is to master adding 1, 2, and 3 first.

21	32=	43	54
31	42	53	42
41=	52	42	54
51	31	63	14=
41=	62		
61	72		
71	32=		
41=			
81			

12	23=	34	45
14=	24	35	41=
13	14=	35	23=
14=	25	14=	
15	26	32	
16	27		
17			
14=			
18			

14 answers

51

Ho Math Chess 何数棋谜　妈! 我会棋谜式加法啦!
Mom! I Learn Addition Using Math-Chess-Puzzles Connection

Contents include both traditional and Math-Chess-Puzzles combined methods. Extra strength
©2008 — 2018 Frank Ho, Amanda Ho　All rights reserved. www.homathchess.com

Student Name _____ Date _____

Adding to 6 凑 6

Add digits of each number then write each sum beside that number.
Circle the following numbers when the sum of their digits is 6.

Think 1 + 5 = 5 + 1 = 6, 2 + 4 = 4 + 2 = 6, and double 3 = 6

If the students have difficulties in doing adding 4, then just let them skip the questions such as 5 + 4. The goal is to master adding 1, 2, and 3 first. Some students may have problems to work backwards, in this case just skip to doubles or adding 10 strategies but ask students to try hard before letting them skip. Work with these students on the problem by problem individually.

21	32	43	54
31	42	53	42
51	52	42	
41	62	63	
51	42		
61	72		
71			
51			
81			

12	23	34	45
13	24=	35	42=
15=	25	35	
14	26		
15=	24=		
16	27		
17	24		
18			

11 answers

Ho Math Chess 何数棋谜　妈！我会棋谜式加法啦！
Mom! I Learn Addition Using Math-Chess-Puzzles Connection

Contents include both traditional and Math-Chess-Puzzles combined methods. Extra strength

©2008 – 2018 Frank Ho, Amanda Ho　　All rights reserved. www.homathchess.com

Student Name _____ Date _____

Adding to 7 凑 7

Add digits of each number then write each sum beside that number.
Circle the following numbers when the sum of their digits is **7**.

Think 1 + 6 = 6 + 1, 2 + 5 = 5 + 2, 3 + 4 = 4 + 3, and 1 + 6 = 6 + 1

21	32	43=	54
31	52=	53	54
61	42	63	52=
41	52=	43=	
51	62		
61=	72		
71			
81			

12	23	34=	45
13	24	35	46
16=	25=	35	43=
14	26	34=	
15	25=		
16=	27		
17	25=		
16=			
18			

Adding to 8 凑 8

Add digits of each number then write each sum beside that number.
Circle the following numbers when the sum of their digits is **8**.

Think 1 + 7 = 7 + 1, 2 + 6 = 6 + 2, 3 + 5 = 5 + 3, double 4 = 8

21	32	43	54
31	42	53=	43
41	62=	63	54
71=	52	43	
51	62=		
61	72		
71=			
81			

12	23	34	45
13	24	35=	34
16	25	35	45
14	26=	34	
15	27		
16	25		
17=			
18			

10 answers

Mom! I Learn Addition Using Math-Chess-Puzzles Connection

Contents include both traditional and Math-Chess-Puzzles combined methods. Extra strength

©2008 – 2018 Frank Ho, Amanda Ho   All rights reserved. www.homathchess.com

Student Name_____ Date_____

Adding to 9 凑 9

Add digits of each number then write each sum beside that number.
Circle the following numbers when the sum of their digits is **9**.

Think 1 + 8 = 8 + 1, 2 + 7 = 7 + 2, 3 + 6 = 6, 4 + 5 = 5 + 4 = double 4 + 1

21	32	43	54
31	42	53	53
81=	62	53	54=
41	52	63=	
51	62	53	
8	72=		
61			
71			
81=			

12	23	34	63=
13	24	35	45=
14	26	35	35
17	45		
15	26		
36	27=		
17			
18=			

8 answers

Ho Math Chess 何数棋谜   妈！我会棋谜式加法啦！

Mom! I Learn Addition Using Math-Chess-Puzzles Connection

Contents include both traditional and Math-Chess-Puzzles combined methods. Extra strength

©2008 − 2018 Frank Ho, Amanda Ho   All rights reserved. www.homathchess.com

Student Name _____ Date _____

Partitioning numbers

Partition 3 拆 3

I have **three** apples and my mom wants me to put them in 2 baskets. How many ways can I put my apples into the two baskets? Write the number of apples in each basket.

Basket A	Basket B

03 30 12 21

Mom! I Learn Addition Using Math-Chess-Puzzles Connection

Contents include both traditional and Math-Chess-Puzzles combined methods. Extra strength
©2008 – 2018 Frank Ho, Amanda Ho   All rights reserved. www.homathchess.com

Student Name _____ Date _____

Partition 4 拆 4

I have **four** apples and my mom wants me to put them in 2 baskets. How many ways can I put my apples into the two baskets? Write the number of apples in each basket.

Basket A	Basket B
🧺	🧺
🧺	🧺
🧺	🧺
🧺	🧺
🧺	🧺

04 40 13 31 22

Partition 5 拆 5

I have **five** apples and my mom wants me to put them in 2 baskets. How many ways can I put my apples into the two baskets? Write the number of apples in each basket.

Basket A	Basket B

05 50 14 41 23 32

Partition 6 拆 6

I have **six** apples and my mom wants me to put them in 2 baskets. How many ways can I put my apples into the two baskets? Write the number of apples in each basket.

Basket A	Basket B

06 60 15 51 24 42 33

Ho Math Chess 何数棋谜　妈！我会棋谜式加法啦！
Mom! I Learn Addition Using Math-Chess-Puzzles Connection
Contents include both traditional and Math-Chess-Puzzles combined methods. Extra strength
©2008 – 2018 Frank Ho, Amanda Ho　All rights reserved. www.homathchess.com

Student Name_____ Date_____

Partition 7 拆 7

I have **seven** apples and my mom wants me to put them in 2 baskets. How many ways can I put my apples into the two baskets? Write the number of apples in each basket.

Basket A	Basket B	Basket A	Basket B

07 70 16 61 25 52 3 43

60

Partition 8 拆 8

I have **eight** apples and my mom wants me to put them in 2 baskets. How many ways can I put my apples into the two baskets? Write the number of apples in each basket.

08 80 17 71 26 62 35 53 44

Ho Math Chess 何数棋谜　妈！我会棋谜式加法啦！
Mom! I Learn Addition Using Math-Chess-Puzzles Connection
Contents include both traditional and Math-Chess-Puzzles combined methods. Extra strength
©2008 – 2018 Frank Ho, Amanda Ho   All rights reserved. www.homathchess.com

Student Name _____ Date _____

Partition 9 拆 9

I have **nine** apples and my mom wants me to put them in 2 baskets. How many ways can I put my apples into the two baskets? Write the number of apples in each basket.

Basket A	Basket B	Basket A	Basket B

09 90 18 81 27 72 36 63 45 54

Partition 10 拆 10

I have **ten** apples and my mom wants me to put them in 2 baskets. How many ways can I put my apples into the two baskets? Write the number of apples in each basket.

Basket A	Basket B	Basket A	Basket B

0, 10; 10, 0; 1, 9; 9, 1; 2, 8; 8, 2; 3, 7; 7, 3; 4, 6; 6, 4; 5, 5

Number Puzzles

Replace each ? with a number.

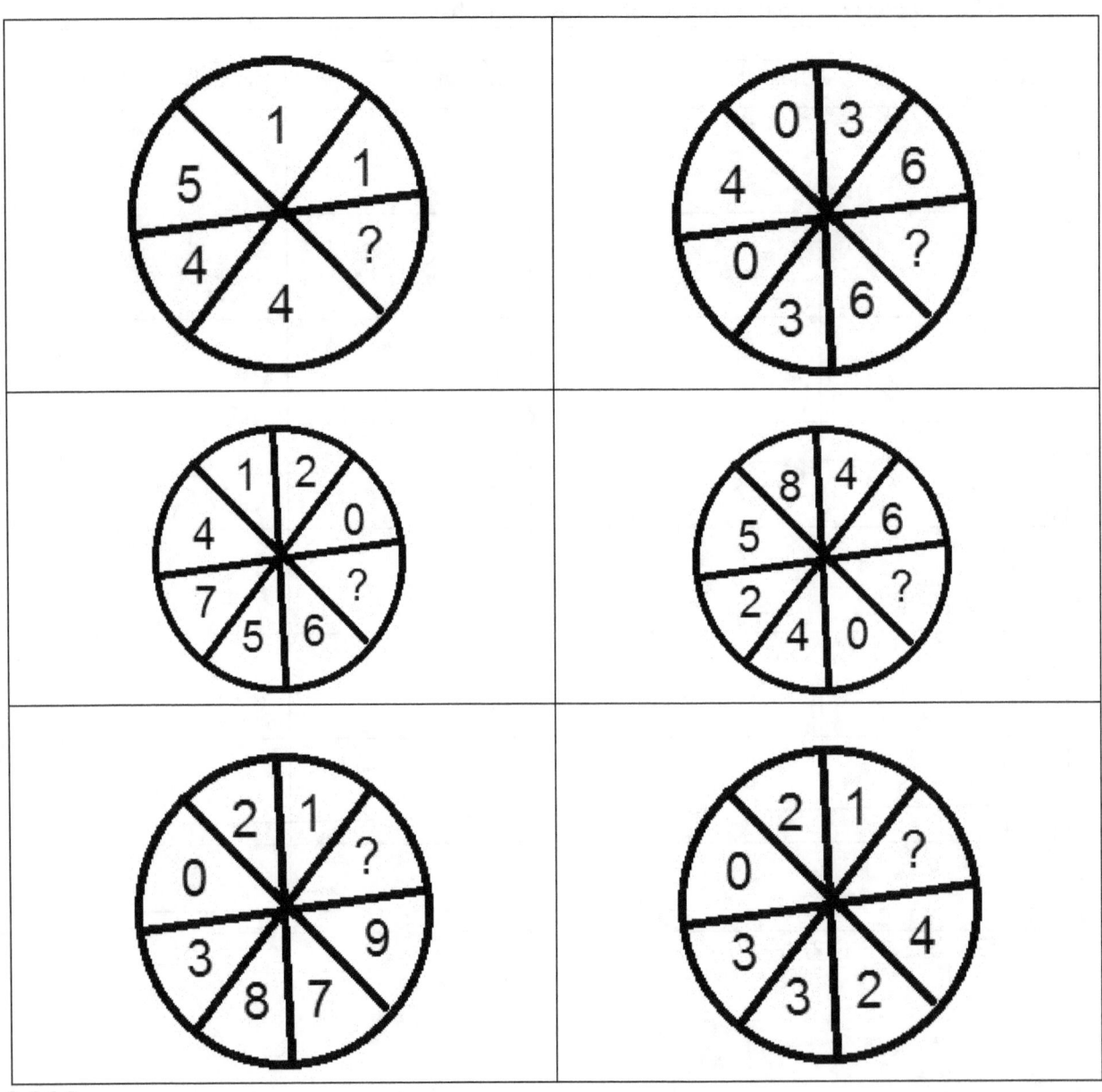

0 3 6; 2 3 5

Ho Math Chess 何数棋谜 妈！我会棋谜式加法啦！
Mom! I Learn Addition Using Math-Chess-Puzzles Connection
Contents include both traditional and Math-Chess-Puzzles combined methods. Extra strength
©2008 – 2018 Frank Ho, Amanda Ho   All rights reserved. www.homathchess.com

Student Name _____ Date _____

Addition strategy: adding 1, 2, 3 加 1, 2, 3

Adding d + 1 (the larger number in counting)

```
  2 5 3 2 5 4 3 3 4 1 2 3 6 2 7 8 3 4 1 7 3
+ 1 1 1 1 1 1 1 1 1 1 1 1 1 1 1 1 1 1 1 1 1
  3 6 4 3 6 5 4 4 5 2 3 4 7 3 8 9 4 5 2 8 4
```

```
  3 6 4 5 7 3 2 1 3 4 5 3 5 1 3 6 3 7 3 8 2
+ 1 1 1 1 1 1 1 1 1 1 1 1 1 1 1 1 1 1 1 1 1
  4 7 5 6 8 4 3 2 4 5 6 4 6 2 4 7 4 8 4 9 3
```

```
  5 2 7 1 8 4 3 6 5 4 7 5 2 7 2 4 8 3 6 3 1
+ 1 1 1 1 1 1 1 1 1 1 1 1 1 1 1 1 1 1 1 1 1
  6 3 8 2 9 5 4 7 6 5 8 6 3 8 3 5 9 4 7 4 2
```

```
  6 4 5 8 2 7 3 5 7 4 6 3 8 5 3 2 4 8 5 3 7
+ 1 1 1 1 1 1 1 1 1 1 1 1 1 1 1 1 1 1 1 1 1
  7 5 6 9 3 8 4 6 8 5 7 4 9 6 4 3 5 9 6 4 9
```

```
  5 7 3 7 4 8 5 3 2 5 7 1 4 6 3 8 4 2 4 8 6
+ 1 1 1 1 1 1 1 1 1 1 1 1 1 1 1 1 1 1 1 1 1
  6 8 4 8 5 9 6 4 3 6 8 2 5 7 4 9 5 3 5 9 7
```

65

Adding d + 1 and 1 + d

```
  0 8 0 8 0 1 0 1 0 1 0 1 0 1 0 1 0 1 0 1
+ 0 1 0 1 0 7 0 5 0 6 0 3 0 4 0 1 0 2 0 6 0 7
  0 9 0 9 0 8 0 6 0 7 0 4 0 5 0 2 0 3 0 7 0 8
```

```
  0 1 0 1 0 1 0 1 0 1 0 1 0 1 0 1 0 1 0 1
+ 0 3 0 7 0 4 0 6 0 7 0 8 0 5 0 2 0 3 0 7 0 6
  0 4 0 8 0 5 0 7 0 8 0 9 0 6 0 3 0 4 0 8 0 7
```

```
  0 3 0 1 0 4 0 6 0 7 0 8 0 1 0 3 0 4 0 6 0 6
+ 0 1 0 8 0 1 0 1 0 1 0 1 0 1 0 1 0 1 0 1 0 1
  0 4 0 9 0 5 0 7 0 8 0 9 0 2 0 4 0 5 0 7 0 7
```

```
  0 7 0 8 0 1 0 1 0 1 0 1 0 1 0 1 0 1 0 1
+ 0 1 0 1 0 7 0 5 0 6 0 3 0 4 0 1 0 2 0 6 0 7
  0 8 0 9 0 8 0 6 0 7 0 4 0 5 0 2 0 3 0 7 0 8
```

```
  0 1 0 1 0 1 0 1 0 1 0 1 0 1 0 1 0 1 0 1
+ 0 3 0 7 0 4 0 6 0 7 0 8 0 5 0 2 0 3 0 7 0 6
  0 4 0 8 0 5 0 7 0 8 0 9 0 6 0 3 0 4 0 8 0 7
```

2. Add d + 2 and 2 + d.

```
  0 2 0 2 0 2 0 2 0 2 0 2 0 2 0 2 0 2 0 2 0 2 0 2
+ 0 3 0 5 0 6 0 7 0 1 0 4 0 5 0 3 0 2 0 5 0 7
  0 5 0 7 0 8 0 9 0 3 0 6 0 7 0 5 0 4 0 7 0 9
```

```
  0 2 0 2 0 2 0 2 0 2 0 2 0 2 0 2 0 2 0 2 0 2 0 2
+ 0 6 0 4 0 6 0 7 0 6 0 7 0 2 0 3 0 7 0 6 0 3
  0 8 0 6 0 8 0 9 0 8 0 9 0 4 0 5 0 9 0 8 0 5
```

```
  0 3 0 4 0 6 0 7 0 5 0 1 0 3 0 4 0 6 0 6 0 5
+ 0 2 0 2 0 2 0 2 0 2 0 2 0 2 0 2 0 2 0 2 0 2 0 2
  0 5 0 6 0 8 0 9 0 7 0 3 0 5 0 6 0 8 0 8 0 7
```

```
  0 2 0 2 0 2 0 2 0 2 0 2 0 2 0 2 0 2 0 2 0 2 0 2
+ 0 7 0 5 0 7 0 5 0 4 0 6 0 4 0 3 0 7 0 6 0 5
  0 9 0 7 0 9 0 7 0 6 0 8 0 6 0 5 0 9 0 8 0 7
```

```
  0 2 0 3 0 7 0 6 0 5 0 4 0 6 0 7 0 6 0 3 0 4
+ 0 2 0 2 0 2 0 2 0 2 0 2 0 2 0 2 0 2 0 2 0 2 0 2
  0 4 0 5 0 9 0 8 0 7 0 6 0 8 0 9 0 8 0 5 0 6
```

67

Adding d + 3 and 3 + d.

```
  0 3 0 3 0 3 0 3 0 3 0 3 0 3 0 3 0 3 0 3
+ 0 3 0 6 0 1 0 4 0 5 0 3 0 2 0 5 0 4 0 6 0 5
  0 6 0 9 0 4 0 7 0 8 0 6 0 5 0 8 0 7 0 9 0 8

  0 3 0 3 0 3 0 3 0 3 0 3 0 3 0 3 0 3 0 3
+ 0 6 0 5 0 4 0 6 0 5 0 5 0 6 0 2 0 3 0 5 0 6
  0 9 0 8 0 7 0 9 0 8 0 8 0 9 0 5 0 6 0 8 0 9

  0 2 0 6 0 6 0 5 0 1 0 3 0 4 0 6 0 6 0 5 0 3
+ 0 3 0 3 0 3 0 3 0 3 0 3 0 3 0 3 0 3 0 3
  0 5 0 9 0 9 0 8 0 4 0 6 0 7 0 9 0 9 0 8 0 6

  0 4 0 6 0 3 0 5 0 2 0 3 0 6 0 4 0 6 0 1 0 1
+ 0 3 0 3 0 3 0 3 0 3 0 3 0 3 0 3 0 3 0 3
  0 7 0 9 0 6 0 8 0 5 0 6 0 9 0 7 0 9 0 4 0 4

  0 3 0 6 0 2 0 5 0 5 0 4 0 2 0 5 0 6 0 3 0 5
+ 0 3 0 3 0 3 0 3 0 3 0 3 0 3 0 3 0 3 0 3
  0 6 0 9 0 5 0 8 0 8 0 7 0 5 0 8 0 9 0 6 0 8

  0 6 0 3 0 5 0 4 0 6 0 3 0 5 0 3 0 3 0 1 0 3
+ 0 3 0 6 0 3 0 3 0 3 0 4 0 3 0 6 0 5 0 3 0 5
  0 9 0 9 0 8 0 7 0 0 0 7 0 8 0 9 0 8 0 4 0 8
```

Ho Math Chess 何数棋谜　妈！我会棋谜式加法啦！
Mom! I Learn Addition Using Math-Chess-Puzzles Connection

Contents include both traditional and Math-Chess-Puzzles combined methods. Extra strength

©2008 – 2018 Frank Ho, Amanda Ho　All rights reserved. www.homathchess.com

d + d with missing operator or number
Fill in each __ with a number or + operator.

6 + 3 = ___ 9	3 + 6 = __ 9
5 + 3 = ___ 8	3 + 5 = ___ 8
4 ___ + 3 = 7	3 __ + 4 = 7
3 __ + 3 = 6	3 + __ 3 = 6
2 + __ 3 = 5	3 + __ 2 = 5
1 + __ 3 = 4	3 + __ 1 = 4
2 __ + 3 = 3 + 2 = ___	3 __ + 2 = 2 + 3 = ___
3 __ + 3 = 2 + 4 = ___	3 + __ 3 = 1 + 5 = ___
4 __ + 3 = 3 + 4 = ___	3 __ + 4 = 2 + 5 = ___
5 __ + 3 = 2 + 6 = ___	3 __ + 5 = 1 + 7 = ___
6 + __ 3 = 1 + 8 = ___	3 + __ 6 = 6 + 3 = ___
5 + __ 3 = 1 + 7 = ___	3 + __ 5 = 2 + 6 = ___
four __ + three = 3 + 4 = ___	three ____ + four = 5 + 2 = ___
five + ____ three = 6 + 3 = ___	three + ____ five = 6 + 2 = ___
four __ + five = 8 + 1 = ___	five __ + four = 9 = 7 + 2 = ___

Ho Math Chess 何数棋谜　妈！我会棋谜式加法啦！
Mom! I Learn Addition Using Math-Chess-Puzzles Connection
Contents include both traditional and Math-Chess-Puzzles combined methods. Extra strength
©2008 – 2018 Frank Ho, Amanda Ho　　All rights reserved. www.homathchess.com

Student Name _____ Date _____

7 days mastering computation skills

After finishing the previous worksheets, the students reach a critical point. The tutor can start to teach or review the following strategies in about 7 days of homework (finish in one week at home or in class.).

The following 7 days of work is very important because it serves as a benchmark for students to see if they could master all addition strategies in a very short time. The tutor must be careful in monitoring a student's progress if after finishing this 7-day of work but still is not proficient in addition strategies. It means that it may take a bit longer for the student than an average student to master these skills, so the tutor must be more patient.

When working on this 7-day of worksheets, the tutor or parent needs to check the student' work after each problem, not to wait until the entire worksheet is finished . Watch how the student works out the solution step by step to make sure that the student understands the reasoning behind each step.

The 7-day strategy short course is only a guide, so if a student could not finish in 7 days, be prepared to give students more time to do it. It is important to test students after they finish at the end of each day to make sure they understand.

If you do not see the students while they are doing it, then you need to teach each day's work before they work on them and make them do a few questions of each day's work, so they do not get stuck while they try to finish them alone.

Ho Math Chess 何数棋谜   妈！我会棋谜式加法啦！
Mom! I Learn Addition Using Math-Chess-Puzzles Connection
Contents include both traditional and Math-Chess-Puzzles combined methods. Extra strength
©2008 – 2018 Frank Ho, Amanda Ho   All rights reserved. www.homathchess.com

Student Name _____ Date _____

Day 1 of doubling 1, 2, or 3

Day 1	1 + 1 = ☐
	2 + 1 = ☐ 2 + 2 = ☐ = double 2 = ☐
	3 + 1 = ☐ 3 + 2 = 2 + ☐ = ☐ 3 + 3 = ☐ = double 3 = 3 + 3 = ☐

Day 2 of 2 adding 1, 2, or 3, adding to 10, doubling 2

2 + 2 = 4 = double 2 = 4
d + d with missing operator or number
Fill in each __ with a number or + operator.

2 + 2 = __ 4	2 + 2 = __ 4
2 + 2 = __ 4	2 + 2 = __ 4
2 __ + 2 = 4	2 __ + 2 = 4
2 __ + 2 = 4	2 __ + 2 = 4
2 + __ 2 = 4	2 + __ 2 = 4
2 + __ 2 = 4	2 + __ 2 = 4
2 __ + 2 = 4	2 __ + 2 = 4
2 __ +2 = 4	2 __ + 2 = 4
2 __ + 2 = 4	2 __ + 2 = 4
2 __ + 2 = 4	2 __ + 2 = 4

Ho Math Chess 何数棋谜 妈！我会棋谜式加法啦！
Mom! I Learn Addition Using Math-Chess-Puzzles Connection

Contents include both traditional and Math-Chess-Puzzles combined methods. Extra strength

©2008 – 2018 Frank Ho, Amanda Ho   All rights reserved. www.homathchess.com

Student Name _____ Date _____

2 + __ 2 = 4	2 + __ 2 = 4
2 + __ 2 = 4	2 + __ 2 = 4
two __ + two = 4	two __ + two = 4
two __ + two = 4	two __ + two = 4
two __ + two = 4	two __ + two = 4
two + __ two = 4	two + __ two = 4
two + __ __ two = 4	Two + __ two = 4
two + __ __ two = 4	two + __ two = 4

Day 3 of 3 adding 1, 2, or 3, adding to 10, doubling 3

3 + 3 = 6 = double 3 = 6

d + d with missing operator or number
Fill in each __ with a number or + operator.

3 + 3 = __ 6	3 + 3 = __ 6
3 + 3 = __ 6	3 + 3 = __ 6
3 __ + 3 = 6	3 __ + 3 = 6
3 __ + 3 = 6	3 __ + 3 = 6
3 + __ 3 = 6	3 + __ 3 = 6
3 + __ 3 = 6	3 + __ 3 = 6
3 __ + 3 = 6	3 __ + 3 = 6
3 __ +3 = 6	3 __ +3 = 6

Ho Math Chess 何数棋谜   妈！我会棋谜式加法啦！
Mom! I Learn Addition Using Math-Chess-Puzzles Connection

Contents include both traditional and Math-Chess-Puzzles combined methods. Extra strength

©2008 — 2018 Frank Ho, Amanda Ho   All rights reserved. www.homathchess.com

Student Name _____   Date _____

3 __ + 3 = 6	3 __ + 3 = 6
3 __ + 3 = 6	3 __ + 3 = 6
3 + __ 3 = 6	3 + __ 3 = 6
3 + __ 3 = 6	3 + __ 3 = 6

three __ + three = 6	three __ + three = 6
three + three = 6	three __ + three = 6
three __ + three = 6	three __ + three = 6
three + __ three = 6	three + __ three = 6
three + __ __ three = 6	three + __ three = 6
three + __ __ three = 6	three + __ three = 6

Ho Math Chess 何数棋谜　妈！我会棋谜式加法啦！
Mom! I Learn Addition Using Math-Chess-Puzzles Connection

Contents include both traditional and Math-Chess-Puzzles combined methods. Extra strength
©2008 – 2018 Frank Ho, Amanda Ho   All rights reserved. www.homathchess.com

Student Name_____ Date_____

Day 4 of 4 adding 1, 2, or 3, adding to 10, doubling 4

4 + 4 = 8 = double 4 = 8

d + d with missing operator or number
Fill in each __ with a number or + operator.

4 + 4 = 8	4 + 4 = __ 8
4 + 4 = __ 8	4 + 4 = __ 8
4 __ + 4 = 8	4 __ + 4 = 8
4 __ + 4 = 8	4 __ + 4 = 6
4 + __ 4 = 8	4 + __ 4 = 8
4 + __ 4 = 8	4 + __ 4 = 8
4 __ + 4 = 8	4 __ + 4 = 8
4 __ +4 = 8	4 __ +4 = 8
4 __ + 4 = 8	4 __ + 4 = 8
4 __ + 4 = 8	4 __ + 4 = 8
4 + __ 4 = 8	4 + __ 4 = 8
4 + __ 4 = 8	4 + __ 4 = 8
four __ + four = 8	four __ + four = 8
four __ + four = 8	four + __ four = 8
four __ + four = 8	four __ + four = 8

Ho Math Chess 何数棋谜　妈！我会棋谜式加法啦！
Mom! I Learn Addition Using Math-Chess-Puzzles Connection
Contents include both traditional and Math-Chess-Puzzles combined methods. Extra strength
©2008－2018 Frank Ho, Amanda Ho　All rights reserved. www.homathchess.com

Student Name _____ Date _____

Day 5 of 5 adding 1, 2, or 3, adding to 10, doubling 5

Day 5

5 + 1 = 1 + 5 = ☐
5 + 2 = 2 + 5 = ☐
5 + 3 = 3 + 5 = ☐ = double ☐ + 2 = ☐
5 + 4 = 1 + ☐ + 4 = double 4 + ☐ = ☐
5 + 5 = double 5 = ☐

1 + 9 = ☐
2 + 8 = ☐
3 + 7 = ☐
4 + 6 = ☐
5 + 5 = ☐

Complete the following pattern. For example,

19 = 91

19 = ☐
28 = ☐
__ = ☐
__ = ☐
__ = ☐

9 + 1 = ☐
8 + 2 = ☐
7 + 3 = ☐
6 + 4 = ☐
5 + 5 = ☐

Complete the following pattern. For example,

19 = 91

__ = ☐
82 = ☐
73 = ☐
__ = ☐
55 = ☐

5 + 5 = 10 = double 5 = 10

d + d with missing operator or number
Fill in each __ with a number or + operator.

5 + 5 = 10	4 + 4 = __ 10
5 + 5 = ___ 10	5 + 5 = ___ 10
5 ___ + 5 = 10	5 __ + 5 = 10
5 __ + 5 = 10	5 __ + 5 = 6

77

Ho Math Chess 何数棋谜 妈！我会棋谜式加法啦！
Mom! I Learn Addition Using Math-Chess-Puzzles Connection

Contents include both traditional and Math-Chess-Puzzles combined methods. Extra strength
©2008 − 2018 Frank Ho, Amanda Ho   All rights reserved. www.homathchess.com

Student Name _____ Date _____

5 + __ 5 = 10	5 + __ 5 = 10
5 + __ 5 = 10	5 + __ 5 = 10
5 __ + 5 = 10	5 __ + 5 = 10
5 __ + 5 = 10	5 __ +5 = 10
5 __ + 5 = 10	5 __ + 5 = 10
5 __ + 5 = 10	5 __ + 5 = 10
5 + __ 5= 10	5 + __ 5 = 10
5 + __ 5 = 10	5 + __ 5 = 10
five __ + five = 10	five __ + five = 10
five __ + five = 10	five __ + five = 10
five __ + five = 10	five __ + five = 10

Mom! I Learn Addition Using Math-Chess-Puzzles Connection

Day 6 of 6 adding 1, 2, or 3, adding to 10, doubling 6

| Day 6 | 6 + 1 = 1 + 6 = ☐
 6 + 2 = 2 + 6 = ☐
 6 + 3 = 3 + 6 = ☐
 6 + 4 = 4 + 6 = ☐
 6 + 5 = 5 + 6 = double 5 + ☐ = ☐
 6 + 6 = double 6 = ☐ | Complete the following pattern.

 19, 28, 37, ___, ___

 ___, 28, 37, ___, ___

 ___, ___, 37, 46, ___

 ___, ___, ___, 46, 55

 ___, 28, ___, 46, ___

 ___, ___, 37, ___, 55

 ___, 28, ___, 46, ___

 19, ___, ___, ___, 55 |

Mom! I Learn Addition Using Math-Chess-Puzzles Connection

6 + 6 = 12 = double 6 = 12

d + d with missing operator or number

Fill in each __ with a number or + operator.

6 + 6 = __ 12	6 + 6 = __ 12
6 + 6 = __ 12	6 + 6 = __ 12
6 __ + 6 = 12	6 __ + 6 = 12
6 __ + 6 = 12	6 __ + 6 = 12
6 + __ 6 = 12	6 + __ 6 = 12
6 + __ 6 = 12	6 + __ 6 = 12
6 __ + 6 = 12	6 __ __ + 6 = 12
6 __ __ + 6 = 12	6 __ __ + 6 = 12
6 __ + 6 = 12	6 __ + 6 = 12
6 __ + 6 = 12	6 __ + 6 = 12
6 + __ __ 6 = 12	6 + __ __ 6 = 12
6 + __ __ 6 = 12	6 + __ __ 6 = 12
six __ + six = 12	six __ + six = 12
six __ + six = 12	six __ + six = 12
six + __ 6 = 12	six + __ 6 = 12
six __ __ + six = 12	six __ + six = 12
six __ + six = 12	six + __ six = 12
six + __ __ six = 12	six + __ six = 12

Mom! I Learn Addition Using Math-Chess-Puzzles Connection

d + d with missing operator or number

Fill in each __ with a number or + operator.

7 + 6 = __ 13	7 + 6 = __ 13
6 + 7 = __ 13	6 + 7 = __ 13
7 __ + 6 = 13	7 __ + 6 = 13
6 __ + 7 = 13	6 __ + 7 = 13
6 + __ 7 = 13	6 + __ 7 = 13
7 + __ 6 = 13	7 + __ 6 = 13
6 __ __ +7 = 13	6 __ __ + 7 = 13
7 __ __ + 6= 13	7 __ __ + 6 = 13
7 __ + 6 = 13	7 __+ 6 = 13
6 __ + 7 = 13	6 __+ 7 = 13
7 + __ __ 6 = 13	7 + __ __ 6 = 13
6 + __ __ 7 = 13	6 + __ __ 7 = 13
six __ + seven = 13	six __ + seven = 13
seven __ + six = 13	seven __+ six = 13
seven __ __ +6 = 13	seven __ __ + 6 = 13
six __ __ +7 = 13	six __ __ + 7 = 13
6+ __ __ seven = 13	6 + __ __ seven = 13
7+ __ __ six = 13	7 + __ __ six = 13

d + d with missing operator or number

Fill in each __ with a number or a + operator.

6 + 8 = __ 14	6 + 8 = __ 14
8 + 6 = __ 14	8 + 6 = __ 14
6 __ + 8 = 14	6 __ + 8 = 14
8 __ + 6 = 14	8 __ = 6 = 14
8 + __ 6 = 14	8 + __ 6 = 14
6 + __ 8 = 14	6 + __ 8 = 14
8 __ __ + 6 = 14	8 __ __ + 6 = 14
6 __ __ + 8 = 14	6 __ __ + 8 = 14
6 __ +8 = 14	6 __ + 8 = 14
8 __ + 6 = 14	8 __ + 6 = 14
8 + __ __ 6 = 14	8+ __ __ 6 = 14
6 + __ __ 8 = 14	8+ __ __ 8 = 14
six __ + eight = 14	six __ + eight = 14
Eight __ + six = 14	Eight __ + six = 14
six __ +8 = 14	six __ __ +8 = 14
eight __ +6= 14	eight __ __ +6= 14
6+ __ __ eight = 14	6+ __ __ eight = 14
8+ __ __ six =14	8+ __ __ six =14

Mom! I Learn Addition Using Math-Chess-Puzzles Connection

Day 6 of 7 adding 1, 2, or 3, adding to 10, doubling 7

Day 6	
7 + 1 = 1 + 7 = ☐ 7 + 2 = 2 + 7 = ☐ 7 + 3 = 3 + 7 = ☐ 7 + 4 = 4 + 7 = ☐ + 1 = ☐ 7 + 5 = 5 + 7 = double 5 + ☐ = ☐ 7 + 6 = double 6 + ☐ = ☐	Complete the following pattern. 91, 82, 73, 64, ___ ___, ___, ___, 64, 55 91, ___, 73, ___, 55 ___, 82, ___, 64, ___ ___, ___, 73, 64, 55 ___, ___, ___, 64, 55 91, ___, ___, ___, 55

7 + 7 = 14 = double 7 = 14

d + d with missing operator or number

Fill in each __ with a number or + operator.

7 + 7 = __14	7 + 7 = __14
7 + 7 = __ 14	7 + 7 = __ 14
7 __ + 7 = 14	7 __ + 7 = 14
7 __ + 7 = 14	7 __ + 7 = 14
7 + __7 = 14	7 + __ 7 = 14
7 + __ 7 = 14	7 + __ 7 = 14
7 __ __ + 7 = 14	7 __ __ + 7 = 14
7 __ __ + 7 = 14	7 __ __ + 7 = 14
7 __+ 7 = 14	7 __+ 7 = 14
7 __+ 7 = 14	7 __+ 7 = 14
7 + __ __ 6 = 14	7 + __ __ 7 = 14
6 + __ __ 7 = 14	7 + __ __ 7 = 14
seven __ + seven = 14	seven __+ seven = 14
seven __ + seven = 14	seven __+ seven = 14
seven __ __ + seven = 14	seven __ __ + seven = 14
seven __ __ + 7 = 14	seven __ __ +7 = 14
seven + __ __ seven = 14	seven + __ __ seven = 14
7 + __ __ seven = 14	7 + __ __ seven = 14

Day 7 of 8 adding 1, 2, or 3, adding to 10, doubling 8

Day 7	$8 + 1 = 1 + 8 = \square$	Complete the following pattern.
	Making 10	91, 82, 73, 64, ___
	$8 + 2 = 2 + 8 = \square$	___, ___, ___, 64, 55
	$8 + 3 = 3 + 8 = \square + 2 + 1 = \square + 1 = \square$	91, ___, 73, ___, 55
	$8 + 4 = 4 + 8 = \square + 2 + 2 = \square + 2 = \square$	___, 82, ___, 64, ___
	Doubling	___, ___, 73, 64, 55
	$8 + 5 = 5 + 8 =$ double 5 $+ \square = \square$	___, ___, ___, 64, 55
	$8 + 6 =$ double 6 $+ \square = \square$	
	$8 + 7 =$ double 7 $+ \square = \square$	91, ___, ___, ___, 55
	$8 + 8 =$ double 8 $= \square$	

d + d with missing operator or number

Fill in each __ with a number or a + operator.

5 + 8 = __ 13	♖ + 8 = __ 13
8 + ♖ = __ 13	8 + 5 = __ 13
5 __ +8 = 13	5 __ + 8 = 13
8 __ + 5 = 13	8 __ + ♖ = 13
8 + __ 5 = 13	8 + __ 5 = 13
♖ + __ 8 = 13	5 + __ 8 = 13
8 __ __ + 5 = 13	8 __ __ + 5 = 13
5 __ __ + 8 = 13	♖ __ __ + 8 = 13
♖ __ __ + 8 = 13	5 __ __ + 8 = 13
8 __ + 5 = 13	8 __ + 5 = 13
8 + __ __ 5 = 13	8 + __ __ ♖ = 13
5 + __ __ 8 = 13	5 + __ __ 8 = 13
five __ + eight = 13	five __ + eight = 13
Eight __ + five = 13	Eight __ + five = 13
five __ __ + 8 = 13	five __ __ + 8 = 13
eight __ __ + 5 = 13	eight __ __ +5 = 13
5 + __ __ eight = 13	5 + __ __ eight = 13
8 + __ __ five = 13	8 + __ __ five = 13

d + d with missing operator or number

Fill in each __ with a number or + operator.

7 + 8 = __ 15	7 + 8 = __ 15
8 + 7 = __ 15	8 + 7 = __ 15
7 __ + 8 = 15	7 __ + 8 = 15
8 __ + 7 = 15	8 __ + 7 = 15
8 + __ 7 = 15	8 + __ 7 = 15
7 + __ 8 = 15	7 + __ 8 = 15
8 __ +7 = 15	8 __ +7 = 15
7 __ +8= 15	7 __ +8 = 15
7 __ + 8 = 15	7 __ + 8 = 15
8 __ + 7 = 15	8 __ + 7 = 15
8 + __ 7 = 15	8 + __ 7 = 15
7 + __ 8 = 15	7 + __ 8 = 15
seven __+ eight = 15	seven __+ eight = 15
eight __+ seven = 15	eight __+ seven = 15
seven __ +8 = 15	seven __ + 8 = 15
eight __ +7 = 15	eight __ + 7 = 15
7 + __ eight = 15	7 + __ eight = 15
8 + __ seven =15	8 + __ seven = 15

Mom! I Learn Addition Using Math-Chess-Puzzles Connection

8 + 8 = 16 = double 8 = 16

d + d with missing operator or number

Fill in each __ with a number or + operator.

8 + 8 = __ 16	7 + 8 = __ 16
8 + 8 = __ 16	8 + 8 = __ 16
8 __ + 8 = 16	8 __ + 8 = 16
8 __ + 8 = 16	8 __ + 8 = 16
8 + __ 8 = 16	8 + __ 8 = 16
8 + __ 8 = 16	8 + __ 8 = 16
8 __ +8 = 16	8 __ + 8 = 16
8 __ +8 = 16	8 __ + 8 = 16
8 __ + 8 = 16	8 __ + 8 = 16
8 __ + 8 = 16	8 __ + 8 = 16
8 + __ 8 = 16	8+ __ 8 = 16
8 + __ 8 = 16	8 + __ 8 = 16
eight __ + eight = 16	eight __ + eight = 16
eight __ + eight = 16	eight __ + eight = 16
eight __ + eight = 16	eight __ +8= 16
eight __ + eight = 16	eight __ + eight = 16
eight + __ eight = 16	eight + __ eight = 16
eight + __ eight = 16	eight + __ eight = 16

9 plus strategy

Day 7		
	$9 + 1 = 1 + 9 = \square$	Complete the following pattern.
	$\mathbf{9 + d = 9 + 1 + (d - 1)}$	91, 82, 73, 64, ___
	$9 + 2 = 2 + 9 = 9 + 1 + \square = \square$	___, ___, ___, 64, 55
	$9 + 3 = 3 + 9 = 9 + 1 + \square = \square$	91, ___, 73, ___, 55
	$9 + 4 = 4 + 9 = 9 + 1 + \square = \square$	___, 82, ___, 64, ___
	$9 + 5 = 5 + 9 = 9 + 1 + \square = \square$	___, ___, 73, 64, 55
	$9 + 6 = 6 + 9 = 9 + 1 + \square = \square$	
	$9 + 7 = 7 + 9 = 9 + 1 + \square = \square$	___, ___, ___, 64, 55
	$9 + 8 = 8 + 9 = 9 + 1 + \square = \square$	91, ___, ___, ___, 55
	$9 + 9 = $ double $9 = \square$	

Day 7

9 + 1 = 1 + 9 = ☐

$9 + d = 9 + 1 + (d - 1)$

9 + 2 = 2 + 9 = 9 + 1 + ☐ = ☐
9 + 3 = 3 + 9 = 9 + 1 + ☐ = ☐
9 + 4 = 4 + 9 = 9 + 1 + ☐ = ☐
9 + 5 = 5 + 9 = 9 + 1 + ☐ = ☐
9 + 6 = 6 + 9 = 9 + 1 + ☐ = ☐
9 + 7 = 7 + 9 = 9 + 1 + ☐ = ☐
9 + 8 = 8 + 9 = 9 + 1 + ☐ = ☐

9 + 9 = double 9 = ☐

Complete the following pattern.

91, 82, 73, 64, ___

___, ___, ___, 64, 55

91, ___, 73, ___, 55

___, 82, ___, 64, ___

___, ___, 73, 64, 55

___, ___, ___, 64, 55

91, ___, ___, ___, 55

Mom! I Learn Addition Using Math-Chess-Puzzles Connection

d + d with missing operator or number

Fill in each __ with a number or + operator.

9 + 9 = __ 18	9 + 9 = __ 18
9 + 9 = __ 18	9 + 9 = __ 18
9 __ + 9 = 18	9 __ + 9 = 18
9 __ + 9 = 18	9 __ + 9 = 18
9 + __ 9 = 18	9 + __ 9 = 18
9 + __ 9 = 18	9 + __ 9 = 18
9 __ + 9 = 18	9 __ + 9 = 18
9 __ + 9 = 18	9 __ + 9 = 18
9 __ + 9 = 18	9 __ + 9 = 18
9 __ + 9 = 18	9 __ + 9 = 18
9 + __ 9 = 18	9 + __ 9 = 18
9 + __ 9 = 18	9 + __ 9 = 18
nine __ + nine = 18	nine __ + nine = 18
nine __ + nine = 18	nine __ + nine = 18
nine __ + nine = 18	nine __ + nine = 18
nine __ + nine = 18	nine __ + nine = 18
nine + __ nine = 18	nine + __ nine = 18
nine + __ nine = 18	nine + __ nine = 18

Ho Math Chess 何数棋谜 妈！我会棋谜式加法啦！
Mom! I Learn Addition Using Math-Chess-Puzzles Connection
Contents include both traditional and Math-Chess-Puzzles combined methods. Extra strength
©2008 — 2018 Frank Ho, Amanda Ho   All rights reserved. www.homathchess.com

Student Name_____ Date_____

Part 2: More practices of addition strategy : adding 1, 2, 3 加 1, 2, 3

If students are not proficient after finishing Part1, then continue to Part 2 and on.

1 + 1 = ____
2 + 1 = ____
3 + 1 = ____
4 + 1 = ____
5 + 1 = ____
6 + 1 = ____
7 + 1 = ____
8 + 1 = ____

1 + 1 = ____
1+2=2 + 1 = ____
1+3=3 + 1 = ____
1+4 = ____
1+5 = ____
1+6 = ____
1+7 = ____
1+8 = ____

1 + 2 = ____
2 + 2 = ____
3 + 2 = ____
4 + 2 = ____
5 + 2 = ____
6 + 2 = ____
7 + 2 = ____
8 + 2 = ____

Ho Math Chess 何数棋谜　妈！我会棋谜式加法啦！
Mom! I Learn Addition Using Math-Chess-Puzzles Connection

Contents include both traditional and Math-Chess-Puzzles combined methods. Extra strength
©2008 − 2018 Frank Ho, Amanda Ho   All rights reserved. www.homathchess.com

Student Name _____ Date _____

1 + 2 = 2 + 1 = ____

2 + 2 = 2 × 2 = ____

2 + 3 = 3 + 2 = ____

2 + 4 = ____

2 + 5 = ____

2 + 6 = ____

2 + 7 = ____

2 + 8 = ____

1 + 2 = 2 + 1 = ____

2 + 2 = 2 × 2 = ____

2 + 3 = 3 + 2 = ____

2 + 4 = ____

2 + 5 = ____

2 + 6 = ____

2 + 7 = ____

2 + 8 = ____

1 + 3 = ____

2 + 3 = ____

3 + 3 = 2 × 3 = ____

4 + 3 = ____

5 + 3 = ____

6 + 3 = ____

7 + 3 = ____

Ho Math Chess 何数棋谜　妈！我会棋谜式加法啦！
Mom! I Learn Addition Using Math-Chess-Puzzles Connection

Contents include both traditional and Math-Chess-Puzzles combined methods. Extra strength

©2008 – 2018 Frank Ho, Amanda Ho　　All rights reserved. www.homathchess.com

Student Name _____ Date _____

3 + 1 = 2 + 1 = ___

3 + 2 = ___

3 + 3 = 2 × 3 = ___

3 + 4 = ___

3 + 5 = ___

3 + 6 = ___

3 + 7 = ___

1 + 4 = ___

2 + 4 = ___

3 + 4 = ___

4 + 4 = 2 × 4 = _____

5 + 4 = ___

6 + 4 = ___

4 + 1 = ___

4 + 2 = ___

4 + 3 = ___

4 + 4 = 2 × 4 = _____

4 + 5 = ___

4 + 6 = ___

Ho Math Chess 何数棋谜 妈！我会棋谜式加法啦！
Mom! I Learn Addition Using Math-Chess-Puzzles Connection

Contents include both traditional and Math-Chess-Puzzles combined methods. Extra strength

©2008－2018 Frank Ho, Amanda Ho   All rights reserved. www.homathchess.com

Student Name _____   Date _____

1 + 5 = 5 + 1 = ____

2 + 5 = ____

3 + 5 = ____

4 + 5 = ____

5 + 5 = 2 × 5 = _____

5 + 1 = ____

5 + 2 = ____

5 + 3 = ____

5 + 4 = ____

5 + 5 = 2 × 5 = _____

1 + 5 = 5 + 1 = ____

2 + 5 = ____

3 + 5 = ____

4 + 5 = ____

5 + 5 = 2 × 5 = _____

5 + 1 = ____

5 + 2 = ____

5 + 3 = ____

5 + 4 = ____

5 + 5 = 2 × 5 = _____

1 + 7 = 7 + 1 = ____
2 + 7 = ____
3 + 7 = ____

7 + 1 = ____
7 + 2 = ____
7 + 3 = ____

1 + 8 = 8 + 1 = ____
2 + 8 = ____

8 + 1 = ____
8 + 2 = ____

1 + 9 = 9 + 1 = ____

9 + 1 = ____

Mom! I Learn Addition Using Math-Chess-Puzzles Connection

Adding 1, 2, or 3 by counting
Counting 0, 1 forward d + d
(King = 0, Pawn = 1, Bishop = 3, Knight = 3, Rook = 5, Queen = 9)
Do not use fingers to count but recite numbers in brain.

Rook	1	6	7	Queen
+ 0	+ Rook	+ King	+ Pawn	+ Pawn
Pawn	7	Bishop	7	Rook
+ 8	+ Pawn	+ Pawn	+ King	+ 1
`34				
Rook	Pawn	Pawn	Bishop	King
+ Pawn	+ Queen	+ 7	+ Pawn	+ 8
Knight	Rook	Knight	1	Queen
+ 1	+ 1	+ 1	+ Pawn	+ 1
Bishop	Rook	Bishop	2	Bishop
+ 2	+ 1	+ 1	+ Rook	+ 1

97

Ho Math Chess 何数棋谜 妈！我会棋谜式加法啦！
Mom! I Learn Addition Using Math-Chess-Puzzles Connection
Contents include both traditional and Math-Chess-Puzzles combined methods. Extra strength
©2008 − 2018 Frank Ho, Amanda Ho   All rights reserved. www.homathchess.com

Student Name _____ Date_____

Counting 0, 1 forward d + d (Start with bigger number and count on dots. zero property)
(♔ = 0, ♙ = 1, ♗ = 3, ♘ = 3, ♖ = 5, ♕ = 9)

5 + ♔	.1 + 5	6 + 0	7 + .1	8 + .♙
.1 + 8	7 + .1	6 + .♙	7 + ♔	♖ + .1
6 + .1	♙ + 9	6 + .1	7 + .1	♔ + 8
8 + .♙	♖ + .1	6 + .♙	7 + .2	♕ + .♙
.♙ + 8	7 + .1	6 + .♙	.♙ + 6	8 + .1

Mom! I Learn Addition Using Math-Chess-Puzzles Connection

Counting forward 2 for d + d with no carrying

(Start with bigger number and count on dots.) (♔ = 0, ♙ = 1, ♗ = 3, ♘ = 3, ♖ = 5, ♕ = 9)

♖	..2	6	..2	7
+ ..2	+ 5	+ ..2	+ 7	+ ..2
..2	7	..2	7	♖
+ 6	+ ..2	+ 4	+ ..2	+ ..2
6	..2	..2	7	..2
+ ..2	+ 7	+ 2	+ ..2	+ 6
3	..2	7	..2	6
+ ..2	+ ♖	+ ..2	+ 7	+ ..2
..2	..2	6	..2	4
+ 5	+ 7	+ ..2	+ 6	+ ..2

Mom! I Learn Addition Using Math-Chess-Puzzles Connection

Contents include both traditional and Math-Chess-Puzzles combined methods. Extra strength
©2008 – 2018 Frank Ho, Amanda Ho   All rights reserved. www.homathchess.com

Student Name _____   Date _____

Counting forward 3 for d + d with no carrying
(Start with a bigger number and count on dots.) (♔ = 0, ♙ = 1, ♗ = 3, ♘ = 3, ♖ = 5, ♕ = 9)

♗ + …♗	…♗ + 4	6 + …♗	…♗ + ♖	6 + …♗
…♗ + 6	♖ + …♗	…♗ + 4	4 + …3	♖ + …♗
5 + ..2	…3 + 6	♗ + …3	♖ + …3	…♗ + 6
♗ + …3	…3 + ♖	6 + …3	…♗ + 4	6 + …3
♖ + …♘	4 + …♗	6 + …♘	…♙ + ♖	4 + …♗

6, 7, 9, 8, 9
9, 8, 7, 7, 8
7, 9, 6, 8, 7
6, 8, 9, 7, 9
8, 7, 9, 8, 7

100

Ho Math Chess 何数棋谜　妈！我会棋谜式加法啦！
Mom! I Learn Addition Using Math-Chess-Puzzles Connection

Contents include both traditional and Math-Chess-Puzzles combined methods. Extra strength

©2008 － 2018 Frank Ho, Amanda Ho　　All rights reserved. www.homathchess.com

Student Name _____　　Date _____

Adding up to 10 by adding 1

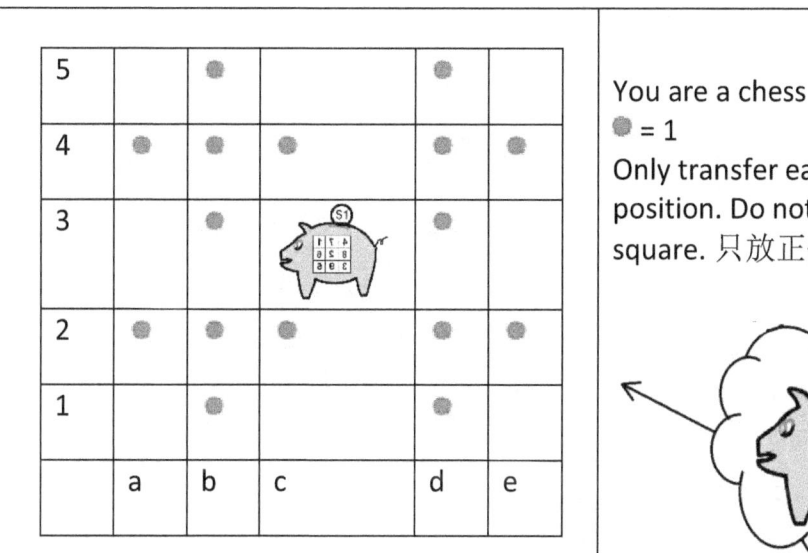

You are a chess piece located at c3.
● = 1
Only transfer each number itself to a correct position. Do not the transfer the entire 3 by 3 square. 只放正個別數即可, 不需放正表格.

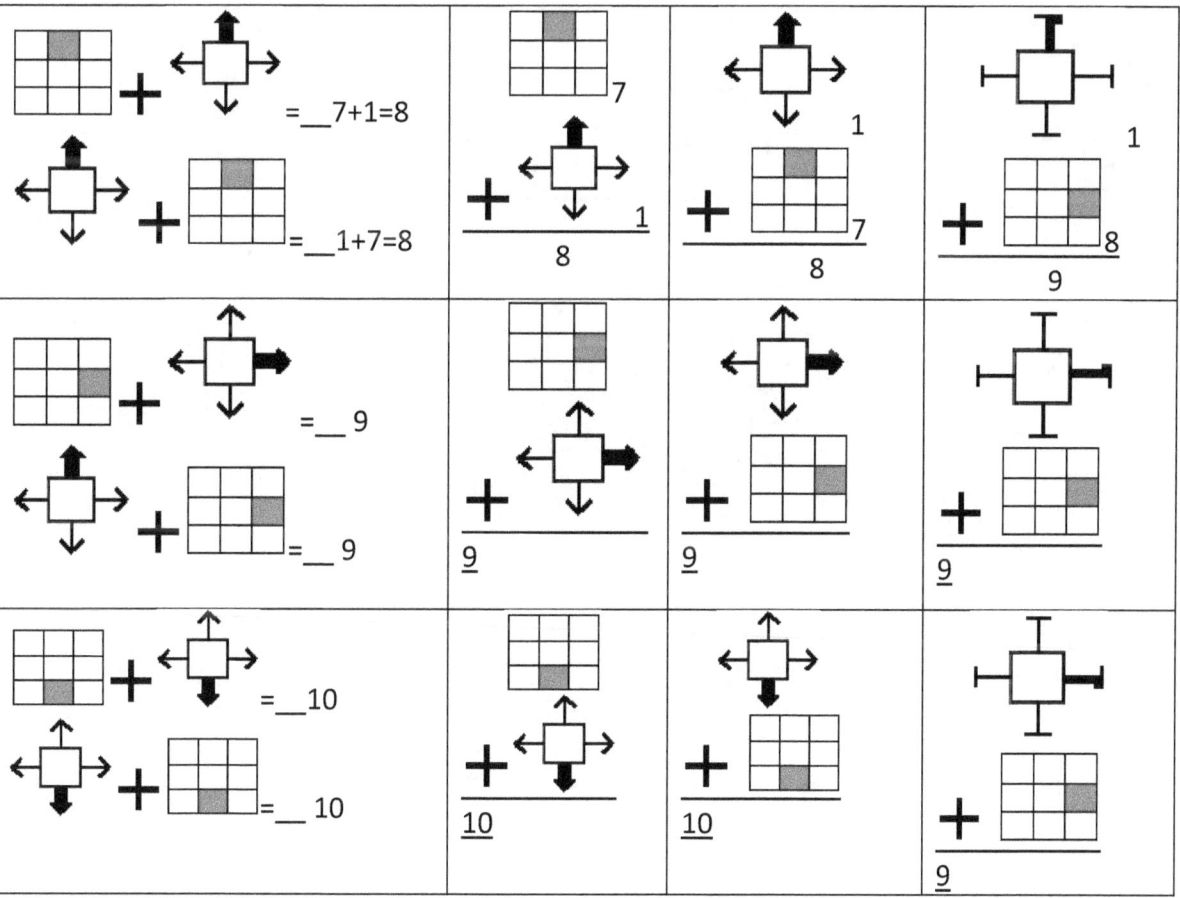

101

Ho Math Chess 何数棋谜 妈！我会棋谜式加法啦！
Mom! I Learn Addition Using Math-Chess-Puzzles Connection
Contents include both traditional and Math-Chess-Puzzles combined methods. Extra strength
©2008 – 2018 Frank Ho, Amanda Ho   All rights reserved. www.homathchess.com

Student Name_____ Date_____

Adding up to 10

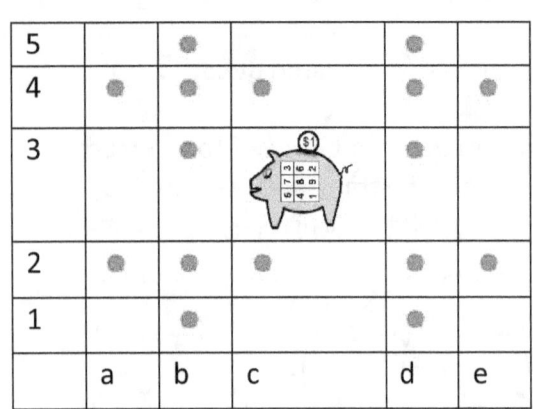

Only transfer each number itself to a correct position. Do not the transfer the entire 3 by 3 square.

You are a chess piece located at c3.
● = 1

Only transfer each number to a correct position. Do not the transfer the 3 by 3 square.

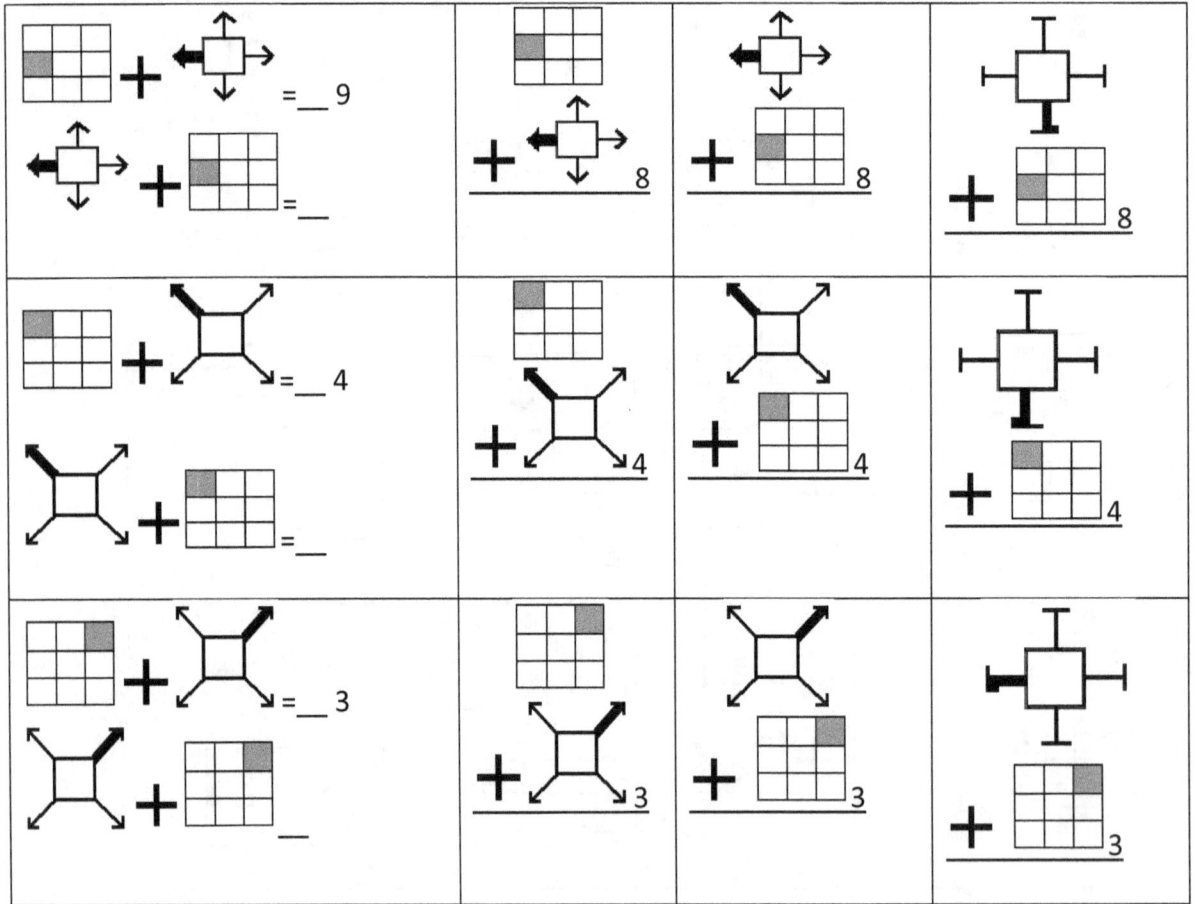

Ho Math Chess 何数棋谜　妈！我会棋谜式加法啦！
Mom! I Learn Addition Using Math-Chess-Puzzles Connection
Contents include both traditional and Math-Chess-Puzzles combined methods. Extra strength
©2008 – 2018 Frank Ho, Amanda Ho　　All rights reserved. www.homathchess.com

Student Name_____ Date_____

Adding up to 10

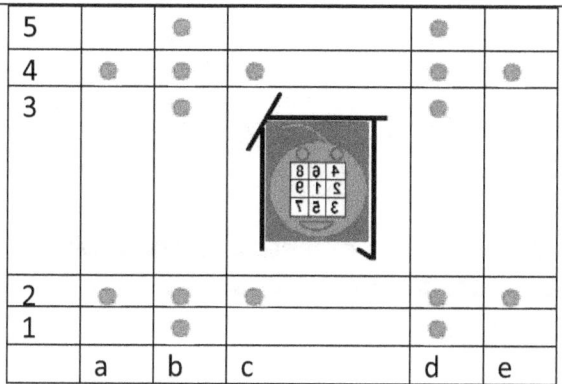

Only transfer each number itself to a correct position. Do not the transfer the entire 3 by 3 square.

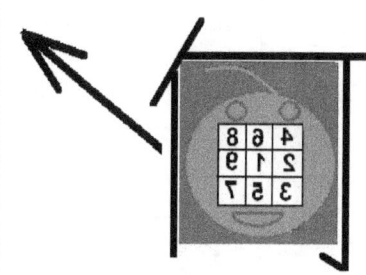

You are a chess piece located at c3.
● = 1

Only transfer each number to a correct position. Do not the transfer the 3 by 3 square.

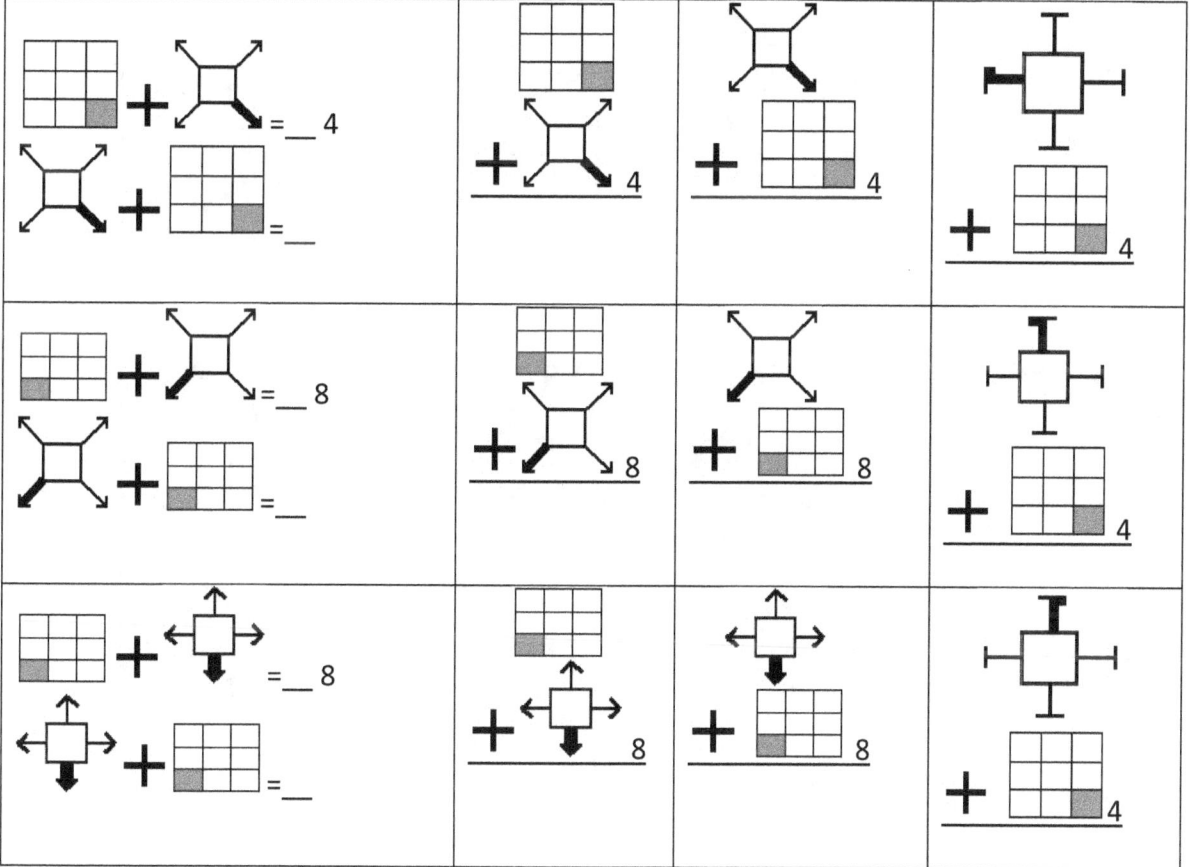

103

Ho Math Chess 何数棋谜   妈！我会棋谜式加法啦！
Mom! I Learn Addition Using Math-Chess-Puzzles Connection
Contents include both traditional and Math-Chess-Puzzles combined methods. Extra strength
©2008 – 2018 Frank Ho, Amanda Ho   All rights reserved. www.homathchess.com

Student Name_____ Date_____

Adding up to 10 by adding 2

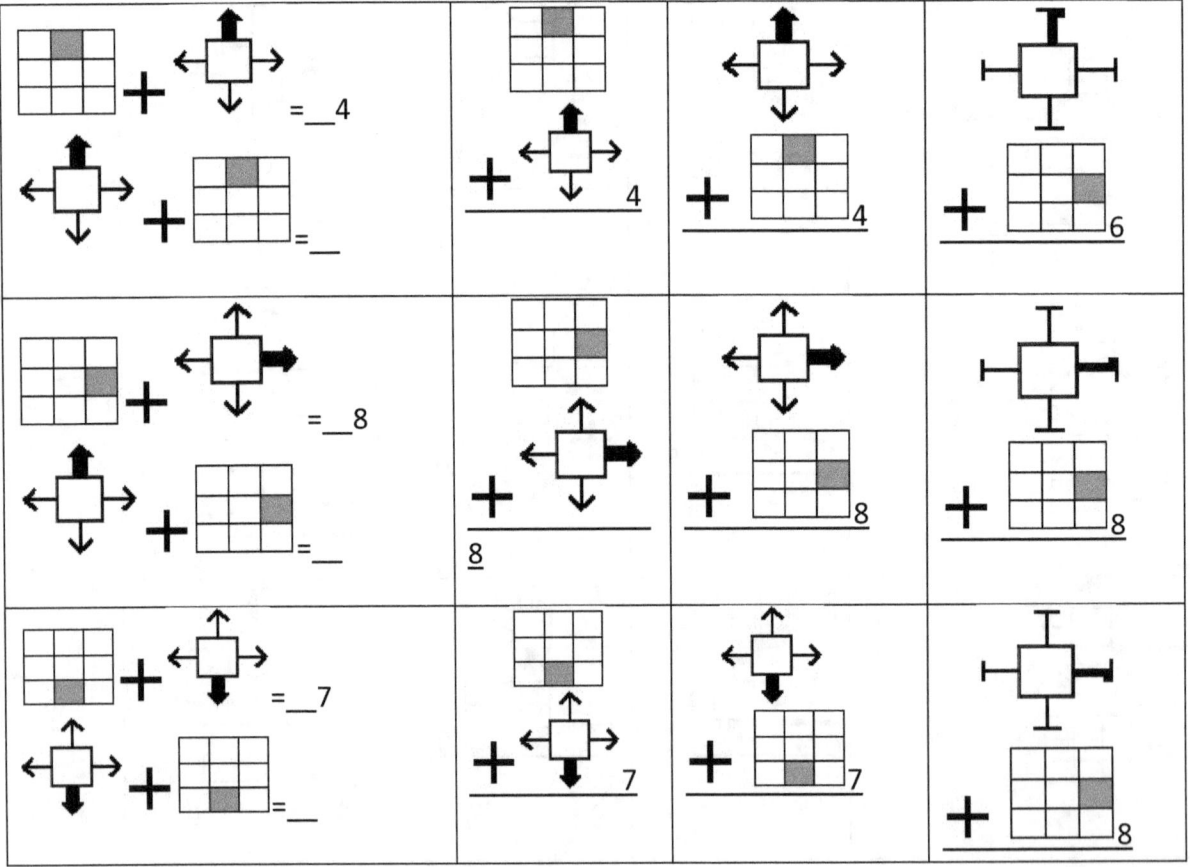

104

Ho Math Chess 何数棋谜　妈！我会棋谜式加法啦！
Mom! I Learn Addition Using Math-Chess-Puzzles Connection

Contents include both traditional and Math-Chess-Puzzles combined methods. Extra strength

©2008 – 2018 Frank Ho, Amanda Ho　　All rights reserved.　www.homathchess.com

Student Name _____　Date _____

Adding up to 10

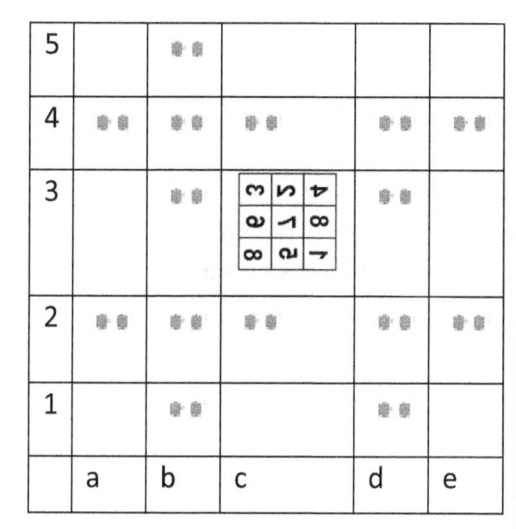

You are a chess piece located at c3.

● = 1

Only transfer each number to a correct position. Do not the transfer the 3 by 3 square. Each number is transformed into 2 operations.

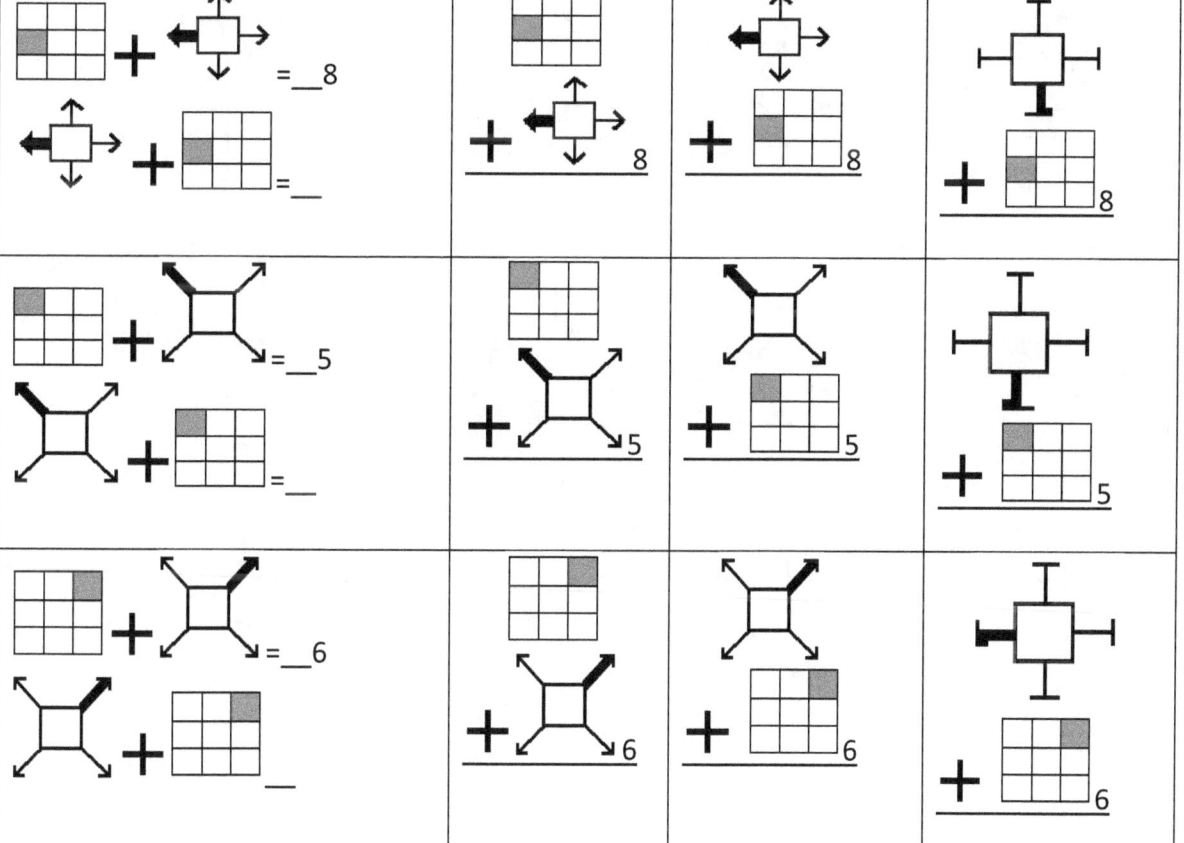

105

… Math Chess 何数棋谜   妈！我会棋谜式加法啦！
Mom! I Learn Addition Using Math-Chess-Puzzles Connection
Contents include both traditional and Math-Chess-Puzzles combined methods. Extra strength
©2008 – 2018 Frank Ho, Amanda Ho   All rights reserved. www.homathchess.com

Student Name _____ Date _____

Adding up to 10

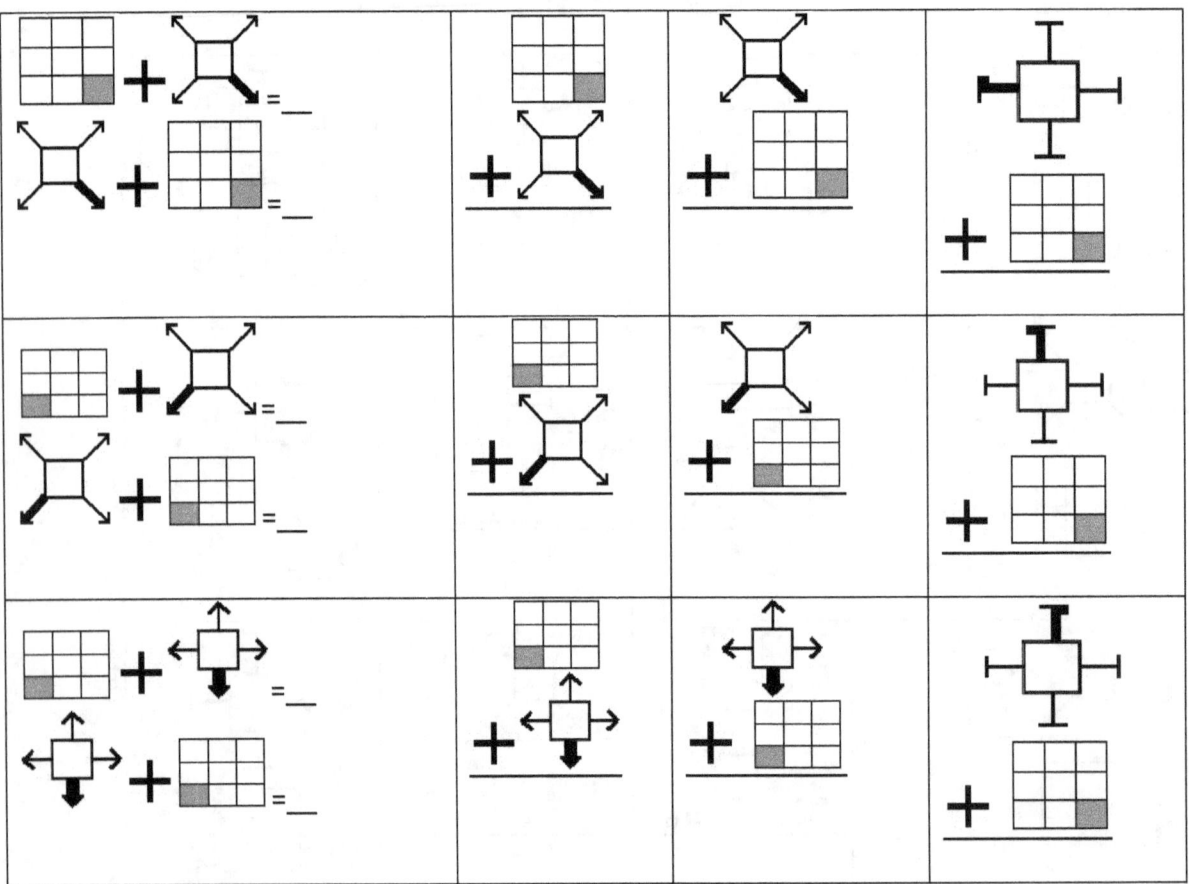

3 3 3 3
6 6 6 3
6 6 6 3

Mom! I Learn Addition Using Math-Chess-Puzzles Connection

Contents include both traditional and Math-Chess-Puzzles combined methods. Extra strength

©2008 – 2018 Frank Ho, Amanda Ho   All rights reserved. www.homathchess.com

Student Name _____ Date _____

Adding up to 10 by adding 3

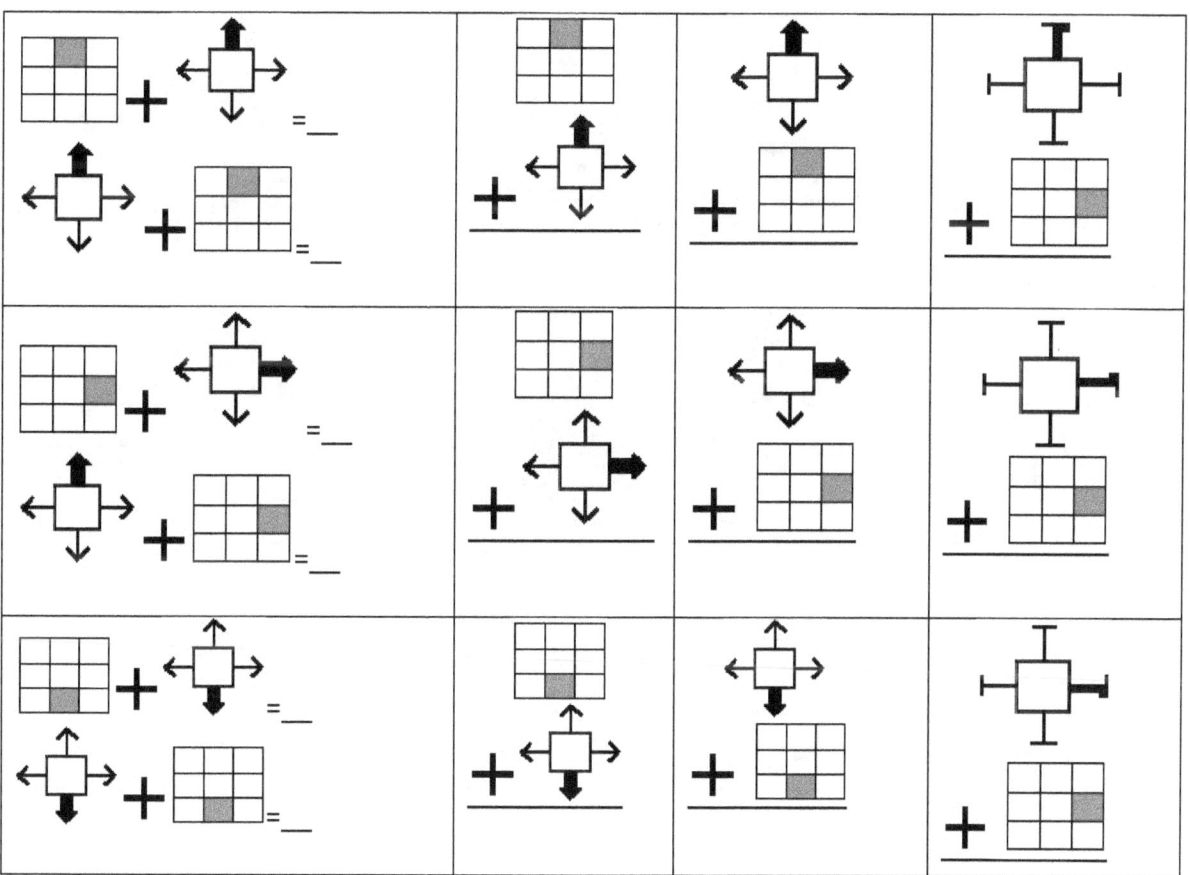

6 6 6 6
6 6 6 6
4 4 4 6

107

Ho Math Chess 何数棋谜　妈！我会棋谜式加法啦！
Mom! I Learn Addition Using Math-Chess-Puzzles Connection
Contents include both traditional and Math-Chess-Puzzles combined methods. Extra strength
©2008 – 2018 Frank Ho, Amanda Ho　　All rights reserved. www.homathchess.com

Student Name_____　Date_____

Adding up to 10

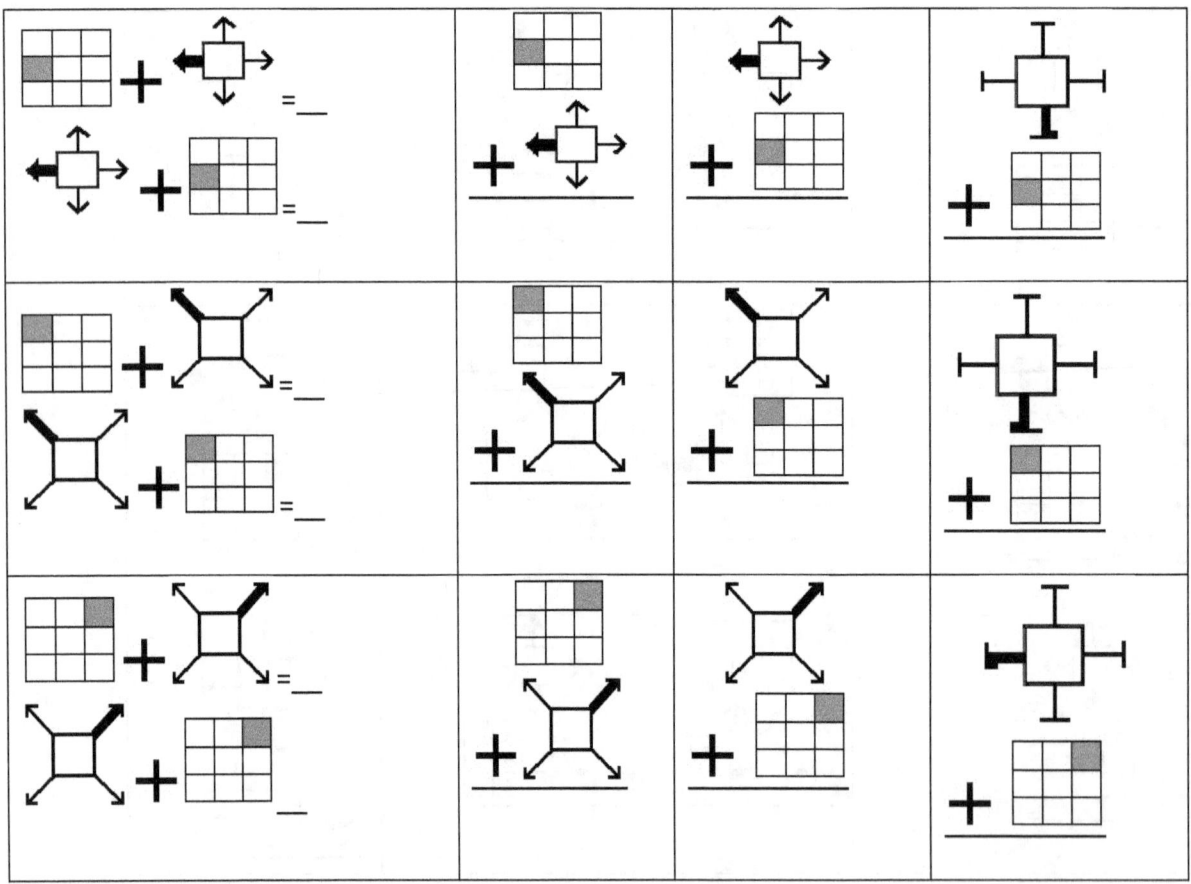

7 7 7 9
4 4 4 4
7 7 7 7

Ho Math Chess 何数棋谜　妈！我会棋谜式加法啦！
Mom! I Learn Addition Using Math-Chess-Puzzles Connection

Contents include both traditional and Math-Chess-Puzzles combined methods. Extra strength

©2008 – 2018 Frank Ho, Amanda Ho　　All rights reserved. www.homathchess.com

Student Name _____ Date _____

Adding up to 10

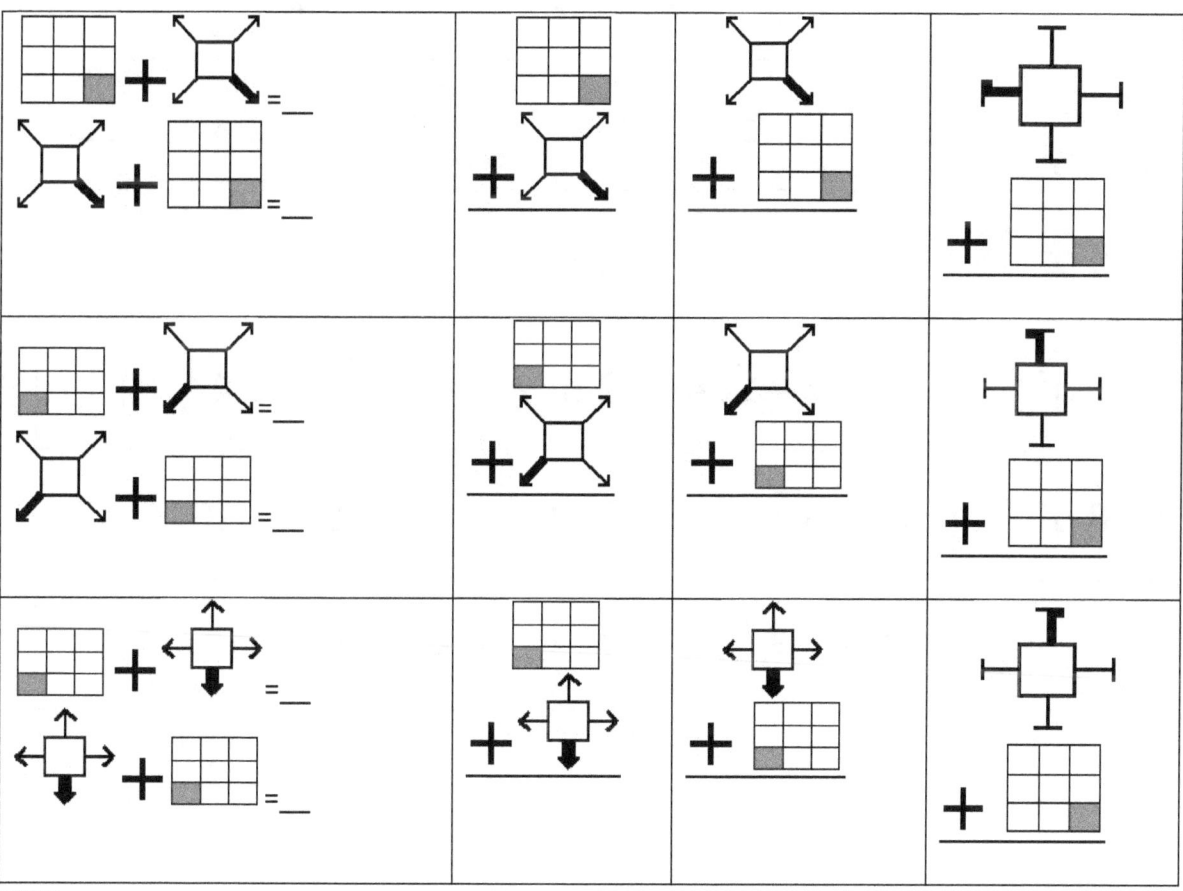

7777
5557
5557

Mom! I Learn Addition Using Math-Chess-Puzzles Connection

Addition strategy: Doubling = 2 times (even numbers) 双数

Fill in each ☐ with a number.

Math and Chess sentence	Number sentence.	Comments
♟♟ + 2	2 + 2 = ☐4	2 of ♟♟ = 2 × 2 = ☐4
♟♟♟ + ♟♟♟	3 + 3 = ☐6	2 of ♟♟♟ = 2 × 3 = ☐6
♟♟♟♟ + ♟♟♟♟	4 + 4 = ☐8	2 of ♟♟♟♟ = 2 × 4 = ☐8
♟♟♟♟♟ + ♟♟♟♟♟	5 + 5 = ☐10	2 of ♟♟♟♟♟ = 2 × 5 = ☐10
6 of ♟ + 6 of ♟	6 + 6 = ☐12	2 of 6 ♟ = 2 × 6 = ☐ 12
7 of ♟ + 7 of ♟	7 + 7 = ☐14	2 of 7 ♟ = 2 × 7 = ☐ 14
8 of ♟ + 8 of ♟	8 + 8 = ☐16	2 of 8 ♟ = 2 × 8 = ☐ 16
9 of ♟s + 9 of ♟	9 + 9 = ☐18	2 of 9 ♟ = 2 × 9 = ☐18

Mom! I Learn Addition Using Math-Chess-Puzzles Connection

Contents include both traditional and Math-Chess-Puzzles combined methods. Extra strength

©2008 – 2018 Frank Ho, Amanda Ho   All rights reserved. www.homathchess.com

Student Name_____ Date_____

Getting even numbers by doubling a number

It is an easy task for some students to understand the concepts of adding 1, 2, 3 or adding to ten or doubling but for some students who sit in the same math class, the work to come up with a correct answer becomes extremely difficult or slow. So here it is we have one more review of doubling concept.

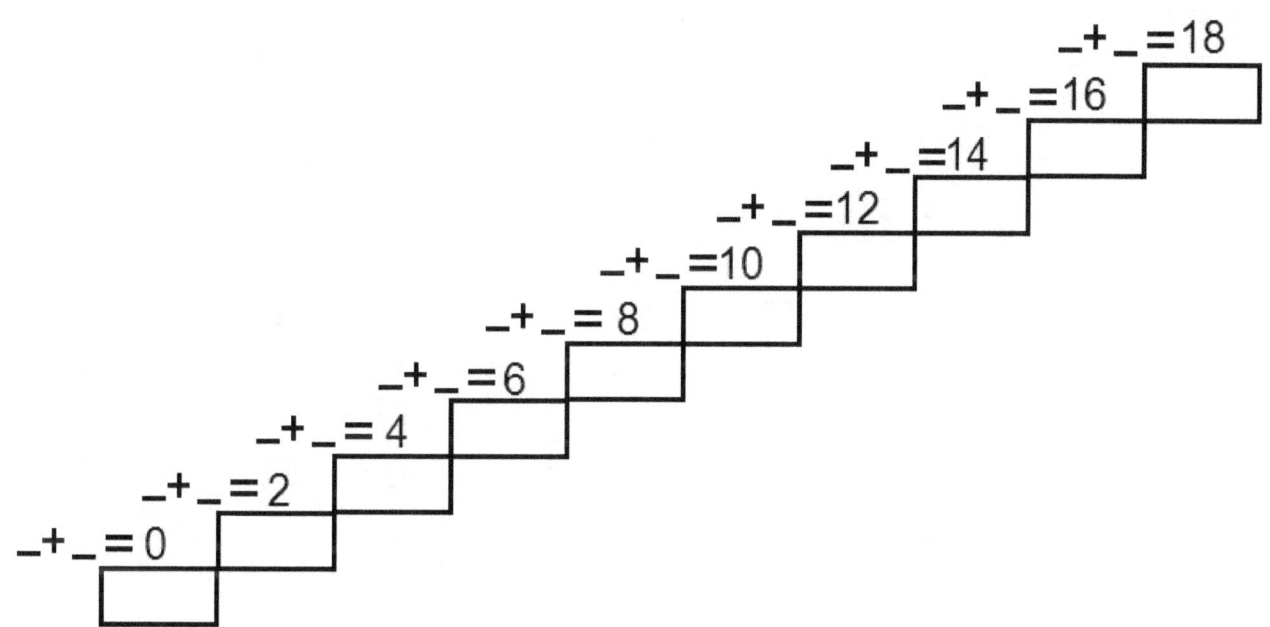

0 0
1 1
2 2
3 3
4 4
5 5
6 6
7 7
8 8
9 9

Student Name _____ Date _____

Doubling = 2 times (even numbers)

This page is not meant to teach multiplication but shows the relationships of doubling.

2 + 2 = ____4	2 + 2 = 2 of 2 = ☐ × 2 = 4	2 plus ___ equals 2 times 2 = ___4
3 + 3 = ____6	3 + 3 = 2 of 3 = ☐ × 3 = 6	3 plus ___ equals 2 times 3 = ___6
4 + 4 = ____8	4 + 4 = 2 of 4 = ☐ × 4 = 8	4 plus ___ equals 2 times 4 = ___8
5 + 5 = ____10	5 + 5 = 2 of 5 = ☐ × 5 = 10	5 plus 5 equals ___ times 5 = ___10
6 + 6 = ____12	6 + 6 = 2 of 6 = ☐ × 6 = 12	6 plus ___ equals 2 times ___ = ___12
7 + 7 = ____14	7 + 7 = 2 of 7 = ☐ × 7 = 14	7 plus 7 equals 2 times ___ = ___14
8 + 8 = ____16	8 + 8 = 2 of 8 = ☐ × 8 = 16	8 plus 8 equals 2 times ___ = ___16
9 + 9 = ____18	9 + 9 = 2 of 9 = ☐ × 9 = 18	9 plus ___ equals 2 times 9 = ___18

Doubling

2	7	4	8	6
+ 2	+ 7	+ 4	+ 8	+ 6
___	___	___	___	___
4	14	8	16	12
♗	8	♕	4	♖
+ ♗	+ 8	+ ♕	+ 4	+ ♖
___	___	___	___	___
6	16	18	8	10
7	8	9	5	3
+7	+ 8	+ 9	+ 5	+3
___	___	___	___	___
14	16	18	10	6
2	7	♕	4	6
+ 2	+ 7	+ ♕	+ 4	+ 6
___	___	___	___	___
4	14	18	8	12

Ho Math Chess 何数棋谜　妈！我会棋谜式加法啦！
Mom! I Learn Addition Using Math-Chess-Puzzles Connection

Contents include both traditional and Math-Chess-Puzzles combined methods. Extra strength
©2008 — 2018 Frank Ho, Amanda Ho　　All rights reserved. www.homathchess.com

Student Name _____ Date _____

Doubling

For each question, fill in each ☐ with the same number.

☐2 + ☐2 = 4	☐ + ☐ = 4	☐ + ☐ = 4	☐ + ☐ = 4	☐ + ☐ = 4
☐3 + ☐3 = 6	☐ + ☐ = 6	☐ + ☐ = 6	☐ + ☐ = 6	☐ + ☐ = 6
☐4 + ☐4 = 8	☐ + ☐ = 8	☐ + ☐ = 8	☐ + ☐ = 8	☐ + ☐ = 8
☐5 + ☐5 = 10	☐ + ☐ = 10	☐ + ☐ = 10	☐ + ☐ = 10	☐ + ☐ = 10

Ho Math Chess 何数棋谜   妈！我会棋谜式加法啦！
Mom! I Learn Addition Using Math-Chess-Puzzles Connection

Contents include both traditional and Math-Chess-Puzzles combined methods. Extra strength
©2008 – 2018 Frank Ho, Amanda Ho   All rights reserved. www.homathchess.com

Student Name _____   Date _____

Doubling

For each question, fill in each ☐ with the same number.

☐6 + ☐6 = 12	☐ + ☐ = 12	☐ + ☐ = 12	☐ + ☐ = 12	☐ + ☐ = 12
☐7 + ☐7 = 14	☐ + ☐ = 14	☐ + ☐ = 14	☐ + ☐ = 14	☐ + ☐ = 14
☐8 + ☐8 = 16	☐ + ☐ = 16	☐ + ☐ = 16	☐ + ☐ = 16	☐ + ☐ = 16
☐9 + ☐9 = 18	☐ + ☐ = 18	☐ + ☐ = 18	☐ + ☐ = 18	☐ + ☐ = 18

Mom! I Learn Addition Using Math-Chess-Puzzles Connection

Doubling

two + two 4	two + two 4	three + three 6	three + three 6	four + four 8
four + four 8	five + five 10	five + five 10	six + six 12	six + six 12
Seven + seven 14	seven + seven 14	eight + eight 16	eight + eight 16	nine + nine 18
nine + nine 18	seven + seven 14	five + five 10	four + four 8	nine + nine 18
seven + seven 14	eight + eight 16	six + six 12	seven + seven 14	six + six 12

Mom! I Learn Addition Using Math-Chess-Puzzles Connection

Doubling

two plus two is □4	2 + 2 = □4
three plus three is □6	3 + 3 = □6
four plus four is □8	4 + 4 = □8
five plus five is □10	5 + 5 = □10
six plus six is □12	6 + 6 = □12
seven plus seven is □14	7 + 7 = □14
eight plus eight is □16	8 + 8 = □16
nine plus nine is □18	9 + 9 = □18
two plus two is □4	2 + 2 = □4
three plus three is □6	3 + 3 = □6
four plus four is □8	4 + 4 = □8
five plus five is □10	5 + 5 = □10
six plus six is □12	6 + 6 = □12
seven plus seven is □14	7 + 7 = □14
eight plus eight is □16	8 + 8 = □16
nine plus nine is □18	9 + 9 = □18

Mom! I Learn Addition Using Math-Chess-Puzzles Connection

d + d with missing operator or number

Fill in each __ with a number or + operator.

9 + 9 = __ 18	9 + 9 = __ 18
9 + 9 = __ 18	9 + 9 = __ 18
9 __ + 9 = 18	9 __ + 9 = 18
9 __ + 9 = 18	9 __ + 9 = 18
9 + __ 9 = 18	9 + __ 9 = 18
9 + __ 9 = 18	9 + __ 9 = 18
9 __ + 9 = 18	9 __ + 9 = 18
9 __ + 9 = 18	9 __ + 9 = 18
9 __ + 9 = 18	9 __ + 9 = 18
9 __ + 9 = 18	9 __ + 9 = 18
9 + __ 9 = 18	9 + __ 9 = 18
9 + __ 9 = 18	9 + __ 9 = 18
nine __ + nine = 18	nine __ + nine = 18
nine __ + nine = 18	nine __ + nine = 18
nine __ + nine = 18	nine __ + nine = 18
nine __ + nine = 18	nine __ + nine = 18
nine + __ nine = 18	nine + __ nine = 18
nine + __ nine = 18	nine + __ nine = 18

Working backwards on doubling

◯ + 0 = 0	4 + ◯ = 8
4 + ◯ = 8	◯ + 0 = 0
◯ + 2 = 4	6 + ◯ = 12
6 + ◯ = 12	◯ + 0 = 0
◯ + 9 = 18	5 + ◯ = 10
5 + ◯ = 10	◯ + 7 = 14
◯ + 3 = 6	8 + ◯ = 16
2 + ◯ = 4	◯ + 0 = 0
◯ + 1 = 2	6 + ◯ = 12
4 + ◯ = 8	◯ + 0 = 0
◯ + 2 = 4	4 + ◯ = 8
7 + ◯ = 14	◯ + 0 = 0
◯ + 9 = 18	5 + ◯ = 10
5 + ◯ = 10	◯ + 9 = 18
◯ + 3 = 6	8 + ◯ = 16
2 + ◯ = 4	◯ + 0 = 0
◯ + 1 = 2	7 + ◯ = 14

All are doubles so just copy the other number to get the answer to each problem.

△ + △ − 1 = 7

△ + △ + 1 = 9

△ + △ − 1 = 13

△ + △ + 1 = 15

△ + △ − 1 = 11

△ + △ + 1 = 13

△ + △ − 1 = 17

△ + △ + 1 = 19

△ + △ − 1 = 3

△ + △ + 1 = 7

△ + △ − 1 = 19

△ + △ + 1 = 23

△ + △ − 1 = 13

△ + △ + 1 = 21

△ + △ − 1 = 1

△ + △ + 1 = 3

Work backwards by using the right-hand number adds or subtracts the left number and then divided by 2.

Mom! I Learn Addition Using Math-Chess-Puzzles Connection

$2 + 4 + \bigcirc = 92$	$\bigcirc + \bigcirc + 5 = 9$
$2 + 2 + \bigcirc = 64$	$\bigcirc - 1 = 8$
$\bigcirc + \bigcirc + \bigcirc + 1 = 10$	$\bigcirc + \bigcirc = 8$
$\bigcirc + \bigcirc + 4 = 8$	$\bigcirc + \bigcirc - 1 = 9$
$\bigcirc + \bigcirc + 1 = 7$	$\bigcirc + \bigcirc - 2 = 8$
$\bigcirc + \bigcirc + 2 = 8$	$\bigcirc + \bigcirc + 2 = 8$
$8 + \bigcirc = 9$	$\bigcirc + \bigcirc - 1 = 7$
$8 + \bigcirc + \bigcirc = 10$	$\bigcirc + \bigcirc + 1 = 7$
$\bigcirc + \bigcirc + \bigcirc = 9$	$\bigcirc + \bigcirc + 3 = 7$
$7 + \bigcirc + 1 = 10$	$\bigcirc + \bigcirc + 5 = 9$
$5 + 2 + \bigcirc = 8$	$\bigcirc + \bigcirc - 1 = 9$
$4 + \bigcirc + \bigcirc = 6$	$\bigcirc + \bigcirc + 2 = 10$
$3 + \bigcirc + \bigcirc = 7$	$\bigcirc + \bigcirc - 2 = 10$
$\bigcirc + \bigcirc + 1 = 5$	$\bigcirc + \bigcirc + 3 = 11$
$\bigcirc + \bigcirc - 1 = 5$	$\bigcirc + \bigcirc - 1 = 11$
$\bigcirc + 3 = 6$	$8 + \bigcirc = 16$
$2 + \bigcirc = 4$	$\bigcirc + 0 = 0$
$\bigcirc + 1 = 2$	$7 + \bigcirc = 14$

Ho Math Chess 何数棋谜   妈！我会棋谜式加法啦！
Mom! I Learn Addition Using Math-Chess-Puzzles Connection
Contents include both traditional and Math-Chess-Puzzles combined methods. Extra strength
©2008 – 2018 Frank Ho, Amanda Ho   All rights reserved. www.homathchess.com

Student Name _____ Date _____

$\bigcirc + \bigcirc + 1 = 5$	2
$\bigcirc + \bigcirc + 2 = 10$	4
$\bigcirc + \bigcirc + 1 = 5$	2
$\bigcirc + \bigcirc + 9 + 1 = 20$	5
$4 + \bigcirc + \bigcirc + 6 = 14$	2
$7 + 3 + \bigcirc + \bigcirc + 1 = 23$	6
$9 + \bigcirc + \bigcirc + 1 = 18$	4
$\triangle + \triangle = 8,\ \triangle + \bigcirc + \bigcirc + 2 = 16$	6 4 4
$\bigcirc + \bigcirc - 1 = 13$	7
$\bigcirc + \bigcirc + 11 - 1 = 28$	9 9
$4 + \bigcirc + \bigcirc - 6 = 14$	8 8
$7 - 3 + \bigcirc + \bigcirc - 1 = 17$	7 7
$\bigcirc + 2 + \triangle = \triangle + 3 + 1$	2
$11 - 5 = \triangle$ $\triangle + \triangle = \bigcirc + \bigcirc + 4$	6 6, 6, 4, 4

122

Ho Math Chess 何数棋谜 妈！我会棋谜式加法啦！
Mom! I Learn Addition Using Math-Chess-Puzzles Connection

Contents include both traditional and Math-Chess-Puzzles combined methods. Extra strength
©2008 – 2018 Frank Ho, Amanda Ho   All rights reserved. www.homathchess.com

Student Name _____ Date _____

$1 + 1 + \bigcirc = 4$	2
$3 + 1 + \bigcirc = 8$	4
$1 + 5 + \bigcirc = 12$	6
$7 + \bigcirc - 1 = 14$	8
$9 + \bigcirc - 1 = 18$	10
$9 - \bigcirc + 8 = 16$	1
$7 + \bigcirc + 8 = 16$	1
$9 + 10 - \bigcirc = 18$	1
$4 - \bigcirc + 3 = 6$	1
$\bigcirc - 1 + 5 = 10$	6
$9 + 1 + 6 + \bigcirc = 20$	4
$8 + 2 + 7 + \bigcirc = 20$	3
$5 + 5 + 5 + \bigcirc = 20$	5
$4 + 6 + 3 + \bigcirc = 20$	7
$2 + 1 + 7 + 2 + 2 + \bigcirc = 20$	6
$10 + \bigcirc + 9 = 20$	1
$3 + 7 + \bigcirc + 6 = 20$	4
$5 + \bigcirc + 3 + 7 = 20$	5
If $\triangle + 3 = 9$, then $\triangle + \triangle + \bigcirc = 16$. What is the value of $\bigcirc$?	6
$\bigcirc + 3 = \bigcirc + 2 + ?$	1
$\bigcirc + \bigcirc + 2 = 8$, What is the value of $\bigcirc$?	3

123

Doubling by relating numbers

8 + 9 uses the double of the smaller number 8. 6 + 7 uses the double of the smaller number 6.

8 + 8 = ____16 8 + 9 = ☐ +1 + 9 = ___17 9 − 2 + 8 = _____15	5 + 5 = ____10 5 + 6 = ____ = 5 + ☐ +1= 11 6 − 2 + 5 = _____9
6+6 = _____12 6 + 7 = ____ = 6 + ☐ + 1 = _13 8−2+7= _____13	9 + 9 = _____18 9 + 8 = 9 + 1 + ☐ = _____17 10 − 2 + 9 = _____17
7 + 7 = _____14 7 + 8 = 7 + ☐ +1 = _____15 9 − 2 + 8 = _____15	3 + 3 = _____6 3 + 4 = _____7 5 − 2 + 4 = _____7
4 + 4= _____8 4 + 5 = _____9 6 − 2 + 5 = _____12	6 + 6 = _____12 7 + 6 =1+ ☐ + 6 = _____13 8 + 7 = _____15 9 + 7 = _____16

Mom! I Learn Addition Using Math-Chess-Puzzles Connection

Using doubles to add

(♔ = 0, ♙ = 1, ♗ = 3, ♘ = 3, ♖ = 5, ♕ = 9)

♙ + ♙ = 2	10 + 10 = 20	2 + 2 = 4	20 + 20 = 40	3 + 3 = 6
♘ + ♘ = 6	30 + 30 = 60	4 + 4 = 8	40 + 40 = 80	♖ + ♖ = 10
50 + 50 = 100	6 + 6 = 12	60 + 60 = 120	7 + 7 = 14	70 + 70 = 140
8 + 8 = 16	80 + 80 = 160	9 + 9 = 18	90 + 90 = 180	10 + 10 = 20
20 + 20 = 40	30 + 30 = 60	40 + 40 = 80	50 + 50 = 100	60 + 60 = 120

Doubling

3	4	6	8
2	10	✴	12
1	14	16	18
	a	b	c

↔↕↨ No
↕↨ Yes

___ + ___ = ___ + ___ = ___ = 2 × ___ = 2 × ___ = ___

___ + ___ = ___ + ___ = ___ = 2 × ___ = 2 × ___ = ___

___ + ___ = ___ + ___ = ___ = 2 × ___ = 2 × ___ = ___

___ + ___ = ___ + ___ = ___ = 2 × ___ = 2 × ___ = ___

___ + ___ = ___ + ___ = ___ = 2 × ___ = 2 × ___ = ___

___ + ___ = ___ + ___ = ___ = 2 × ___ = 2 × ___ = ___

___ + ___ = ___ + ___ = ___ = 2 × ___ = 2 × ___ = ___

___ + ___ = ___ + ___ = ___ = 2 × ___ = 2 × ___ = ___

12
32
20
12
16
28
8
36

Ho Math Chess 何数棋迷   妈！我会棋谜式加法啦！
Mom! I Learn Addition Using Math-Chess-Puzzles Connection

Contents include both traditional and Math-Chess-Puzzles combined methods. Extra strength
©2008 – 2018 Frank Ho, Amanda Ho   All rights reserved. www.homathchess.com

Student Name _____   Date _____

(♔ = 0, ♙ = 1, ♗ = 3, ♘ = 3, ♖ = 5, ♕ = 9)

4 + □4 = 8	□4 + 4 = 8	4 + □4 = 8	□4 + 4 = 8
□4 + 4 = 8	4 + □4 = 8	□4 + 4 = 8	4 + □4 = 8

4 + 4 = 8

four + four = □8

$$\begin{array}{r} 4 \\ +\ 4 \\ \hline □8 \end{array} \qquad \begin{array}{r} 4 \\ +\ 4 \\ \hline □8 \end{array}$$

4 + ♖ = 4 + 4 + 1
    = □8 + 1
    = □9

four + five = □9

$$\begin{array}{r} 4 \\ +\ ♖ \\ \hline □9 \end{array} \qquad \begin{array}{r} 4 \\ 4 \\ +\ 1 \\ \hline □9 \end{array}$$

4 + 7 = 1 + 3 + 7
    = 1 + □10
    = □11

Four + seven = □11

$$\begin{array}{r} 4 \\ +\ 7 \\ \hline 11 \end{array} \qquad \begin{array}{r} 1 \\ ♗ \\ +\ 7 \\ \hline 11 \end{array}$$

4 + □8 = 12	□4 + 8 = 12	8 + □4 = 12	4□ + 8 = 12
□4 + 8 = 12	8 + □4 = 12	4□ + 8 = 12	8 + □4 = 12

Ho Math Chess 何数棋谜 妈！我会棋谜式加法啦！
Mom! I Learn Addition Using Math-Chess-Puzzles Connection
Contents include both traditional and Math-Chess-Puzzles combined methods. Extra strength
©2008 — 2018 Frank Ho, Amanda Ho   All rights reserved. www.homathchess.com

Student Name _____ Date _____

d + d
(♔ = 0, ♙ = 1, ♗ = 3, ♘ = 3, ♖ = 5, ♕ = 9)

5 + □5 = 10	5□ + 5 = 10	5 + □6 = 10	5□ + 5 = 10
5□ + 6 = 11	5 + □6 = 11	5□ + 6 = 11	5 + □6 = 11

5 + 7 = 12   five + seven = □12

♖
+ 7
―――
12

7
+ ♖
―――
12

5 + 8 = 5 + 5 + 3
     = □10 + 3
     = □ 13

five + eight = □13

♖
+ 8
―――
13

♖
♖
+ ♗
―――
13

5 + 9 = 4 + ___ + 9
     = □ 14

9 + d = 9 + 1 + (d−1)

five + nine = □14

♖
+ 9
―――
14

4
1
+ 9
―――
14

9 + □5 = 14	9□ + 5 = 14	9 + □5 = 14	9□ + 5 = 14
6□ + 9 = 15	5 + □9 = 14	5□ + 9 = 14	5 + □9 = 14

128

Ho Math Chess 何数棋谜　妈！我会棋谜式加法啦！
Mom! I Learn Addition Using Math-Chess-Puzzles Connection

Contents include both traditional and Math-Chess-Puzzles combined methods. Extra strength
©2008－2018 Frank Ho, Amanda Ho　　All rights reserved. www.homathchess.com

Student Name _____ Date _____

d + d

6 + □6 = 12	6□ + 6 = 12	6 + □6 = 12	6□ + 6 = 12
6□ + 7 = 13	7 + □6 = 13	6□ + 7 = 13	7 + □6 = 13

6 + 8 = 14   six + eight = □ 14

```
   6        8
 + 8      + 6
  14       14
```

8 + 6 = 6 + 6 + 2   eight + six = □ 14
　　 = □12 + 2
　　 = □ 14

```
   8        2
 + 6        6
  14      + 6
           14
```

6 + 9 = 1 + 5 + 9   six + nine = □ 15
　　 = 10 + □ 5
　　 = □ 15

Split the small
number to be 1 and
a smaller number.

```
   6        1
 + 9        5
  15      + 9
           15
```

9 + □ 6 = 15	9□ + 6 = 15	9 + □6 = 15	9□ + 6 = 15
6□ + 8 = 14	8 + □6 = 14	6□ + 8 = 14	8 + □6 = 14

Ho Math Chess 何数棋谜 妈！我会棋谜式加法啦！
Mom! I Learn Addition Using Math-Chess-Puzzles Connection
Contents include both traditional and Math-Chess-Puzzles combined methods. Extra strength
©2008 – 2018 Frank Ho, Amanda Ho   All rights reserved. www.homathchess.com

Student Name_____ Date_____

d + d

7 + □7 = 14	□7 + 7 = 14	7 + □7 = 14	□7 + 7 = 14
□7 + 7 = 14	7 + □7 = 14	□7 + 7 = 14	7 + □7 = 14

7 + 7 = 14   seven + seven = □14

$$\begin{array}{r} 7 \\ +\ 7 \\ \hline 14 \end{array} \qquad \begin{array}{r} 7 \\ +\ 7 \\ \hline 14 \end{array}$$

7 + 8 = 7 + 7 + 1   seven + eight = □15
    = □ 14 + 1
    = □ 15

$$\begin{array}{r} 7 \\ +\ 8 \\ \hline 15 \end{array} \qquad \begin{array}{r} 7 \\ 7 \\ +\ ♙ \\ \hline 15 \end{array}$$

8 + 7 = 1 + 7 + 7   eight + seven = □15
    = 1 + □14
    = □ 15

$$\begin{array}{r} 8 \\ +\ 7 \\ \hline 15 \end{array} \qquad \begin{array}{r} ♙ \\ 7 \\ +\ 7 \\ \hline 15 \end{array}$$

7 + □8 = 15	8□ + 7 = 15	7 + □7 = 14	8□ + 7 = 15
7□ + 8 = 15	8 + □6 = 14	6□ + 8 = 14	8 + □6 = 14

d + d

8 + □8 = 16	8□ + 8 = 16	8 + □8 = 16	8□ + 8 = 16
8□ + 9 = 17	8 + □9 = 17	8□ + 9 = 17	8 + □9 = 17

8 + 8 = 16   eight + eight = □16

$$\begin{array}{r} 8 \\ +\ 8 \\ \hline 16 \end{array} \qquad \begin{array}{r} 8 \\ +\ 8 \\ \hline 16 \end{array}$$

8 + 9 = 7 + 1 + 9   seven + nine = □16
   = □7 + 10
   = □ 17

$$\begin{array}{r} 8 \\ +\ ♛ \\ \hline 17 \end{array} \qquad \begin{array}{r} 1 \\ 7 \\ +\ ♛ \\ \hline 17 \end{array}$$

9 + 8 = 9 + 1 + 7   nine + eight = □17
   = 10 + □7
   = □ 17

$$\begin{array}{r} ♛ \\ +\ 8 \\ \hline 17 \end{array} \qquad \begin{array}{r} ♛ \\ 1 \\ +\ 7 \\ \hline 17 \end{array}$$

8 + □9 = 17	8□ + 9 = 17	8 + □9 = 17	8□ + 9 = 17
8□ + 9 = 17	8 + □9 = 17	8□ + 9 = 17	8 + □9 = 17

Mom! I Learn Addition Using Math-Chess-Puzzles Connection

Contents include both traditional and Math-Chess-Puzzles combined methods. Extra strength
©2008 – 2018 Frank Ho, Amanda Ho   All rights reserved. www.homathchess.com

Student Name _____ Date _____

d + d

7 + 7 14	7 + 7 14	7 + 7 14	7 + 8 15	8 + 7 15
7 + 8 15	8 + 7 15	7 + 8 15	8 + 7 15	8 + 7 15
8 + 7 15	7 + 8 15	8 + 7 15	7 + 8 15	8 + 7 15
7 + 8 15	8 + 7 15	7 + 8 15	8 + 7 15	8 + 7 15

Mom! I Learn Addition Using Math-Chess-Puzzles Connection

d + d

6	7	6	7	6
+ 6	+ 6	+ 7	+ 6	+ 7
12	13	13	13	13
7	6	7	6	6
+ 6	+ 7	+ 6	+ 7	+ 7
1 3	1 3	13	13	1 3
6	6	9	♛	6
+ ♛	+ 9	+ 6	+ 6	+ ♛
15	15	15	15	15
8	6	6	8	6
+ 6	+ 8	+ 8	+ 6	+ 8
14	14	14	14	14

Ho Math Chess 何数棋谜 妈！我会棋谜式加法啦！
Mom! I Learn Addition Using Math-Chess-Puzzles Connection
Contents include both traditional and Math-Chess-Puzzles combined methods. Extra strength
©2008 − 2018 Frank Ho, Amanda Ho   All rights reserved. www.homathchess.com

Student Name _____ Date _____

d + d

8	8	8	8	8
+ 8	+ 8	+ 8	+ 8	+ 8
16	16	16	16	16
8	9	♛	8	8
+ 9	+ 8	+ 8	+ ♛	+ ♛
17	17	17	17	17
8	♛	8	9	8
+ ♛	+ 8	+ ♛	+ 8	+ 9
17	17	17	17	17
♛	8	♛	8	9
+ 8	+ 9	+ 8	+ ♛	+ 8
17	17	17	17	17

134

d + d

6 + 6 = 12	6 + 6 = 12	6 + 6 = 12	7 + 6 = 13	6 + 7 = 13
7 + 6 = 13	6 + 7 = 13	7 + 6 = 13	6 + 7 = 13	6 + 7 = 13
8 + 6 = 14	6 + 8 = 14	6 + 8 = 14	6 + 8 = 14	8 + 6 = 14
♛ + 6 = 15	6 + 9 = 15	♛ + 6 = 15	6 + ♛ = 15	6 + 9 = 15

d + d

♖ + ♖ = 10	5 + 5 = 10	5 + 6 = 11	6 + ♖ = 11	♖ + 6 = 11
8 + 5 = 13	5 + 8 = 13	♖ + 8 = 13	8 + 5 = 13	5 + 8 = 13
7 + 5 = 12	7 + 5 = 12	5 + 7 = 12	♖ + 7 = 12	7 + ♖ = 12
♖ + 9 = 14	9 + 5 = 14	♖ + 9 = 14	9 + ♖ = 14	5 + 9 = 14

d + d with missing operator or number

Fill in □ with a number or + operator

7 + 7 = □14	7 + 7 = □14
7 + 7 = □14	7 + 7 = □14
7□ + 7 = 14	7□ + 7 = 14
7□ + 7 = 14	7□ + 7 = 14
7 + □7 = 14	7 + □ 7 = 14
7 + □7 = 14	7 + □ 7 = 14
7 □+ 7 = 14	7 □+7 = 14
7 □+7 = 14	7 □+7 = 14
7 □+ 7 = 14	7 □+ 7 = 14
7 □+ 7 = 14	7 □+ 7 = 14
7+□ 7 = 14	7+□ 7 = 14
7+□ 7 = 14	6+□ 8 = 14
seven □+ seven = 14	seven □+ seven = 14
seven □+ seven = 14	seven □+ seven = 14
seven □ +7 = 14	seven □+7 = 14
seven +□7 = 14	seven +□7= 14
7+□□ seven = 14	7+□ seven = 14
7+□□ Seven =14	7+□ Seven =14

d + d with missing operator or number

Fill in □ with a number or + operator

7 + 8 = □15	7 + 8 = □15
8 + 7 = □15	8 + 7 = □15
7□ + 8 = 15	7□ + 8 = 15
8□ + 7 = 15	8□ + 7 = 15
8 + □ 7 = 15	8 + □7 = 15
7 + □ 8 = 15	7 + □8 = 15
8 □+7 = 15	8 □+7 = 15
7 □ +8= 15	7 □+8 = 15
7 □+ 8 = 15	7 □+ 8 = 15
8 □+ 7 = 15	8 □+ 7 = 15
8+□ 7 = 15	8+□ 7 = 15
7+□ 8 = 15	7+□ 8 = 15
Seven □+ eight = 15	Seven □+ eight = 15
Eight □+ seven = 15	Eight □+ seven = 15
Seven □+8 = 15	Seven □ +8= 15
eight □ +7 = 15	eight □ +7= 15
7+□ eight = 15	7+□ eight = 15
8+□ Seven =15	8+□Seven =15

d + d with missing operator or number

Fill in □ with a number or + operator

7 + 6 = □13	7 + 6 = □13
6 + 7 = □13	6 + 7 = □13
7□ + 6 = 13	7□ + 6 = 13
6□ + 7 = 13	6□ + 7 = 13
6 + □7 = 13	6 + □ 7 = 13
7 + □+6 = 13	7 + □ 6 = 13
6 □□+7 = 13	6 □□+7 = 13
7 □□ +6= 13	7 □□+6 = 13
7 □+ 6 = 13	7 □+ 6 = 13
6 □+ 7 = 13	6 □+ 7 = 13
7+ □□ 6 = 13	7+□□ 6 = 13
6+ □□ 7 = 13	6+□□ 7 = 13
six □ + seven = 13	six □+ seven = 13
seven □ + six = 13	seven □+ six = 13
seven □ □ +6 = 13	seven □ □ +6 = 13
six □ □ +7 = 13	six □ □ +7= 13
6+ □□ seven = 13	6+ □□ seven = 13
7+ □□ six=13	7+ □□ six=13

d + d with missing operator or number

Fill in □ with a number or a + operator

6 + 8 = □14	6 + 8 = □14
8 + 6 = □14	8 + 6 = □14
6□ + 8 = 14	6□+ 8 = 14
8□ + 6 = 14	8□ + 6 = 14
8 + □6 = 14	8 + □ 6 = 14
6 + □ 8 = 14	6 + □ 8 = 14
8 □□ +6= 14	8 □ □ +6 = 14
6 □□ +8 = 14	6 □□ +8 = 14
6 □ +8 = 14	6 □ + 8 = 14
8 □+ 6 = 14	8 □ + 6 = 14
8+ □□ 6 = 14	8+ □□ 6 = 14
6+ □□ 8 = 14	8+ □□ 8 = 14
six □ + eight = 14	six □+ eight = 14
Eight □ + six = 14	Eight □+ six = 14
six □ +8 = 14	six □ □ +8 = 14
eight □ +6= 14	eight □ □ +6= 14
6+ □□ eight = 14	6+ □□ eight = 14
8+ □□ six =14	8+ □□ six =14

d + d with missing operator or number

Fill in □ with a number or a + operator

5 + 8 = □13	♖ + 8 = □13
8 + ♖ = □13	8 + 5= □13
5□ + 8 = 13	5□ + 8 = 13
8□ + 5= 13	8□ + ♖ = 13
8 + □5 = 13	8 + □5 = 13
♖ + □ 8 = 13	5+ □8 = 13
8 □□ +5 = 13	8 □□ +5= 13
5 □□ +8= 13	♖ □□ +8= 13
♖ □□ + 8 = 13	5 □□ +8 = 13
8 □+ 5 = 13	8 □ +5 = 13
8+ □□ 5= 13	8+ □□ ♖ = 13
5+ □□ 8 = 13	5+ □□ 8 = 13
five □ + eight = 13	five □ + eight = 13
Eight □ +five = 13	Eight □ + five = 13
five □ □ +8= 13	five □ □ +8 = 13
eight □ □ +5= 13	eight □ □ +5= 13
5+ □□ eight = 13	5+ □□ eight = 13
8+ □□ five =13	8+ □□ five =13

Ho Math Chess 何数棋谜　妈！我会棋谜式加法啦！
Mom! I Learn Addition Using Math-Chess-Puzzles Connection
Contents include both traditional and Math-Chess-Puzzles combined methods. Extra strength
©2008 — 2018 Frank Ho, Amanda Ho　　All rights reserved. www.homathchess.com

Student Name _____ Date_____

Doubling, up, and down

5					
4	8	9	8	7	6
3		5	5	5	
2	3	4	3	2	1
1					
	a	b	c	d	e

You are at c3 = ☐ (fill in a number.).

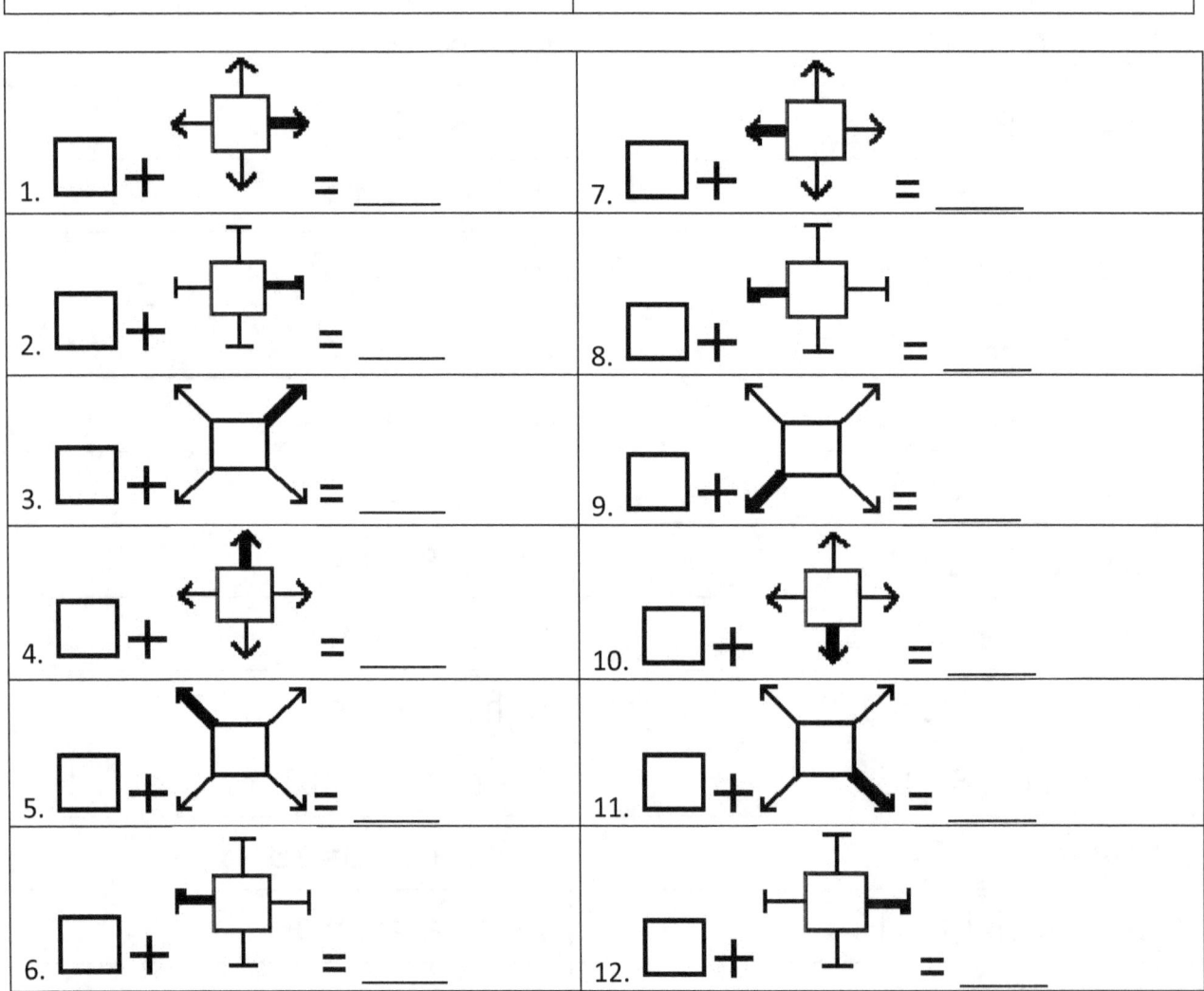

10 10   1 8   2 9   13 8
14 7
13 6

142

Ho Math Chess 何数棋谜 妈！我会棋谜式加法啦！
Mom! I Learn Addition Using Math-Chess-Puzzles Connection
Contents include both traditional and Math-Chess-Puzzles combined methods. Extra strength
©2008 — 2018 Frank Ho, Amanda Ho   All rights reserved. www.homathchess.com

Student Name _____ Date _____

Doubling, up, and down

5					
4	8	9	8	7	9
3		6	6	6	
2	5	4	3	2	1
1					
	a	b	c	d	e

You are at c3 = ☐ (fill in a number.).

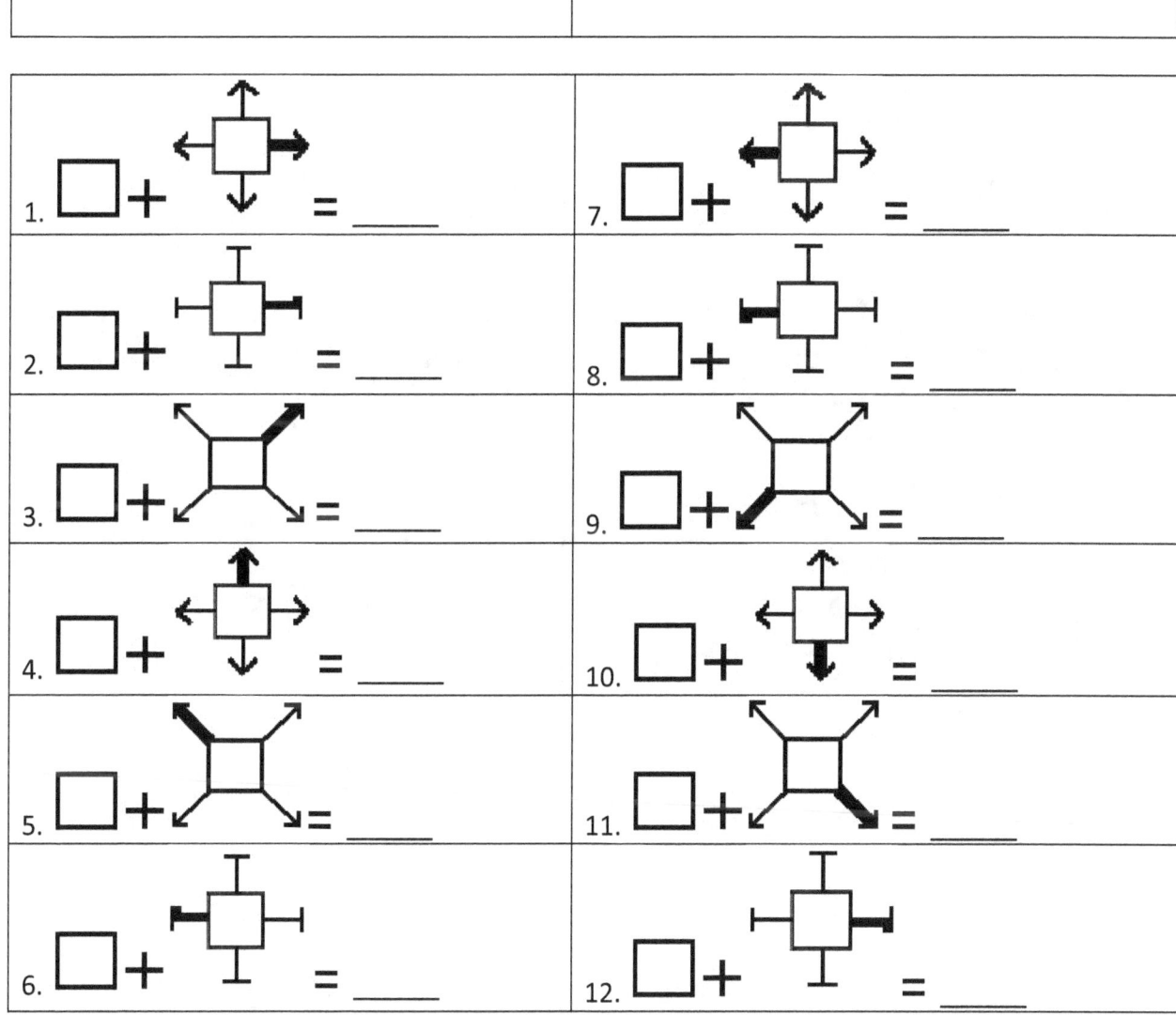

12, 12   15, 11   13, 10   14, 9   15, 8   14, 7

143

Ho Math Chess 何数棋谜   妈！我会棋谜式加法啦！
Mom! I Learn Addition Using Math-Chess-Puzzles Connection
Contents include both traditional and Math-Chess-Puzzles combined methods. Extra strength
©2008 – 2018 Frank Ho, Amanda Ho   All rights reserved. www.homathchess.com

Student Name _____   Date _____

Doubling, up, and down

5					
4	9	8	9	8	9
3		7	7	7	
2	6	5	4	3	2
1					
	a	b	c	d	e

You are at c3 = ☐ (fill in a number.).

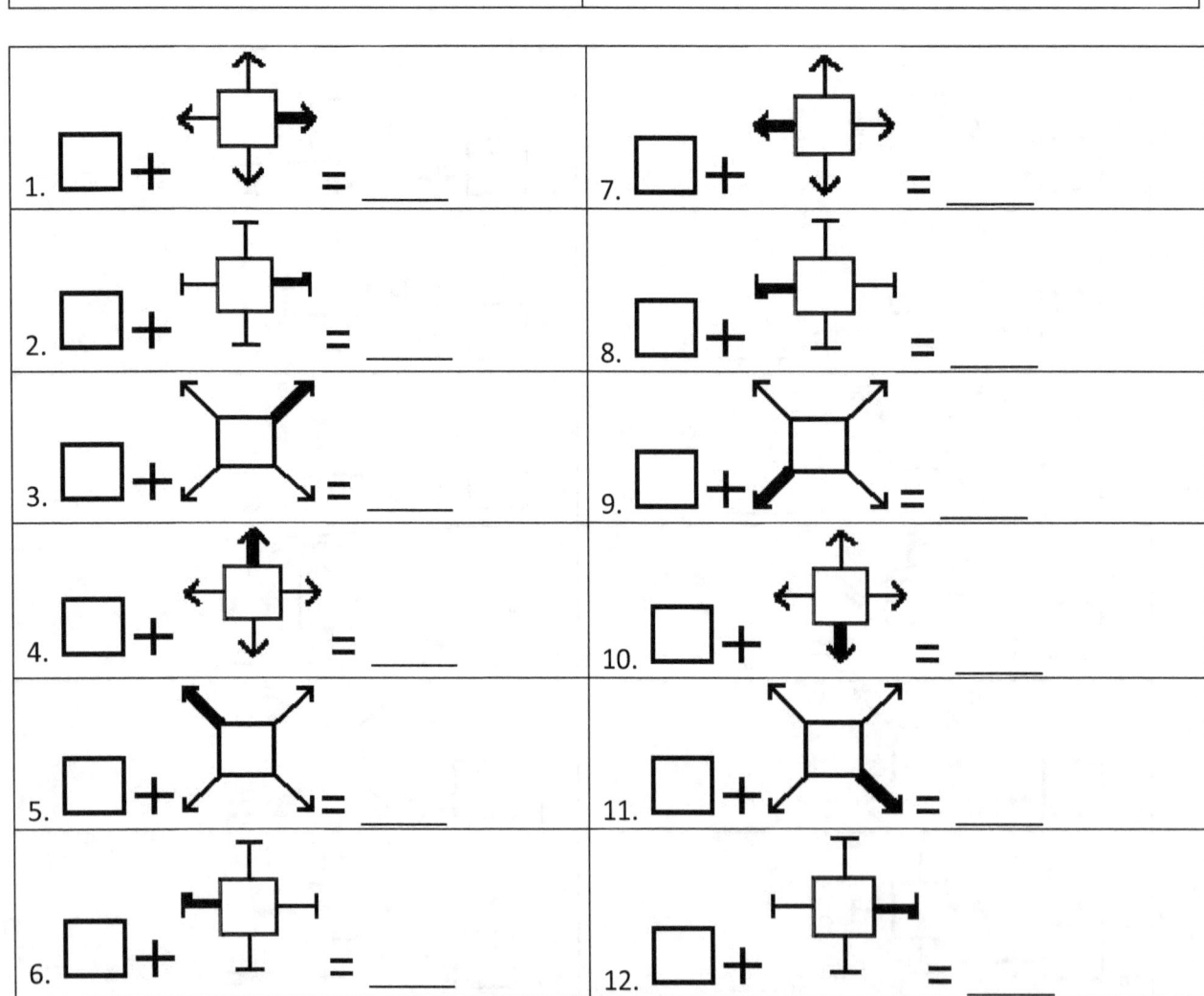

14, 14   16, 13   15, 12   16, 11   15, 10   16, 9

144

Ho Math Chess 何数棋谜   妈！我会棋谜式加法啦！
Mom! I Learn Addition Using Math-Chess-Puzzles Connection

Contents include both traditional and Math-Chess-Puzzles combined methods. Extra strength
©2008 – 2018 Frank Ho, Amanda Ho   All rights reserved. www.homathchess.com

Student Name_____ Date_____

Doubling

5	4		7		2
4		4	7	2	
3	9	9	1	8	8
2		3	6	5	
1	3		6		5
	a	b	c	d	e

You are a chess piece located at c3.

⤢+⤢ =____ + ____ = ____
5 + 5 =10

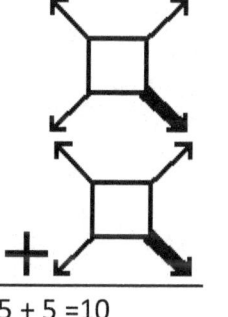
5 + 5 =10

⤢+⤢ =____ + ____ = ____
3 + 3 = 6

3 + 3 = 6

Doubling

	a	b	c	d	e
5	4		7		2
4		4	7	2	
3	9	9	1	8	8
2		3	6	5	
1	3		6		5

You are a chess piece located at c3.

◨ + ◨ = ___ + ___ = ___

4 + 4 = 8

4 + 4 = 8

◨ + ◨ = ___ + ___ = ___

2 + 2 = 4

2 + 2 = 4

Doubling

5	4		7		2
4		4	7	2	
3	9	9	1	8	8
2		3	6	5	
1	3		6		5
	a	b	c	d	e

You are a chess piece located at c3.

8+8=16

8+8=16

6 + 6 = 12

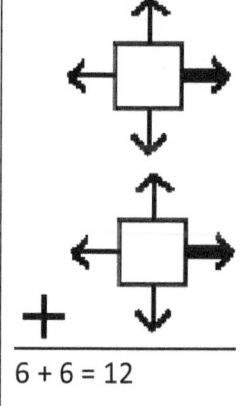

6 + 6 = 12

Adding strategy: adding to 10 凑十

Memorizing the following making up to 10. Replace each ? with a digit.

Remember 1 +9 = 2 + 8 = 3 + 7 = 4 + 6 = 5 + 5 = 10 (19, 28, 37, 46, 55)

For each column, the left digit + the right digit = 10					
9 1	**1 9**	9 1	1 9	9 1	? ?
8 2	**2 8**	8 2	? ?	8 2	2 ?
7 3	**3 7**	7 3	3 7	7 3	7 ?
6 4	**4 6**	6 4	? ?	6 4	? ?
5 5	**5 5**	5 5	? ?	5 5	? ?

9 1	? ?	? ?	1 9	9 1	1 ?
? ?	2 8	8 2	? ?	8 2	2 ?
7 3	? ?	? ?	3 7	7 3	? 7
? ?	4 6	6 4	? ?	6 4	? 6
? ?	5 5	? ?	5 5	5 5	? 5

? 1	1 ?	? ?	? 9	9 ?	? ?
8 ?	? 8	8 ?	? ?	? 2	2 ?
? 3	3 ?	? 3	3 ?	7 ?	? ?
6 ?	? 6	? ?	? 6	6 4	? 6
? 5	5 ?	5 5	? ?	? 5	5 ?

Ho Math Chess 何数棋谜 妈！我会棋谜式加法啦！
Mom! I Learn Addition Using Math-Chess-Puzzles Connection

Contents include both traditional and Math-Chess-Puzzles combined methods. Extra strength

©2008 – 2018 Frank Ho, Amanda Ho   All rights reserved. www.homathchess.com

Student Name _____ Date _____

Counting a pair of numbers to make 10.

9 8 7 6 5 5 3 2 1 4
1 2 3 4 5 5 6 7 8 9
5 4 6 7 8 3 2 1 9 5
7 4 3 2 1 9 6 5 8 5
8 3 4 6 5 2 1 9 7 5
4 5 7 6 3 1 9 2 8 5
3 6 8 7 2 9 1 5 4
8+1, 4+1, 7+1, 5+1, 6+1, 2+1, 1+1, 0+1, 4+1, 3+1
6-1, 9, 8-1, 9-1, 6-1, 7-1, 4-1, 5-1, 3-1, 1
2-1, 7+1, 4-1, 3+1, 6-1, 1+1, 8-1, 5+1, 8+1, 6-1

Ho Math Chess 何数棋谜 妈！我会棋谜式加法啦！
Mom! I Learn Addition Using Math-Chess-Puzzles Connection

Contents include both traditional and Math-Chess-Puzzles combined methods. Extra strength

©2008 – 2018 Frank Ho, Amanda Ho   All rights reserved. www.homathchess.com

Student Name _____ Date _____

Circle all tens.

9 + 1=	5 + 5=	9 + 2=
8 + 3=	4 + 6=	2 + 8=
7 + 4=	3 + 7=	8 + 2=
5 + 6=	6 + 4=	1 + 9=
8 + 3=	2 + 8=	7 + 2=
5 + 6=	2 + 9=	8 + 1=
4 + 6=	6 + 3=	3 + 6=
11 - 1=	7 + 2=	6 + 4=
12 - 2=	1 + 9=	5 + 5=
13 - 3=	3 + 8=	5 + 4=
14 - 4=	2 + 8=	6 + 5=
15 - 5=	4 + 5=	7 + 4=
16 - 6=	4 + 6=	8 + 2=
17 - 7=	7 + 2=	9 + 1=
18 - 8=	9 + 2=	9 + 2=
19 - 9=	8 + 3=	6 + 4=

Matching 2 digits in columns A an B such that their sum is 10.

A	B
1	7
3	9
5	3
7	5
2	6
4	8
8	2
6	4
9	1

A	B
7	2
9	1
3	5
5	8
6	7
8	6
2	3
4	4
1	9

A	B
5	6
6	2
7	5
2	3
3	9
1	4
4	7
8	8
9	1

A	B
2	6
4	4
6	5
7	7
9	3
1	2
3	1
5	8
8	9

Ho Math Chess 何数棋谜   妈！我会棋谜式加法啦！
Mom! I Learn Addition Using Math-Chess-Puzzles Connection

Contents include both traditional and Math-Chess-Puzzles combined methods. Extra strength
©2008 – 2018 Frank Ho, Amanda Ho   All rights reserved. www.homathchess.com

Student Name _____   Date _____

Filling each square box by a digit.

8 + 2 = ☐	2 + 8 = ☐
8 + ☐ = 10	2 + ☐ = 10
☐ + 2 = 10	☐ + 8 = 10
8 ☐ 2 = 10	2 ☐ 8 = 10
8 ☐ ☐ = 10	2 ☐ ☐ = 10
7 + 3 = ☐	3 + 7 = ☐
7 + ☐ = 10	3 + ☐ = 10
☐ + 3 = 10	☐ + 7 = 10
7 ☐ 3 = 10	3 ☐ 7 = 10
7 ☐ ☐ = 10	3 ☐ ☐ = 10

Mom! I Learn Addition Using Math-Chess-Puzzles Connection

Filling each square box by a digit.

9 + 1 = ☐	1 + 9 = ☐
9 + ☐ = 10	1 + ☐ = 10
☐ + 1 = 10	☐ + 9 = 10
9 ☐ 1 = 10	1 ☐ = 10
9 ☐ ☐ = 10	1 ☐ ☐
6 + 4 = ☐	4 + 6 = ☐
6 + ☐ = 10	4 + ☐ = 10
☐ + 4 = 10	☐ + 6 = 10
6 ☐ 4 = 10	4 ☐ 6 = 10
6 ☐ ☐ = 10	4 ☐ ☐ = 10

Ho Math Chess 何数棋谜 妈！我会棋谜式加法啦！
Mom! I Learn Addition Using Math-Chess-Puzzles Connection
Contents include both traditional and Math-Chess-Puzzles combined methods. Extra strength
©2008 – 2018 Frank Ho, Amanda Ho   All rights reserved. www.homathchess.com

Student Name _____ Date _____

each ☐ with a number.

Math and Chess sentence	Number sentence.	Comments
5 + ♙♙♙♙♙	$5 + 5 = 10$	
2 + ♙♙♙♙♙♙♙♙	2+8=10	Count right to left to make a ten. (Add the bigger number to the smaller number.)
♙♙♙♙♙♙♙ + 3	7+3=10	Count left to right to make a ten.
1 + ♙♙♙♙♙♙♙♙♙	3+7=10	Count right to left to make a ten.
♙♙♙♙♙♙♙♙♙ + 2	9+1=10	Count left to right to make a ten.
4 + ♙♙♙♙♙♙	1+9=10	Count right to left to make a ten.
♙♙♙♙♙♙♙♙ + 3	8+2=10	
2 + ♙♙♙♙♙♙♙♙	2+8=10	
6 + ♙♙♙♙	1+9=10	
♙♙♙♙♙ + 5	9+1=10	
4 + ♙♙♙♙♙♙	4+6=10	
6 + ♙♙♙♙	6+4=10	

10's complement

10's complement	10's complement	Make a ten
10 − 4 = 6	10 − 6 = 4	Four plus six is 10
10 − 1 = 9	10 − 9 = 1	One plus nine is 10
10 − 3 = 7	10 − 7 = 3	Three plus seven is 10
10 − 5 = 5	10 − 5 = 5	Five plus five is 10
10 − 7 = 3	10 − 3 = 7	Seven plus three is 10
10 − 9 = 1	10 − 1 = 9	Nine plus one is 10
10 − 6 = 4	10 − 4 = 6	Six plus four is 10
10 − 2 = 8	10 − 8 = 2	Two plus eight is 10
10 − 1 = 9	10 − 9 = 1	One plus nine is 10
10 − 8 = 2	10 − 2 = 8	Eight plus two is 10

Mom! I Learn Addition Using Math-Chess-Puzzles Connection

Contents include both traditional and Math-Chess-Puzzles combined methods. Extra strength

©2008 − 2018 Frank Ho, Amanda Ho   All rights reserved. www.homathchess.com

Student Name _____   Date _____

♚ = 0, ♙ = 1, ♗ = 3, ♘ = 3, ♖ = 5, ♕ = 9

♙ ♘ ♖ ♕ ♗ ♕

9	♙	2	8	3
+ 1	+ 9	+ 8	+ 2	+ 7
10	10	10	10	10
7	6	4	♖	♗
+ 3	+ 4	+ 6	+ 5	+ 7
10	10	10	10	10
7	2	8	♙	9
+ ♗	+ 8	+ 2	+ 9	+ ♙
10	10	10	10	10
8	2	7	♗	4
+ 2	+ 8	+ ♗	+ 7	+ 6
10	10	10	10	10
♖	4	6	♗	7
+ 5	+ 6	+ 4	+ 7	+ ♗
10	10	10	10	10

156

Ho Math Chess 何数棋谜　妈！我会棋谜式加法啦！
Mom! I Learn Addition Using Math-Chess-Puzzles Connection

Contents include both traditional and Math-Chess-Puzzles combined methods. Extra strength
©2008－2018 Frank Ho, Amanda Ho　　All rights reserved. www.homathchess.com

Making ten

♔ = 0, ♙(white) = 1, ♗(black) = 3, ♘ = 3, ♖ = 5, ♕ = 9

7	6	4	♖	♗
+ ♗	+ 4	+ 6	+ 5	+ 7
10	10	10	10	10
7	2	8	♙	9
+ 3	+ 8	+ 2	+ 9	+ 1
10	10	10	10	10
8	2	7	♘	4
+ 2	+ 8	+ 3	+ 7	+ 6
10	10	10	10	10
5	4	6	♗	7
+ ♖	+ 6	+ 4	+ 7	+ ♘
10	10	10	10	10
9	♙	2	8	♘
+ 1	+ 9	+ 8	+ 2	+ 7
10	10	10	10	10

Mom! I Learn Addition Using Math-Chess-Puzzles Connection

Associative property 結合律

(a + b) + c = a + (b + c) = (a + c) + b

1 + 2 + 3 = (1 + 2) + 3 = ☐3 + 3 = ☐ 6

1 + 2 + 3 = 1 + (2 + 3) = 1 + ☐5 = ☐6

1 + 2 + 3 = (1 + 3) + 2 = ☐4 + 2 = ☐6

❄ King = 0, ♙ Pawn = 1, ♝ Bishop = 3, ♞ Knight = 3, ♖ Rook = 5, ♕ Queen = 9

7	6	4	♖	3
2	♝	4	4	♖
+ 3	+ 4	+ 6	+ 5	+ 7
12	13	14	15	16

9	6	4	♖	♝
2	♖	♝	2	4
+ 1	+ 4	+ 7	+ 8	+ 6
12	15	14	15	13

Memorization of adding to 10

Fill in □ with a number.

♔=0, ♙=1, ♗=3, ♘=3, ♖=5, ♕=9

♙ ♖ ♘ ♔ ♗ ♕

1		2		♘		4		♖	
□9	♕	□8		□7	7	□6		□5	6
10	□1	10	8	10	□3	10	6	10	□4
	10		□2		10		□4		10
4		♘	10	2		♗	10	4	
□6	7	□7		□8	♕	□7		□6	♘
10	□3	10	3	10	□1	10	8	10	□7
	10		□7		10		□2		10
♗		2	10	♘		4	10	♖	
□7	9	□8		□7	7	□6		□5	6
10	□1	10	4	10	□3	10	6	10	□7
	10		□		10		□4		10
4		♘	10	2		♗	10	6	
□6	7	□7		□8	♕	□7		□4	♘
10	□3	10	5	10	□1	10	8	10	□7
	10		□5		10		□2		10
			10				10		

159

Mom! I Learn Addition Using Math-Chess-Puzzles Connection

Contents include both traditional and Math-Chess-Puzzles combined methods. Extra strength

©2008 – 2018 Frank Ho, Amanda Ho   All rights reserved. www.homathchess.com

Student Name _____   Date _____

Memorization of adding to 10

Fill in □ with a number.

✻ ♚ = 0, ♖ ♙ = 1, ♗ = 3, ♘ = 3, ♜ ♖ = 5, ✻ ♕ = 9

	2		5		6		2		7
1	□8	2	□5	♘	□4	4	□8	♖	□3
□9	10	□8	10	□7	10	□6	10	□5	10
10		10		10		10		10	
	6		♕		♘		4		8
4	□4	♘	□1	2	□7	1	□6	6	□2
□6	10	□7	10	□8	10	□9	10	□4	10
10		10		10		10		10	
	♕		6		2		4		6
7	□1	2	□4	♗	□8	4	□6	♜	□4
□3	10	□8	10	□7	10	□6	10	□5	10
10		10		10		10		10	
	6		4		8		3		7
4	□4	3	□6	2	□2	1	□7	6	□3
□6	10	□7	10	□8	10	□9	10	□4	10
10		10		10		10		10	
	♗		7		6		4		♕
	□7		□3		□4		□6		□1
	10		10		10		10		10

160

Ho Math Chess 何数棋谜   妈！我会棋谜式加法啦！
Mom! I Learn Addition Using Math-Chess-Puzzles Connection

Contents include both traditional and Math-Chess-Puzzles combined methods. Extra strength

©2008 – 2018 Frank Ho, Amanda Ho   All rights reserved. www.homathchess.com

Student Name _____ Date _____

Grouping 10

Fill in the following the ☐ with a number.

✻ ♔ = 0, ♙ = 1, ♗ = 3, ♘ = 3, ♖ = 5, ✻ ♕ = 9

6 + 4 + 2 + ♖ + 3 + ♗ + 7 = ☐30

3 + 2 + 7 + 8 + 2 = ☐22

♙ + 2 + 6 + 9 + 4 = ☐22

8 + 4 + ♖ + 6 + 5 + 2 = ☐30

♙ + 6 + 9 + 4 + 9 = ☐29

2 + ☐ + 8 + 7 + 5 = ☐25

9 + 6 + 3 + 7 + 4 + ♙ + 5 + ♖ = ☐40

2 + ☐ + 6 + 7 + 4 + 8 + 1 + 9 + 3 = ☐43

Addition strategy: 9 plus 1 (9 + d = 9 + 1 + (d−1))   (9 加 1, 9 + d = 9 + 1 + (d−1))

9 + 6 = 9 + 1 + 5 = 10 + 5 = 15

♔ = 0, ♙ = 1, ♗ = 3, ♘ = 3, ♖ = 5, ♕ = 9

The pre-requisite to use the above strategy is that the students know how to partition a number into a sum of 1 + a smaller number such as 4 = 1 + 3, 5 = 1+ 4 etc. I have taught some students who got confused on this strategy, one of the reasons is they do not understand 9 can be 1 + 8.

4 = 1 + ____
6 = 1 + ____
5 = 1 + ____
7 = 1 + ____
8 = 1 + ____
9 = 1 + ____
3 = 1 + ____
2 = 1 + ____
4 = ____ + 1
5 = ____ + 1
7 = ____ + 1
6 = ____ + 1
8 = ____ + 1
9 = ____ + 1
4 = ____ + 3
6 = ____ + 5
5 = ____ + 4

If 10 + 1 = 11, then 9 + 2 = 9 + 1 + ___ 1 = 10 + ___ 1= ____ 11

If 10 + 2 = 12, then 9 + 3 = 9 + 1 + ___ = 10 + ___ = ____

If 10 + 3 = 13, then 9 + 4 = 9 + 1 + ___ = 10 + ___ = ____

If 10 + 4 = 14, then 9 + 5 = 9 + 1 + ___ = 10 + ___ = ____

If 10 + 5 = 15, then 9 + 6 = 9 + 1 + ___ = 10 + ___ = ____

If 10 + 6 = ___16, then 9 + 7 = 9 + 1 + ___ = 10 + ___ = ____

If 10 + 7 = ___17, then 9 + 8 = 9 + 1 + ___ = 10 + ___ = ____

If 10 + 8 = ___18, then 9 + 9 = double 9 = = ____

Ho Math Chess 何数棋谜 妈！我会棋谜式加法啦！
Mom! I Learn Addition Using Math-Chess-Puzzles Connection
Contents include both traditional and Math-Chess-Puzzles combined methods. Extra strength
©2008 – 2018 Frank Ho, Amanda Ho   All rights reserved. www.homathchess.com

Student Name_____   Date_____

9	6	9	2	9
+ 5	+ 9	+ 2	+ 9	+ 3
14	15	11	11	12
5	9	9	8	9
+ 9	+ 4	+ 7	+ 9	+ 5
14	13	16	17	14
9	3	9	4	5
+ 6	+ 9	+ 6	+ 9	+ 9
15	12	15	13	14
9	6	7	9	9
+ 8	+ 9	+ 9	+ 8	+ 8
17	15	16	17	17
8	7	6	9	3
+ 9	+ 9	+ 9	+ 5	+ 9
17	16	15	14	12

Mom! I Learn Addition Using Math-Chess-Puzzles Connection

Contents include both traditional and Math-Chess-Puzzles combined methods. Extra strength
©2008 — 2018 Frank Ho, Amanda Ho   All rights reserved. www.homathchess.com

Student Name_____ Date_____

♔ = 0, ♙ = 1, ♗ = 3, ♘ = 3, ♖ = 5, ♕ = 9

3 + ♕ 12	10 + 4 14	9 + 4 13	4 + 9 13	10 + 5 15
9 + ♖ 14	♖ + 9 14	10 + 6 16	♕ + 6 15	6 + 9 15
10 + 7 17	♕ + 7 16	7 + 9 16	10 + 8 18	9 + 8 17
8 + 9 17	10 + 9 19	♕ + 9 18	10 + 7 17	♕ + 7 16

165

Mom! I Learn Addition Using Math-Chess-Puzzles Connection

Adding 9 by relating

Adding 9 +d by splitting the other number d to be 1 and d − 1

✲ ♔ = 0, ♕ ♙ = 1, ⛥ ♝ = 3, ✜ ♞ = 3, ✦ ♖ = 5, ❆ ♛ = 9

⌐+ 3 +⌐	⌐+ 5 +⌐	⌐+ 4 +⌐
10 + 9	10 + 9	10 + ♛
↓ ♛ ↓	↓ ♛ ↓	↓ ♛ ↓
◯ ◯ ◯	◯ ◯ ◯	◯ ◯ ◯
13 12 12	15 14 14	14 13 13
⌐+ 6 +⌐	⌐+ 8 +⌐	⌐+ ♛ +⌐
10 + 9	10 + 9	10 + 9
↓ ♛ ↓	↓ ♛ ↓	↓ ♛ ↓
◯ ◯ ◯	◯ ◯ ◯	◯ ◯ ◯
16 15 15	18 17 17	19 18 18
⌐+ ♖ +⌐	⌐+ ♝ +⌐	⌐+ ♙ +⌐
8 + 9	8 + 9	8 + 9
↓ ♛ ↓	↓ ♛ ↓	↓ ♛ ↓
◯ ◯ ◯	◯ ◯ ◯	◯ ◯ ◯
13 14 14	11 12 12	9 10 10
⌐+ 2 +⌐	⌐+ ♝ +⌐	⌐+ 2 +⌐
10 + 9	7 + 8	8 + 10
↓ ♛ ↓	↓ ♛ ↓	↓ ♛ ↓
◯ ◯ ◯	◯ ◯ ◯	◯ ◯ ◯
12 11 11	10 12 11	10 11 12

Mom! I Learn Addition Using Math-Chess-Puzzles Connection

Contents include both traditional and Math-Chess-Puzzles combined methods. Extra strength

©2008 – 2018 Frank Ho, Amanda Ho   All rights reserved. www.homathchess.com

Student Name _____ Date _____

Adding 9 by relating

Adding 9 +d by splitting the other number d to be 1 and d − 1

(♔ = 0, ♙ = 1, ♗ = 3, ♘ = 3, ♖ = 5, ♕ = 9)

↱+ ♕ +↲ ♙ + 2 □ ♙ □ + ♙ □ 10, 11, 11	↱+ ♕ +↲ 2 + ♗ □ 2 □ + ♙ □ 11, 12, 12	↱+ ♕ +↲ ♗ + 4 □ ♗ □ + ♙ □ 12, 13, 13
↱+ 9 +↲ 4 + ♖ □ 4 □ + 1 □ 13, 14, 14	↱+ ♕ +↲ ♖ + 6 □ ♖ □ + 1 □ 14, 15, 15	↱+ ♕ +↲ 6 + 7 □ 6 □ + 1 □ 15, 16, 16
↱+ ♕ +↲ 7 + 8 □ 7 □ + 1 □ 16, 17, 17	↱+ 9 +↲ 8 + ♕ □ 8 □ + 1 □ 17, 18, 18	↱+ 8 +↲ 8 + ♕ □ 8 □ + ♙ □ 16, 17, 17

167

Adding 9 by relating

Adding 9 + d by splitting the other number d to be 1 and d − 1

Adding 9 by relating

Adding 9 +d by splitting the other number d to be 1 and d − 1

(♔ = 0, ♙ = 1, ♗ = 3, ♘ = 3, ♖ = 5, ♕ = 9)

♖ + 7 ↓ 6 ↓ ○ ○ ○ 14 15 16	6 + 6 ↓ 7 ↓ ○ ○ ○ 15 16 15	7 + ♕ ↓ 8 ↓ ○ ○ ○ 16 17 18
2 + 4 ↓ ♗ ↓ ○ ○ ○ 11 12 13	♗ + ♖ ↓ 4 ↓ ○ ○ ○ 12 13 14	4 + 6 ↓ ♖ ↓ ○ ○ ○ 13 14 15
♗ + ♖ ↓ 4 ↓ ○ ○ ○ 12 13 14	4 + 6 ↓ ♖ ↓ ○ ○ ○ 13 14 15	♖ + 7 ↓ 6 ↓ ○ ○ ○ 14 15 16
4 + 6 ↓ ♖ ↓ ○ ○ ○ 13 14 15	♖ + 7 ↓ 6 ↓ ○ ○ ○ 14 15 16	6 + 8 ↓ 7 ↓ ○ ○ ○ 15 16 17

Ho Math Chess 何数棋谜   妈！我会棋谜式加法啦！

Mom! I Learn Addition Using Math-Chess-Puzzles Connection

Contents include both traditional and Math-Chess-Puzzles combined methods. Extra strength

©2008 — 2018 Frank Ho, Amanda Ho   All rights reserved. www.homathchess.com

Student Name _____   Date _____

Using 9 plus technique

3	229	345	564
2	765	**999**	989
1	548	888	899
	a	b	c

The original square is at b2 = ☐.

```
  0708090204040503
+ 9909090909090909
  ────────────────
  10617181113151412
```

b2 + ✛ = ☐☐☐ + ○○○ = 1887

☐ + ✕ = ☐☐☐ + ○○○ = 1898

b2 + ✛ = ☐☐☐ + ○○○ = 1344

☐ + ↔ = 1332, ☐☐☐ + ○○○ = 1764

b2 + ✕ = ☐☐☐ + ○○○ = 1547

☐ + ✛ = ☐☐☐ + ○○○ = 1869

b2 + ✕ = ☐☐☐ + ○○○ = 1563

☐ + ✕ = ☐☐☐ + ○○○ = 1228

Ho Math Chess 何数棋谜   妈！我会棋谜式加法啦！
Mom! I Learn Addition Using Math-Chess-Puzzles Connection

Contents include both traditional and Math-Chess-Puzzles combined methods. Extra strength

©2008 – 2018 Frank Ho, Amanda Ho   All rights reserved. www.homathchess.com

Student Name _____ Date _____

9 plus

Adding 9 + d by splitting the other number d to be 1 and d − 1

5		9		9	
4	8	6	1	7	6
3		3	9	2	
2	5	8	4	5	7
1		9		9	
	a	b	c	d	e

You are at square c3

= ☐.

10 16
11 14
13 17
12 15

171

9 plus

Adding 9 + d by splitting the other number d to be 1 and d − 1

5		9		9	
4	8	6	1	7	6
3		3	9	2	
2	5	8	4	5	7
1		9		9	
	a	b	c	d	e

You are at square c3

= ☐.

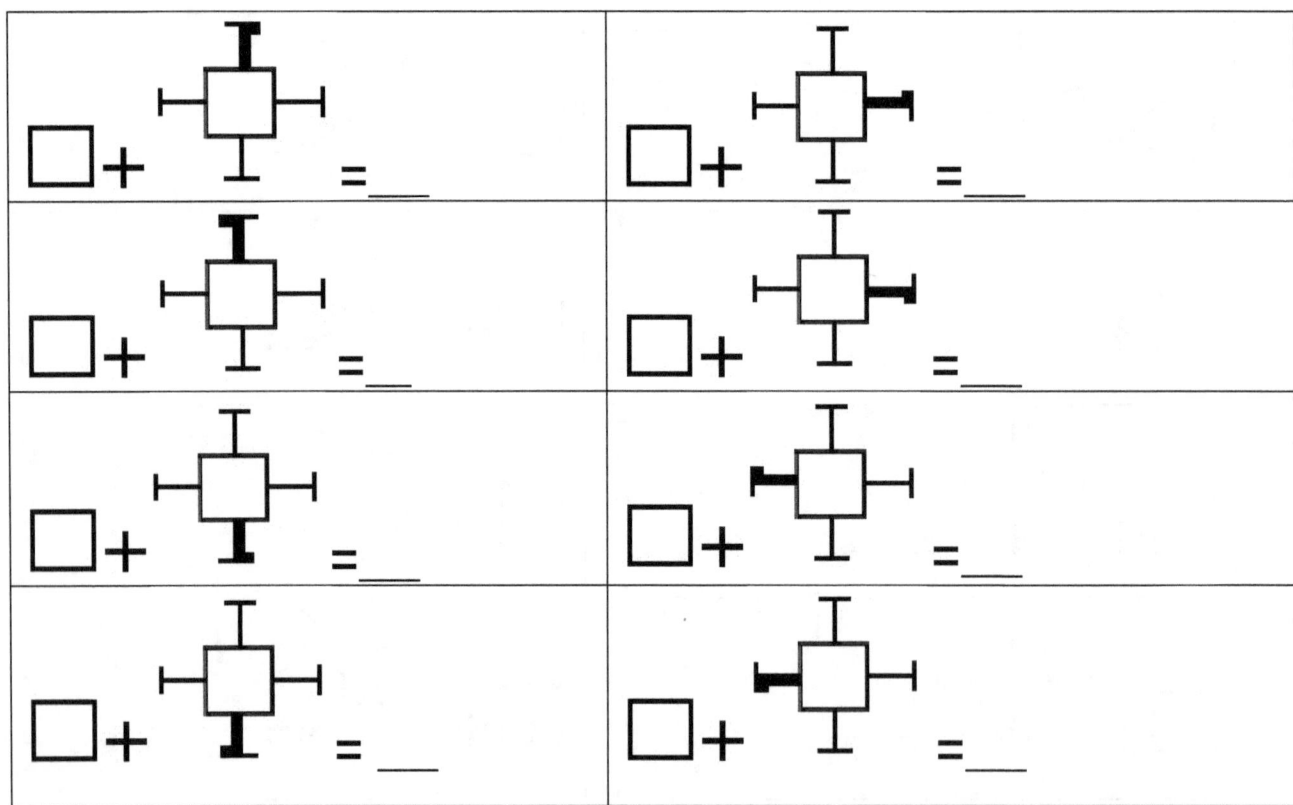

18 15
18 16
18 17
18 14

Mom! I Learn Addition Using Math-Chess-Puzzles Connection

Contents include both traditional and Math-Chess-Puzzles combined methods. Extra strength
©2008 – 2018 Frank Ho, Amanda Ho   All rights reserved. www.homathchess.com

Student Name _____ Date _____

Adding 9 + d by splitting the other number d to be 1 and d − 1

	2	1	2
	1	9	2
	1	3	3

You are a chess piece located at a square indicated by a shaded square.

○ = +

9+1+1=10+1

9+3=9+1+2=12 (Spit 3 to be 12)

10 10 10 11
11 11 11 12
12 12 12 10
10 10 10 11

173

Mom! I Learn Addition Using Math-Chess-Puzzles Connection

Contents include both traditional and Math-Chess-Puzzles combined methods. Extra strength
©2008 − 2018 Frank Ho, Amanda Ho   All rights reserved. www.homathchess.com

Student Name_____   Date_____

Adding 9 +d by splitting the other number d to be 1 and d − 1

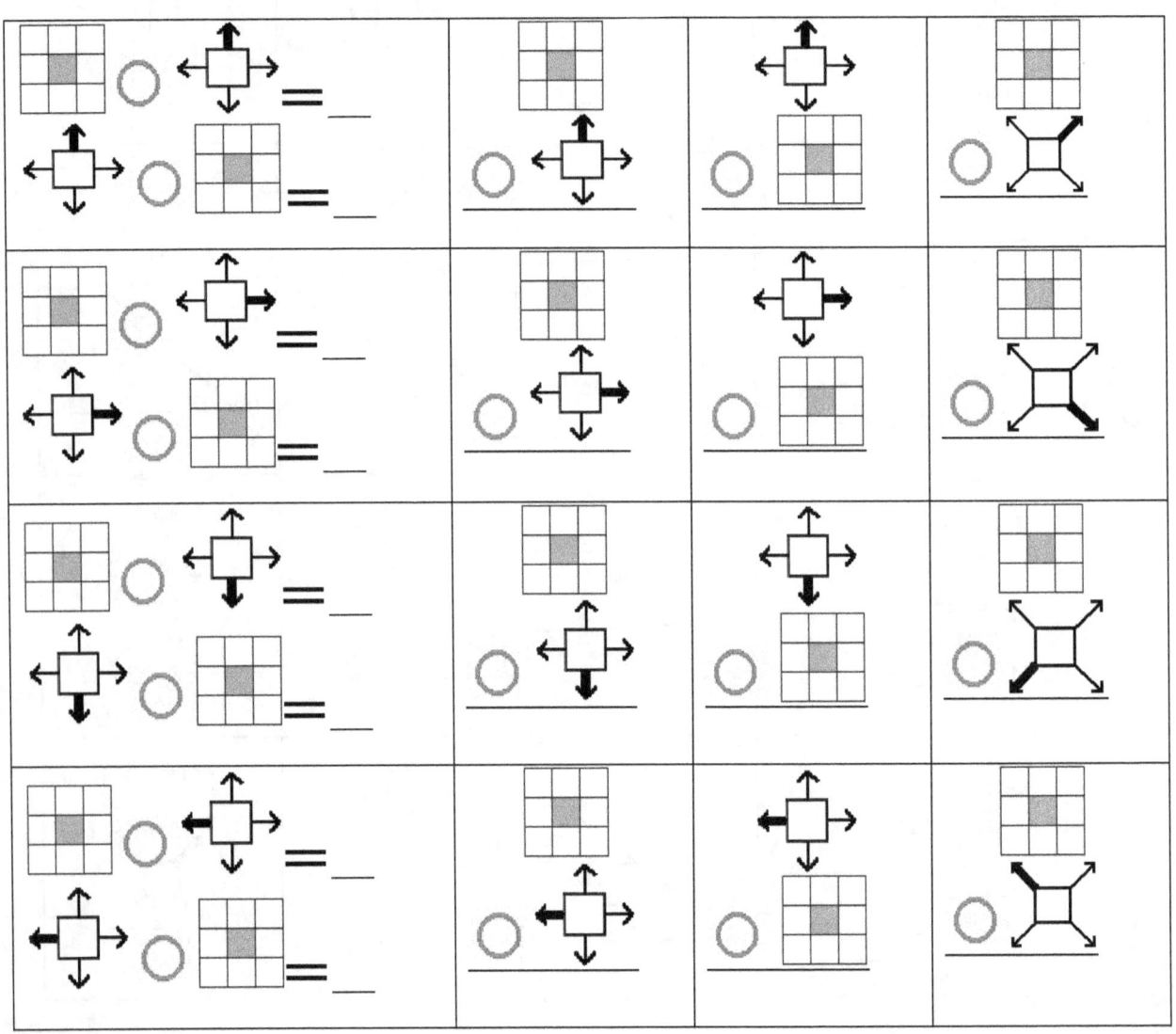

10 10 10 11
10 10 10 11
11 11 11 12
12 12 12 10

174

Ho Math Chess 何数棋谜   妈！我会棋谜式加法啦！
Mom! I Learn Addition Using Math-Chess-Puzzles Connection

Contents include both traditional and Math-Chess-Puzzles combined methods. Extra strength

©2008 – 2018 Frank Ho, Amanda Ho   All rights reserved. www.homathchess.com

Student Name _____   Date _____

Adding 9 + d by splitting the other number d to be 1 and d − 1

You are a chess piece located at a square indicated by a shaded square.

○ = +

12 12 12 10
10 10 10 11
10 10 10 11
11 11 11 12

175

Ho Math Chess 何数棋谜　妈！我会棋谜式加法啦！
Mom! I Learn Addition Using Math-Chess-Puzzles Connection
Contents include both traditional and Math-Chess-Puzzles combined methods. Extra strength
©2008 – 2018 Frank Ho, Amanda Ho   All rights reserved. www.homathchess.com

Student Name_____   Date_____

Adding 9 + d by splitting the other number d to be 1 and d − 1

You are a chess piece located at a square indicated by a shaded square.

○ = +

11 11 11 12
12 12 12 10
10 10 10 11
10 10 10 11

176

Mom! I Learn Addition Using Math-Chess-Puzzles Connection

Contents include both traditional and Math-Chess-Puzzles combined methods. Extra strength

©2008 − 2018 Frank Ho, Amanda Ho   All rights reserved. www.homathchess.com

Student Name _____ Date _____

Adding 9 +d by splitting the other number d to be 1 and d − 1

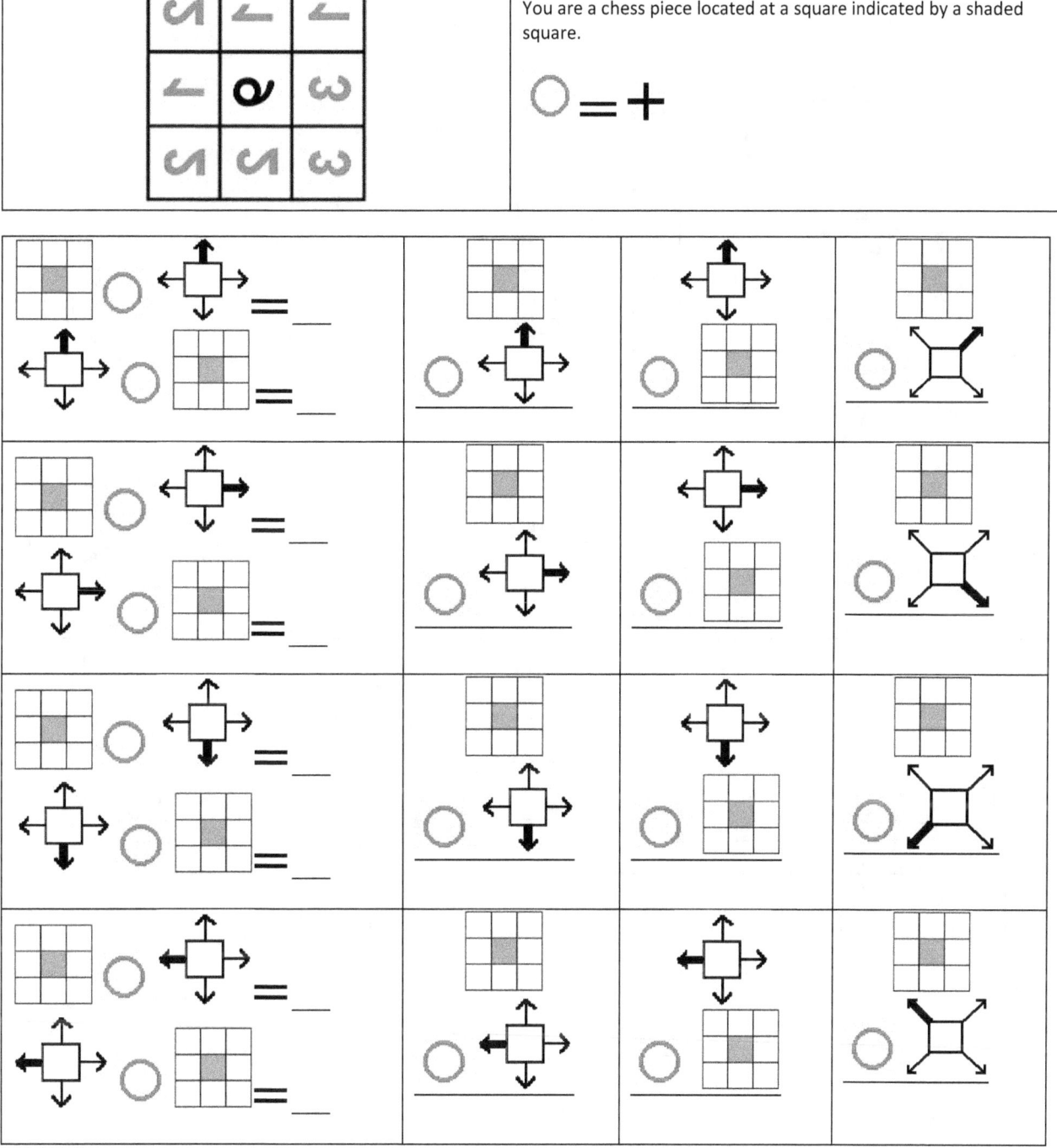

10 10 10 10
12 12 12 1217
11 11 11 11
10 10 10 11

177

Ho Math Chess 何数棋谜 妈！我会棋谜式加法啦！
Mom! I Learn Addition Using Math-Chess-Puzzles Connection
Contents include both traditional and Math-Chess-Puzzles combined methods. Extra strength
©2008 – 2018 Frank Ho, Amanda Ho   All rights reserved. www.homathchess.com

Student Name _____ Date _____

Adding 9 +d by splitting the other number d to be 1 and d − 1

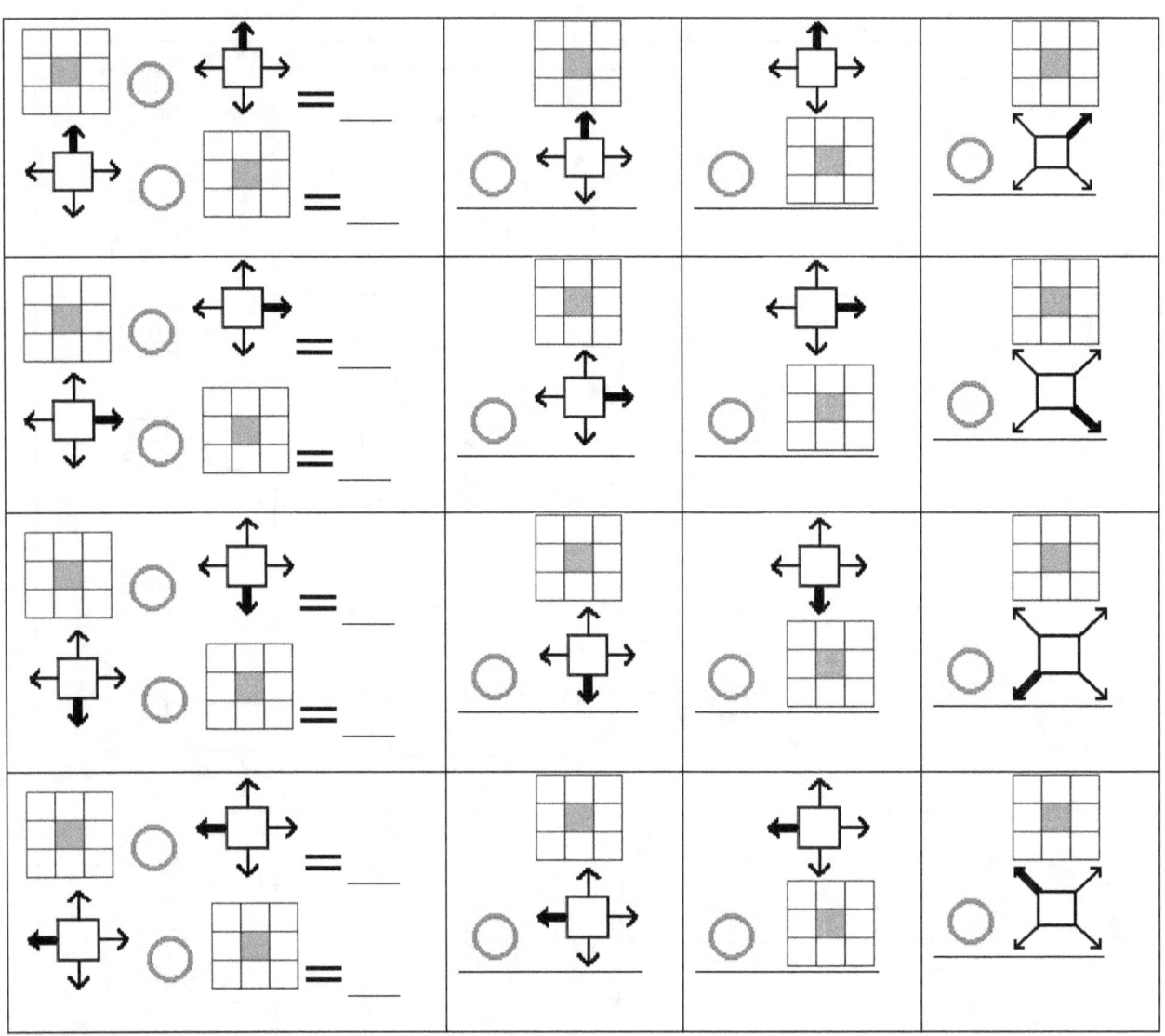

11 11 12 12 13 13 14 14
11 12 13 14
11 12 13 14
18 15 16 17

178

Mom! I Learn Addition Using Math-Chess-Puzzles Connection

Contents include both traditional and Math-Chess-Puzzles combined methods. Extra strength
©2008 – 2018 Frank Ho, Amanda Ho   All rights reserved. www.homathchess.com

Student Name_____   Date_____

Adding 9 + d by splitting the other number d to be 1 and d − 1
Only transfer each number to a correct position. Do not the transfer the 3 by 3 square.

14 14 11 11 12 12 13 13   14 11 12 13   14 11 12 13   17 18 15 16

179

Ho Math Chess 何数棋谜　妈！我会棋谜式加法啦！
Mom! I Learn Addition Using Math-Chess-Puzzles Connection
Contents include both traditional and Math-Chess-Puzzles combined methods. Extra strength
©2008 – 2018 Frank Ho, Amanda Ho　　All rights reserved. www.homathchess.com

Student Name _____　Date _____

Adding 9 + d by splitting the other number d to be 1 and d − 1
Only transfer each number to a correct position. Do not the transfer the 3 by 3 square.

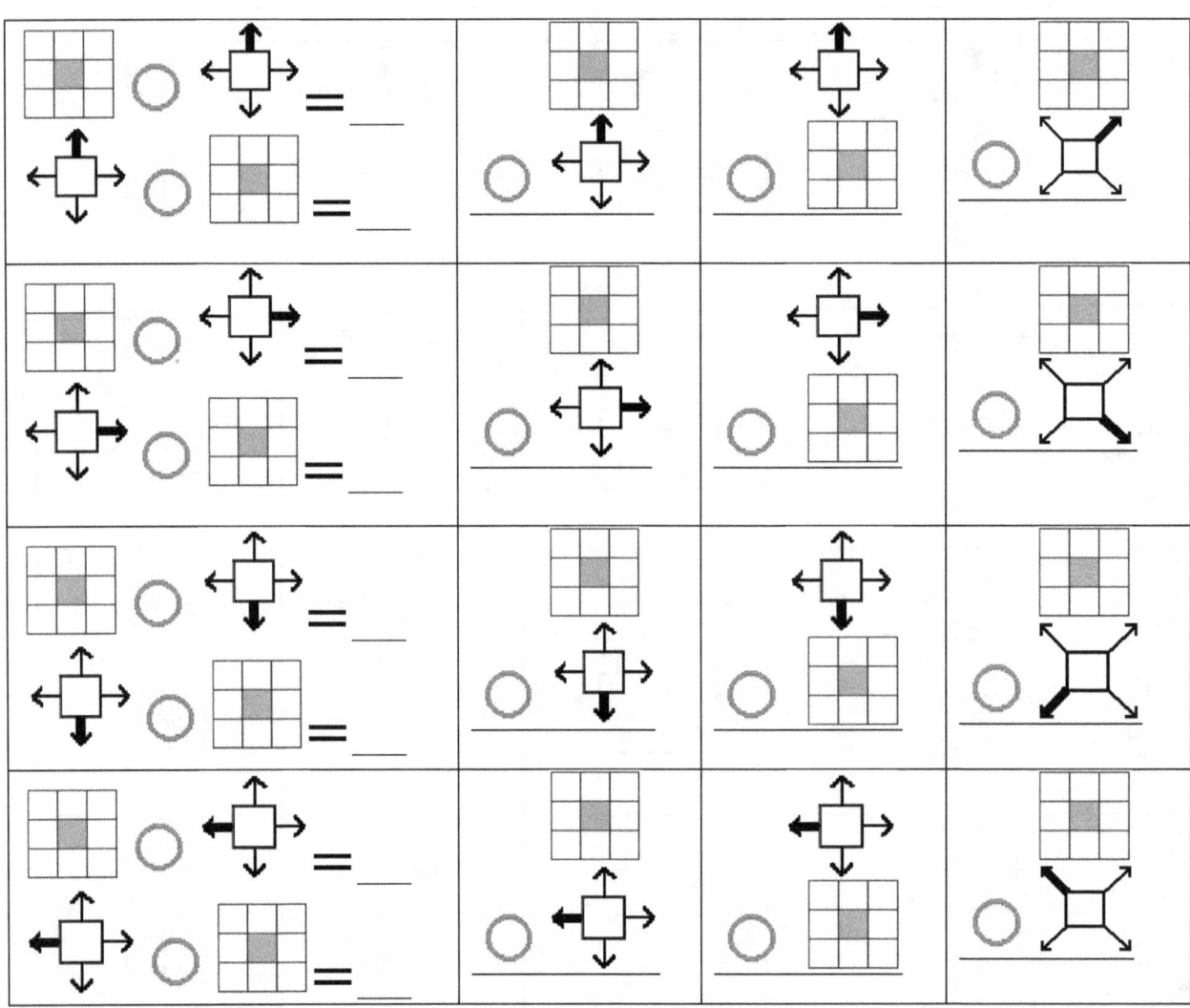

13, 13, 14, 14, 11, 11, 12, 12
13, 14, 11, 12
13, 14, 11, 12
16, 17, 18, 15

Ho Math Chess 何数棋谜　妈！我会棋谜式加法啦！
Mom! I Learn Addition Using Math-Chess-Puzzles Connection
Contents include both traditional and Math-Chess-Puzzles combined methods. Extra strength
©2008 − 2018 Frank Ho, Amanda Ho   All rights reserved. www.homathchess.com

Student Name_____ Date_____

Adding 9 +d by splitting the other number d to be 1 and d − 1
Only transfer each number to a correct position. Do not the transfer the 3 by 3 square.

You are a chess piece located at a square indicated by a shaded square.

1 12 13 13 14 14 11 11
12 13 14 11
12 13 14 11
15 16 17 18

Ho Math Chess 何数棋谜 妈！我会棋谜式加法啦！
Mom! I Learn Addition Using Math-Chess-Puzzles Connection
Contents include both traditional and Math-Chess-Puzzles combined methods. Extra strength
©2008 — 2018 Frank Ho, Amanda Ho All rights reserved. www.homathchess.com

Student Name_____ Date_____

Adding 9 +d by splitting the other number d to be 1 and d − 1
Only transfer each number to a correct position. Do not the transfer the 3 by 3 square.

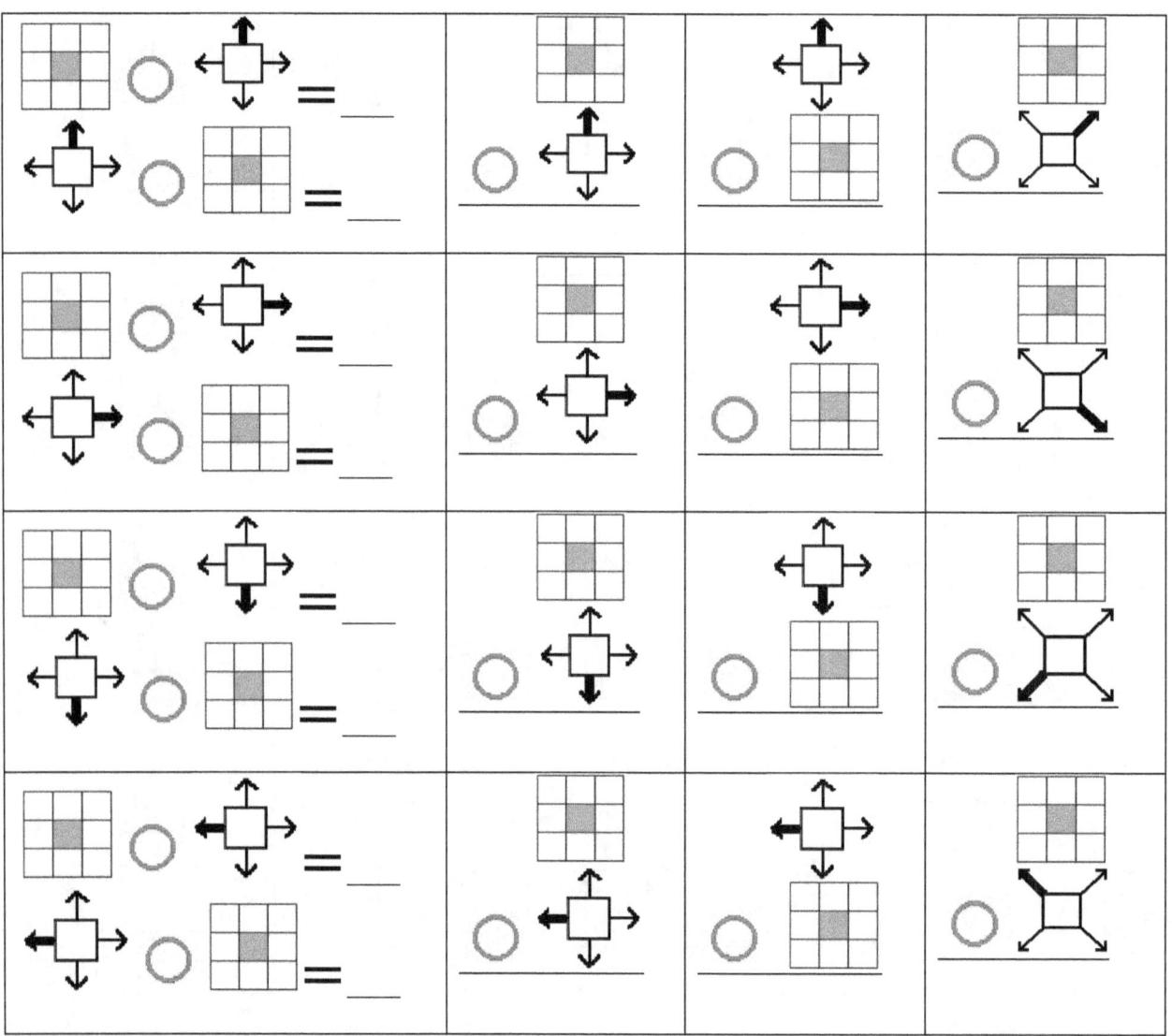

14 14 13 13 12 12 11 11
14 13 12 11
14 13 12 11
16 16 15 18

182

Mom! I Learn Addition Using Math-Chess-Puzzles Connection

Using doubles, 9 + d to add
9 + d = 9 + 1 + (d-1), 9 + 6 = 9 + **1** + **5** = 15

6 + 7 = 13   7 + 8 = 15   5 + 6 = 11

7 + 8 = 15   6 + 8 = 14   6 + 9 = 15

3 + 5 = 8   7 + 9 = 16   8 + 9 = 17

5 + 6 = 11   5 + 6 = 11   5 + 7 = 12

7 + 9 = 16   6 + 8 = 14   5 + 6 = 11
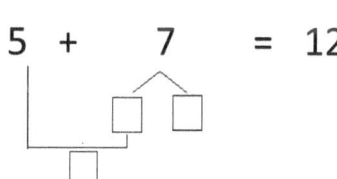

5 + 7 = 12   8 + 9 = 17   7 + 8 = 15
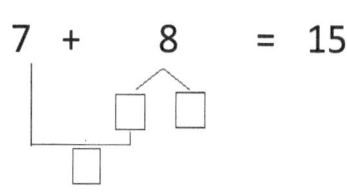

183

Mom! I Learn Addition Using Math-Chess-Puzzles Connection

Using doubles or 9 + d to add
(♔ = 0, ♙ = 1, ♗ = 3, ♘ = 3, ♖ = 5, ♕ = 9)

2 + 2 ☐4	2 + ♘ ☐5	♗ + ♗ ☐6	♗ + 4 ☐7	4 + 4 ☐8
3 + 2 ☐5	♘ + 3 ☐6	4 + 3 ☐7	4 + 4 ☐8	4 + ♖ ☐9
5 + 5 10	♖ + 6 11	6 + 6 12	6 + 7 13	7 + 7 14
6 + 5 11	6 + 6 12	7 + 6 13	7 + 7 14	8 + 7 15
8 + 8 16	8 + 9 17	♕ + ♕ 18	7 + 8 15	♕ + 8 17

Mom! I Learn Addition Using Math-Chess-Puzzles Connection

Using doubles or 9 + d to add
(♔ = 0, ♙ = 1, ♗ = 3, ♘ = 3, ♖ = 5, ♕ = 9)

2	3	4	5	6
+ 2	+ 3	+ 4	+ 5	+ 6
4	6	8	10	12
7	8	9	♘	4
+ 7	+ 8	+ 9	+ ♘	+ 4
14	16	18	6	8
13	13	♖	6	6
+ ♘	+ 4	+ ♖	+ 5	+ 6
16	17	10	11	12
6	7	7	7	8
+ 7	+ 6	+ 7	+ 8	+ 7
13	13	14	15	15
8	8	♕	9	6
+ 8	+ ♕	+ 8	+ 9	+ 5
16	17	17	18	11

Ho Math Chess 何数棋谜   妈！我会棋谜式加法啦！
Mom! I Learn Addition Using Math-Chess-Puzzles Connection

Contents include both traditional and Math-Chess-Puzzles combined methods. Extra strength

©2008 – 2018 Frank Ho, Amanda Ho   All rights reserved. www.homathchess.com

Student Name _____   Date _____

Using doubles or 9 + d to add

(♔ = 0, ♙ = 1, ♗ = 3, ♘ = 3, ♖ = 5, ♕ = 9)

2 + 2 4	♗ + ♗ 6	4 + 4 8	♖ + 5 10	9 + 9 18
7 + 7 14	8 + 8 16	♕ + ♕ 18	2 + 2 4	2 + ♗ 5
8 + 8 16	♖ + 5 10	6 + 5 11	5 + 6 11	4 + 3 7
♕ + 9 18	7 + 6 13	8 + 7 15	♕ + 8 17	8 + 9 17
9 + 8 17	♖ + 6 11	7 + 6 13	6 + 7 13	6 + ♖ 11

Ho Math Chess 何数棋谜   妈！我会棋谜式加法啦！
Mom! I Learn Addition Using Math-Chess-Puzzles Connection
Contents include both traditional and Math-Chess-Puzzles combined methods. Extra strength
©2008 — 2018 Frank Ho, Amanda Ho   All rights reserved. www.homathchess.com

Student Name _____   Date _____

Using doubles or 9 + d to add
(♔ = 0, ♙ = 1, ♗ = 3, ♘ = 3, ♖ = 5, ♕ = 9)

7	8	8	♖	7
+ 6	+ 9	+ 7	+ 8	+ 5
13	17	15	13	12
7	8	8	6	5
+ 8	+ 9	+ 7	+ ♖	+ 7
15	17	15	11	12
♕	5	6	7	7
+ 8	+ 6	+ ♖	+ 8	+ 5
17	11	11	15	12

Using doubles or 9 + d to add

(♔ = 0, ♙ = 1, ♗ = 3, ♘ = 3, ♖ = 5, ♕ = 9)

9 + 8 17	7 + 6 13	8 + 7 15	♕ + 8 17	8 + 6 14
♕ + 8 17	♖ + 6 11	7 + 6 13	6 + 7 13	6 + 5 11

Ho Math Chess 何数棋谜　妈！我会棋谜式加法啦！
Mom! I Learn Addition Using Math-Chess-Puzzles Connection

Contents include both traditional and Math-Chess-Puzzles combined methods. Extra strength

©2008 – 2018 Frank Ho, Amanda Ho　　All rights reserved. www.homathchess.com

Student Name _____ Date _____

Using doubles or 9 + d to add

(♔ = 0, ♙ = 1, ♗ = 3, ♘ = 3, ♖ = 5, ♕ = 9)

7 + 6 13	8 + 9 17	8 + 7 15	♖ + 8 13	7 + 5 12
7 + 8 15	8 + 9 17	8 + 7 15	6 + ♖ 11	5 + 7 12
♕ + 8 17	♖ + 6 11	6 + 5 11	7 + 8 15	7 + 5 12
9 + 8 17	7 + 6 13	8 + 7 15	♕ + 8 17	8 + 6 14

189

Mom! I Learn Addition Using Math-Chess-Puzzles Connection

Contents include both traditional and Math-Chess-Puzzles combined methods. Extra strength
©2008 – 2018 Frank Ho, Amanda Ho   All rights reserved. www.homathchess.com

Student Name _____ Date _____

Adding 5 by doubling and 10's complement

		1		5	
4	6(5+1)	6(5+1)	4	7 (5+2)	3
3		3	5	6(5+1)	
2	8(5+3)	4	5	8(5+3)	7(5+2)
1		9		2	
	a	b	c	d	e

You are at square c3 = ☐ .

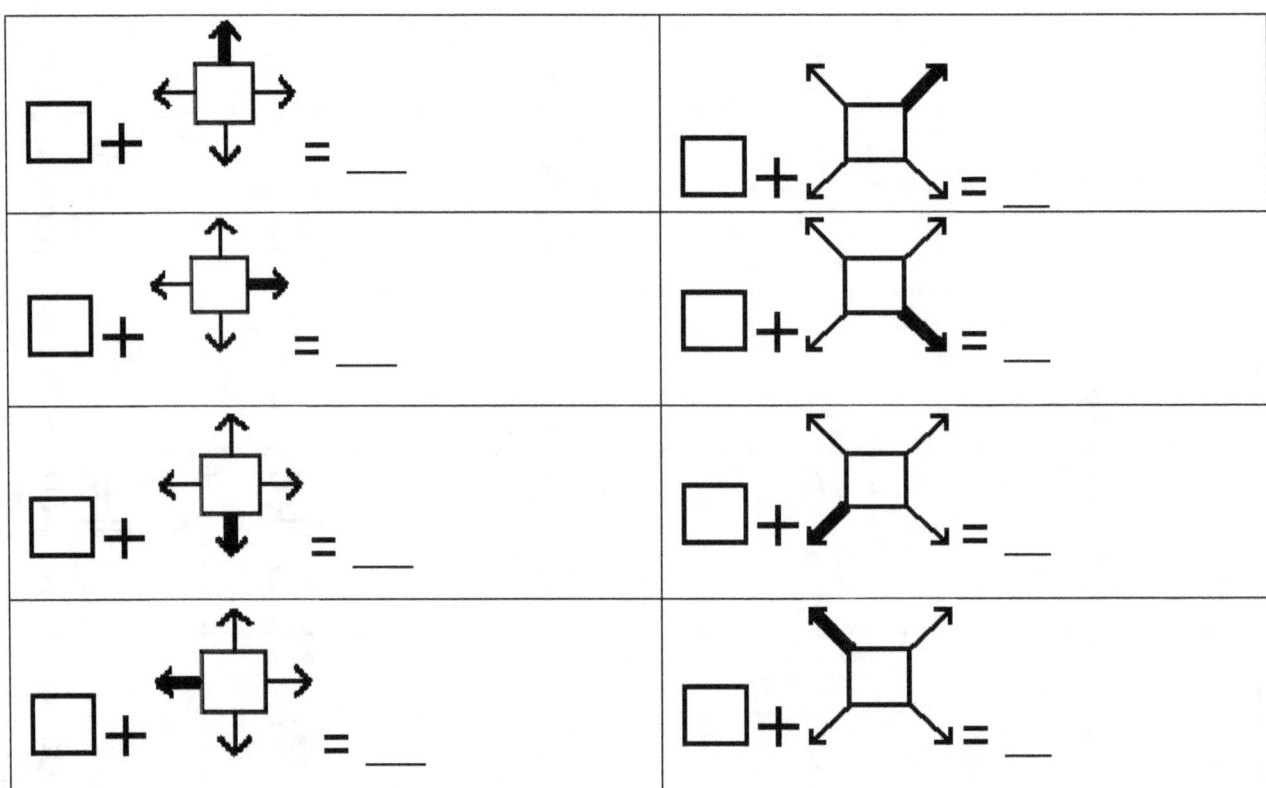

9, 12
11, 13
10, 43
8, 11

190

Mom! I Learn Addition Using Math-Chess-Puzzles Connection

Contents include both traditional and Math-Chess-Puzzles combined methods. Extra strength

©2008 – 2018 Frank Ho, Amanda Ho   All rights reserved. www.homathchess.com

Student Name _____   Date _____

Adding 5 by doubling and 10's complement

5		9		9	
4	8(5+3)	6	1	7(5+2)	6
3		3		2	
2	5	8(5+3)	4	5	7(5+2)
1		9		9	
	a	b	c	d	e

You are at square c3 = ☐.

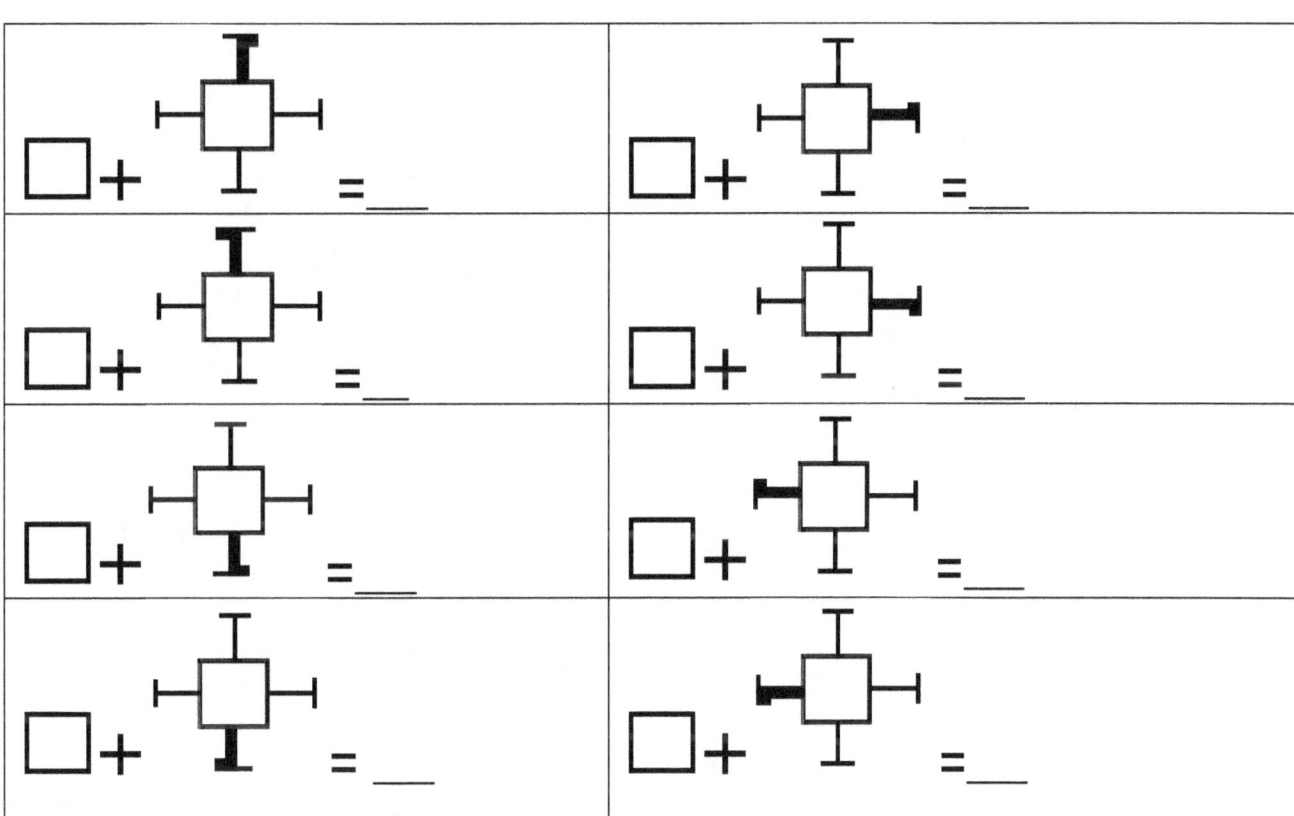

14, 11
14, 12
14, 13
14, 10

Mom! I Learn Addition Using Math-Chess-Puzzles Connection

Contents include both traditional and Math-Chess-Puzzles combined methods. Extra strength
©2008 – 2018 Frank Ho, Amanda Ho   All rights reserved. www.homathchess.com

Student Name _____ Date _____

Adding 6 by doubling and 10's complement

5					
4	8(6+2)	6	1	7(6+1)	6
3		3	6	2	
2	5	8(6+2)	4	5	7(6+1)
1		9		9	
	a	b	c	d	e

(row 1 top values: 9 above b, 9 above d)

You are at square c3

= ☐.

7, 13
8, 11
10, 14
9, 12

192

Mom! I Learn Addition Using Math-Chess-Puzzles Connection

Contents include both traditional and Math-Chess-Puzzles combined methods. Extra strength

©2008 – 2018 Frank Ho, Amanda Ho   All rights reserved. www.homathchess.com

Student Name _____ Date _____

Adding 6 by doubling and 10's complement

5		9		9	
4	8(6+2)	6	1	7	6
3		3	6	2	
2	5	8(6+2)	4	5	7
1		9		9	
	a	b	c	d	e

You are at square c3

= ☐.

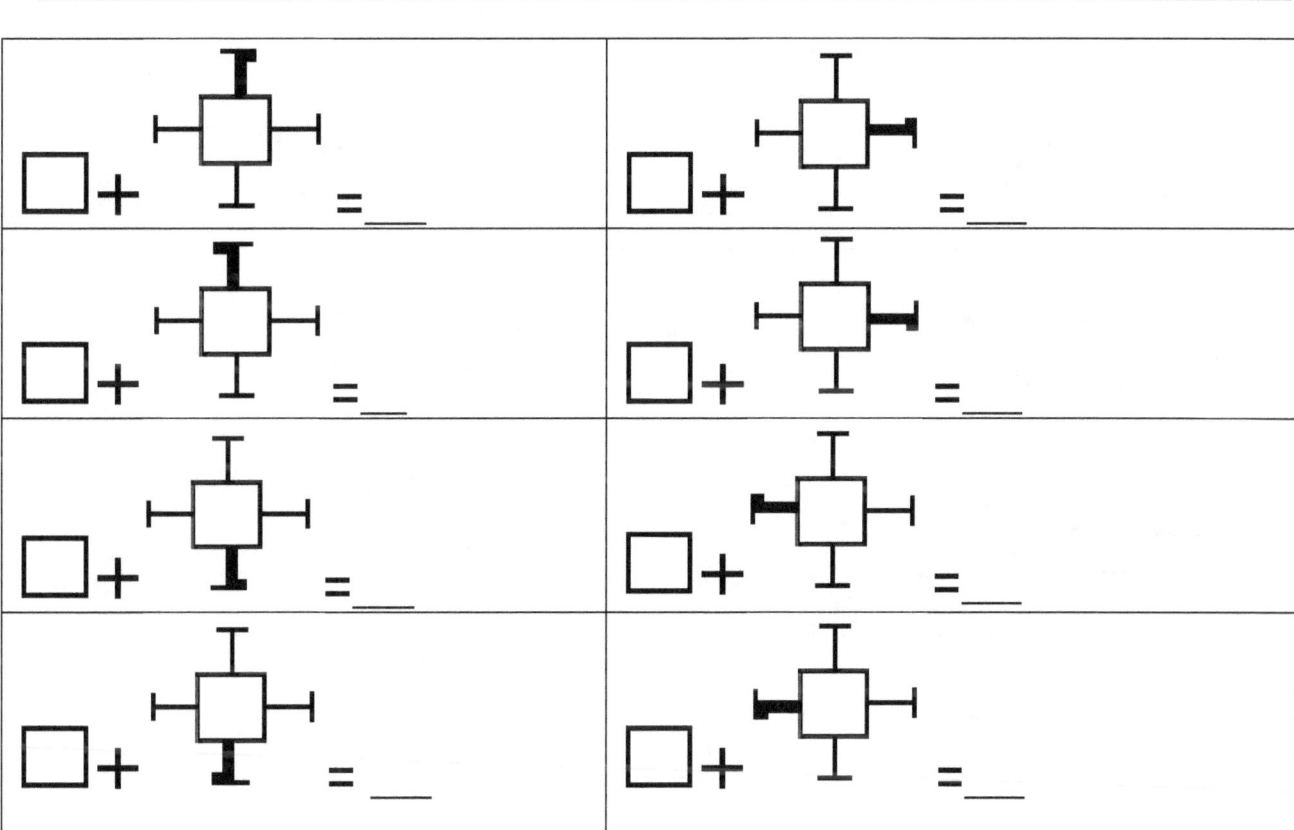

15, 12
15, 13
15, 14
15, 11

193

Mom! I Learn Addition Using Math-Chess-Puzzles Connection

Contents include both traditional and Math-Chess-Puzzles combined methods. Extra strength
©2008 – 2018 Frank Ho, Amanda Ho   All rights reserved. www.homathchess.com

Student Name _____ Date _____

Adding 7 by doubling and 10's complement

5		9		9	
4	8(7+1)	6	1	7	6
3		3	7	2	
2	5	8(7+1)	4	5	7
1		9		9	
	a	b	c	d	e

You are at square c3

= ☐.

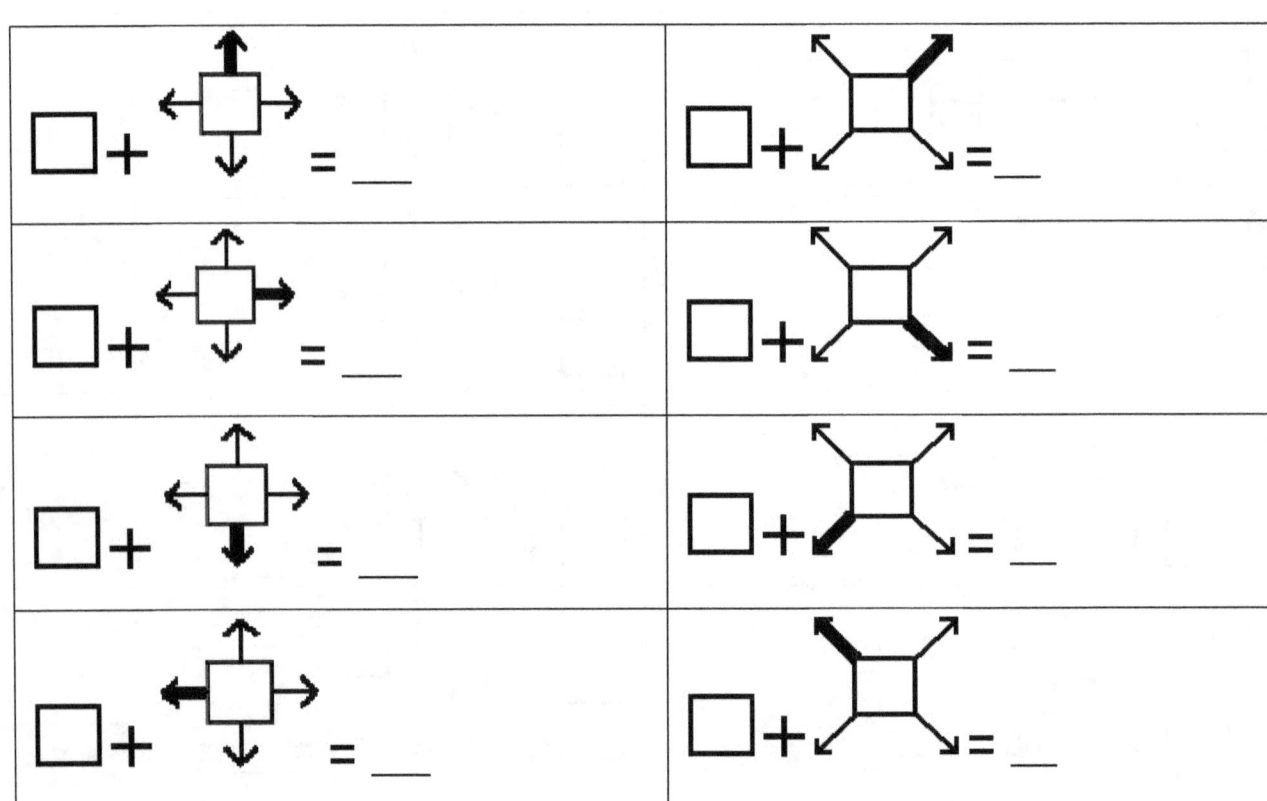

8, 14
9, 12
11, 15
10, 13

194

Adding 7 by doubling and 10's complement

5		9		9	
4	8(7+1)	6	1	7	6
3		3	7	2	
2	5	8(7+1)	4	5	7
1		9		9	
	a	b	c	d	e

You are at square c3

= ▢.

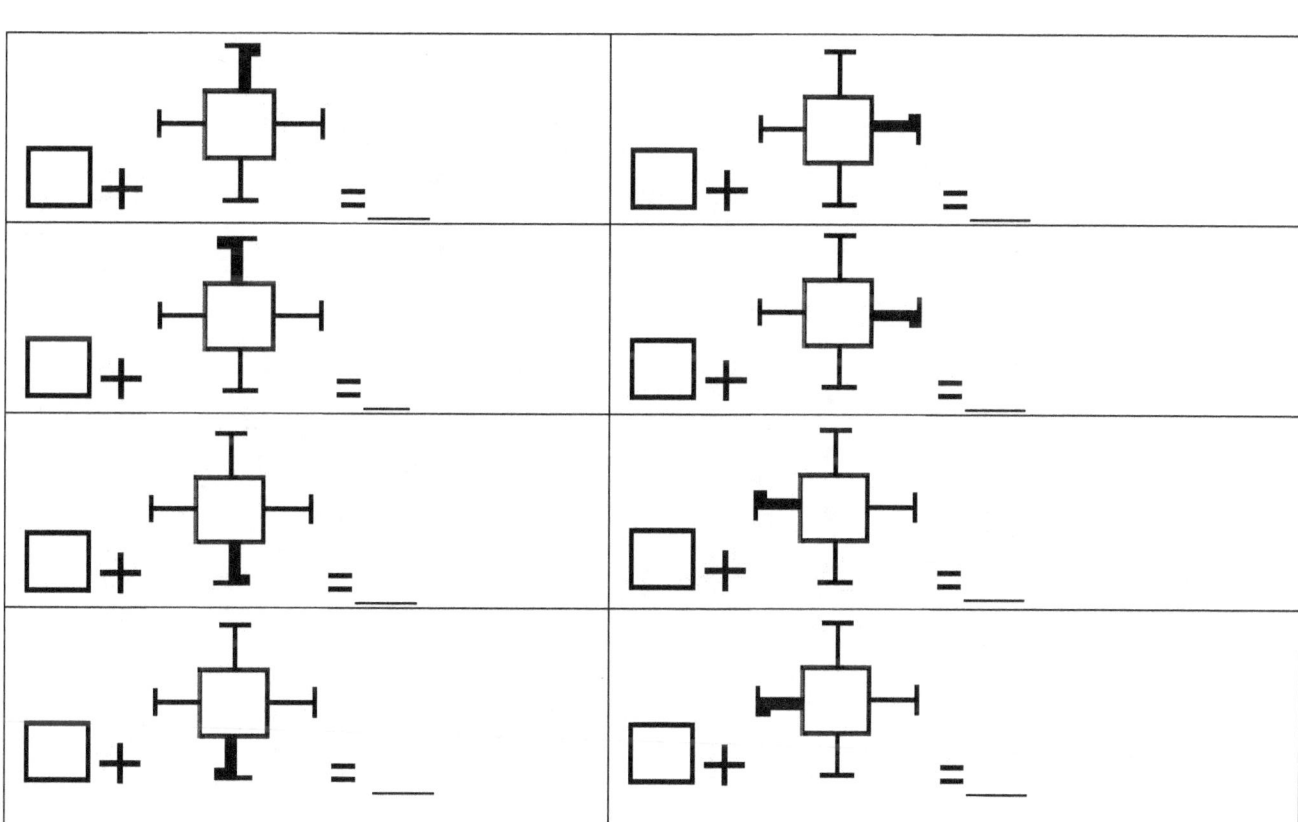

16 13
16 14
16 15
16 12

Ho Math Chess 何数棋谜　妈！我会棋谜式加法啦！
Mom! I Learn Addition Using Math-Chess-Puzzles Connection

Contents include both traditional and Math-Chess-Puzzles combined methods. Extra strength

©2008 － 2018 Frank Ho, Amanda Ho　　All rights reserved. www.homathchess.com

Student Name _____ Date _____

Adding 8 by doubling and 10's complement

5		9		9	
4	8	6	1	7	6
3		3	8	2	
2	5	8	4	5	7
1		9		9	
	a	b	c	d	e

You are at square c3

= ☐.

9　15
10　13
12　16
11　14

196

Adding 8 by doubling

5		9		9	
4	8	6	1	7	6
3		3	8	2	
2	5	8	4	5	7
1		9		9	
	a	b	c	d	e

You are at square c3
= ☐.

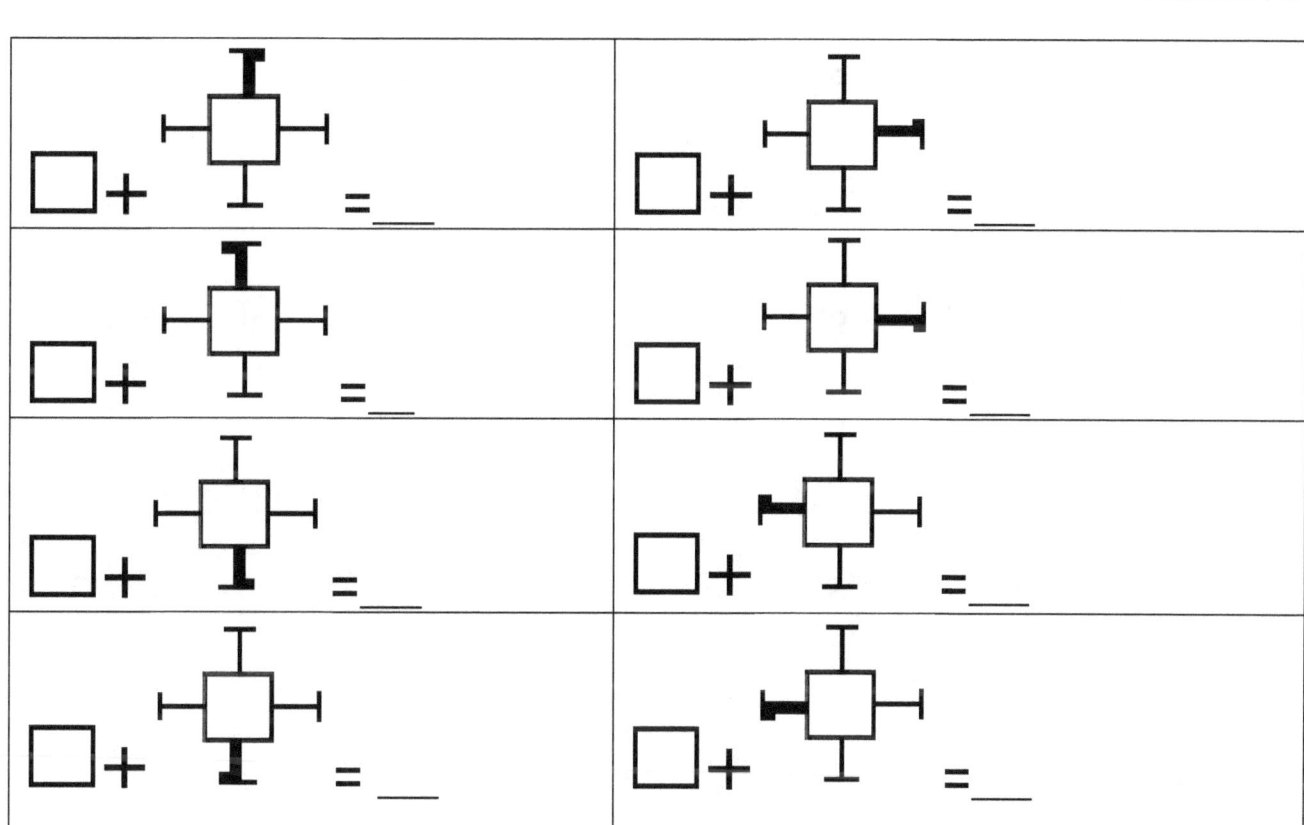

17 14
17 15
17 16
17 13

Ho Math Chess 何数棋谜   妈！我会棋谜式加法啦！
Mom! I Learn Addition Using Math-Chess-Puzzles Connection
Contents include both traditional and Math-Chess-Puzzles combined methods. Extra strength
©2008 – 2018 Frank Ho, Amanda Ho   All rights reserved. www.homathchess.com

Student Name _____   Date _____

Making 10 or adding around 10 or doubling

7 + 5 = 10   9 + 3 = 12   5 + 9 = 14

5 + 8 = 13   7 + 4 = 11   6 + 8 = 14

6 + 5 = 11   8 + 6 = 14   9 + 4 = 13

9 + 4 = 13   6 + 9 = 15   7 + 4 = 11

7 + 8 = 15   6 + 7 = 13   8 + 5 = 13

Mom! I Learn Addition Using Math-Chess-Puzzles Connection

Contents include both traditional and Math-Chess-Puzzles combined methods. Extra strength

©2008 − 2018 Frank Ho, Amanda Ho   All rights reserved. www.homathchess.com

Student Name _____ Date _____

Adding up to 10 or doubling

Use the following strategies:

9 + number to split the number to be 1 + (number −1). 9 + 5 = 9 + 1 + 4
Use doubling whenever you can. 7 + 8 = 7 + 7 + 1 = 14 + 1
Use adding up to 10. 7 + 4 = 7 + 3 + 1 = 10 + 1 = 11

#	Problem
1	7 + 4 = 7 + ( ) + 1 = ( ) + 1 = ( ) 7+3+1=10+1=11
2	7 + 5 = 5 + ( ) + 2 = ( ) + 2 = ( ) 5+5+2=10+2=12
3	7 + 6 = 6 + ( ) + 1 = ( ) + 1 = ( )13
4	7 + 8 = 7 + 7 + ( ) = ( ) + 1 = ( )15
5	7 + 9 = 6 + 1 + ( ) = ( ) + 6 = ( ) 16
6	6 + 5 = 1 + ( ) + 5 = ( ) + 1 = ( ) 11
7	6 + 6 = ( ) 12
8	6 + 7 = 6 + 6 + ( ) = 12 + ( ) = ( ) 13
9	6 + 8 = 6 + 6 + ( ) = 12 + ( ) = ( ) 16
10	6 + 9 = 9 + 1 + ( ) = ( ) + 5 = ( ) 15
11	7 + 8 = 7 + 7 + ( ) = 1 + ( ) = ( ) 15
12	5 + 6 = 5 + 5 + ( ) = ( ) + 1 = ( ) 11
13	5 + 7 = 5 + 5 + ( ) = ( ) + 2 = ( ) 5+5+2=10+2=12

Mom! I Learn Addition Using Math-Chess-Puzzles Connection

Mixed doubling, adding to 10, adding 1, 2, 3

Evaluate the following results.

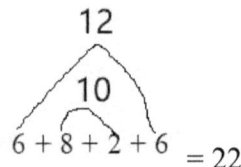 = 22

7 + 8 + 3 + 8
4 + 8 + 2 + 4
5 + 6 + 5 + 4
9 + 3 + 3 + 1
2 + 5 + 5 + 8
4 + 2 + 4 + 8
7 + 1 + 7 + 9
9 + 2 + 9 = 8
4 + 8 + 8 + 6

26, 18, 20, 16, 20, 18, 24, 28, 26

Mixed doubling and adding to 10

9 + 1 + 5 + 5
6 + 2 + 8 + 6
7 + 5 + 3 + 5
2 + 7 + 8 + 7
8 + 8 + 4 + 6
7 + 3 + 7 + 7
5 + 5 + 5 + 5
2 + 8 + 6 + 6
8 + 8 + 1 + 9
7 + 3 + 8 + 8

20, 22, 20, 24, 26, 24, 20, 22, 26, 26

Ho Math Chess 何数棋谜 妈！我会棋谜式加法啦！
Mom! I Learn Addition Using Math-Chess-Puzzles Connection

Contents include both traditional and Math-Chess-Puzzles combined methods. Extra strength
©2008 − 2018 Frank Ho, Amanda Ho   All rights reserved. www.homathchess.com

Student Name _____ Date _____

8 + 8 + 8 + 2
6 + 6 + 7 + 3
7 + 7 + 7 + 3
5 + 5 + 6 + 5
6 + 6 + 4 + 6
4 + 8 + 2 + 4
9 + 6 + 1 + 6
9 + 9 + 9 + 1
1 + 8 + 1 + 2
7 + 3 + 3 + 3

26, 22, 24, 21, 22, 18, 26, 28, 12, 16

Mom! I Learn Addition Using Math-Chess-Puzzles Connection

Mixed doubling, adding to 10, adding 1, 2, 3

Evaluate the following results.

= 24
7 + 8 + 3 + 8 + 1
4 + 8 + 2 + 4 + 2
5 + 6 + 5 + 4 + 3
9 + 3 + 3 + 1 + 2
2 + 5 + 5 + 8 + 1
4 + 2 + 4 + 8 + 3
7 + 1 + 7 + 9 + 3
9 + 2 + 9 + 8 + 2
4 + 8 + 8 + 6 + 1

27, 20, 23, 18, 21, 21, 27, 30, 27

Multi-direction addition of doubling and adding to 10

$6 + ♖ = 5 + 5 + \boxed{}1 = \boxed{}11 = 6 + \boxed{}4 + ♟$
        $\lfloor 10 \rfloor$ $\lfloor 10 \rfloor$

where above: $6\ (5+1) + ♖$

$7 + 6 = 6 + 6 + 1\boxed{} = \boxed{}13 = 3 + \boxed{}7 + ♗$
        $\lfloor 12 \rfloor$ $\lfloor 10 \rfloor$

where above: $5 + ♖$

Ho Math Chess 何数棋谜   妈！我会棋谜式加法啦！
Mom! I Learn Addition Using Math-Chess-Puzzles Connection

Contents include both traditional and Math-Chess-Puzzles combined methods. Extra strength

©2008 — 2018 Frank Ho, Amanda Ho   All rights reserved. www.homathchess.com

Student Name _____ Date _____

10
+ ♗

$8 + ♖ = 5 + 5 + \Box 3 = \Box 13 = 8 + \Box 2 + ♗$
         └ 10 ┘   └ 10 ┘

6
+ ♖

$9 + 2 = 9 + 1 + \Box 1 = \Box 11 = 8 + \Box 2 + ♟$
         └ 10 ┘   └ 10 ┘

Mom! I Learn Addition Using Math-Chess-Puzzles Connection

Contents include both traditional and Math-Chess-Puzzles combined methods. Extra strength

©2008 — 2018 Frank Ho, Amanda Ho   All rights reserved. www.homathchess.com

Student Name _____ Date _____

If-Then for plus (do the left + first then the right +) review

$$10 - ♙ = 9$$
$$+ \ 7 \quad + ♔ = 7 \quad +$$
$$\overline{} \qquad \overline{}$$
$$17 \qquad\qquad 16$$

If 10 + 7 = ☐17, then ♛ + 7 must be ☐16.

If 8 + 8 = ☐16, then ♛ + 7 must be ☐16.

$$10 - ♙ = 9$$
$$+ \ 6 \quad + ♔ = 6 \quad +$$
$$\overline{} \qquad \overline{}$$
$$16 \qquad\qquad 15$$

If 10 + 6 = ☐16, then ♛ + 6 must be ☐15.

If 6 + 6 = ☐12, then ♛ + 6 must be ☐15.

Ho Math Chess 何数棋谜　妈！我会棋谜式加法啦！
Mom! I Learn Addition Using Math-Chess-Puzzles Connection
Contents include both traditional and Math-Chess-Puzzles combined methods. Extra strength
©2008 – 2018 Frank Ho, Amanda Ho　　All rights reserved. www.homathchess.com

Student Name _____ Date _____

$$7 + ♙ = 8$$
$$+\ 7 + ♚ = 7\ +$$
$$\overline{14} \quad \overline{15}$$

If 7 + 7 = □14, then 7 + 8 must be □15.
If 8 + 8 = □16, then 8 + 7 must be □15.

$$10 - ♙ = 9$$
$$+\ 5 + ♚ = 5\ +$$
$$\overline{15} \quad \overline{14}$$

If 10 + ♖ = □15, then ♛ + ♖ must be □14.
If ♖ + 10 = □15, then ♖ + ♛ must be □14.

Mom! I Learn Addition Using Math-Chess-Puzzles Connection

$$\begin{array}{r} 6 + 2 = 8 \\ + \quad 6 + ♔ = 6 \quad + \\ \hline 12 \qquad 14 \end{array}$$

If 6 + 6 = ☐12, then 8 + 6 must be ☐14.

If 6 + 6 = ☐12, then 6 + 8 must be ☐14.

$$\begin{array}{r} 10 - ♙ = 9 \\ + \quad 4 + ♔ = 4 \quad + \\ \hline 6 \qquad 13 \end{array}$$

If 10 + 4 = ☐14, then ♕ + 4 must be ☐13.

If 4 + 10 = ☐14, then 4 + ♕ must be ☐13.

Mom! I Learn Addition Using Math-Chess-Puzzles Connection

Contents include both traditional and Math-Chess-Puzzles combined methods. Extra strength

©2008 – 2018 Frank Ho, Amanda Ho   All rights reserved. www.homathchess.com

Student Name _____ Date _____

$$\begin{array}{c} \text{♖} + \text{♗} = 8 \\ + \quad \text{♖} + \text{♔} = 5 \quad + \\ \hline 10 \qquad\qquad 13 \end{array}$$

If ♖ + ♖ = **10**, then 8 + ♖ must be **13**.

If ♖ + ♖ = **10**, then ♖ + 8 must be **13**.

$$\begin{array}{c} 6 + \text{♙} = 7 \\ + \quad 6 + \text{♔} = 6 \quad + \\ \hline 12 \qquad\qquad 13 \end{array}$$

If 6 + 6 = ☐**12**, then 7 + 6 must be ☐**13**.

If 6 + 6 = ☐**12**, then 6 + 7 must be ☐**13**.

Mom! I Learn Addition Using Math-Chess-Puzzles Connection

Contents include both traditional and Math-Chess-Puzzles combined methods. Extra strength

©2008 – 2018 Frank Ho, Amanda Ho   All rights reserved. www.homathchess.com

Student Name _____   Date _____

```
           10    -   ♙   =   9
       +        ♗   +  ♔   =   3       +
       ─────────────       ─────
            13                12
```

If 10 + ♗ = **13**, then ♕ + ♗ must be **12**.

If ♗ + 10 = **13**, then ♗ + ♕ must be **12**.

```
            4    +   4   =   8
       +        4   +  ♔   =   4       +
       ─────────────       ─────
             8                12
```

If 4 + 4 = ☐**8**, then 8 + 4 must be ☐**12**.

If 4 + 4 = ☐**8**, then 4 + 8 must be ☐**12**.

Ho Math Chess 何数棋谜　妈！我会棋谜式加法啦！
Mom! I Learn Addition Using Math-Chess-Puzzles Connection
Contents include both traditional and Math-Chess-Puzzles combined methods. Extra strength
©2008 – 2018 Frank Ho, Amanda Ho　All rights reserved. www.homathchess.com

Student Name _____ Date _____

$$\begin{array}{c} ♖ + 2 = 7 \\ + \quad ♖ + ♔ = 5 \quad + \\ \hline 10 \qquad 12 \end{array}$$

If ♖ + ♖ = ☐10, then 7 + ♖ must be ☐12.

If ♖ + ♖ = ☐10, then ♖ + 7 must be ☐12.

$$\begin{array}{c} 9 + ♔ = 9 \\ + \quad ♙ + ♙ = 2 \quad + \\ \hline 10 \qquad 11 \end{array}$$

If ♕ + ♙ = ☐10, then ♕ + 2 must be ☐11.

If ♙ + ♕ = ☐10, then 2 + ♕ must be ☐11.

Ho Math Chess 何数棋谜 妈！我会棋谜式加法啦！
Mom! I Learn Addition Using Math-Chess-Puzzles Connection
Contents include both traditional and Math-Chess-Puzzles combined methods. Extra strength
©2008 − 2018 Frank Ho, Amanda Ho   All rights reserved. www.homathchess.com

Student Name_____   Date_____

$$\begin{array}{ccc} & 8 & + \text{♔} = 8 \\ + & 2 & + \text{♙} = 3 \quad + \\ \hline & 10 & \quad\quad 11 \end{array}$$

If 8 + 2 = □10, then 8 + ♗ must be □11.

If 2 + 8 = □10, then ♗ + 8 must be □11.

$$\begin{array}{ccc} & 7 & + \text{♔} = 7 \\ + & 3 & + \text{♙} = 4 \quad + \\ \hline & 10 & \quad\quad 11 \end{array}$$

If 7 + ♗ = □10, then 7 + 4 must be □11.

If ♗ + 7 = □10, then 4 + 7 must be □11.

Mom! I Learn Addition Using Math-Chess-Puzzles Connection

Contents include both traditional and Math-Chess-Puzzles combined methods. Extra strength
©2008 – 2018 Frank Ho, Amanda Ho   All rights reserved. www.homathchess.com

Student Name _____ Date _____

```
            ♖ + ♙ =  6
    +       ♖ + ♔ =  5      +
   _____            _____
     10                11
```

If ♖ + ♖ = ☐10, then 6 + ♖ must be ☐11.

If ♖ + ♖ = ☐10, then ♖ + 6 must be ☐11.

Assessment of 9 + d, doubling, adding to 10

	D	T	123	DT
D	○ + ○ = 10 5, 5	○ + ○ + 6 + 4 = 18 4+4 The following has 5 answers ○ + ○ + 7 + △ = 18 ○ + ○ + 7 + △ = 18 ○ + ○ + 7 + △ = 18 ○ + ○ + 7 + △ = 18 ○ + ○ + 7 + △ = 18 1,1,9 2,2,7 3,3,5 4,4,3 5,5,8	○ + ○ + 3 = 11 4 + 4	
T		6 + 4 + 7 + △ = 20 5	6 + 4 + △ = 13 3 △ + 7 + 3 = 12 2	
123			2 + 4 + ○ = 9 3	△ + △ + 7 + 3 = 26 8+ 8

Part 3: More practices of chess and math integrated problems

Students have finished all addition strategies, so let's have some fun to review them. If students need more practices, then this part offers more practices.

Additions to 10

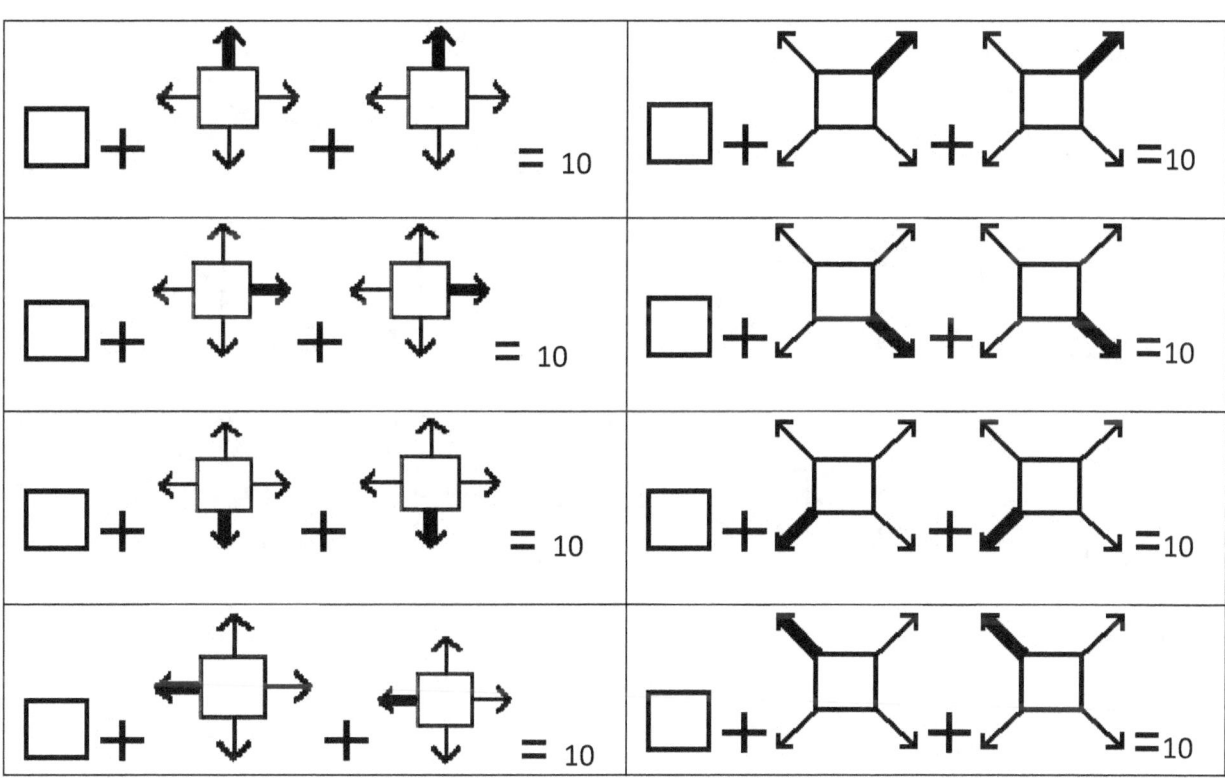

037 028
046 073
055 082
064 091

Adding up to 10

5	?	?	?	?	?
4	2	9	3	2	5
3	?	6	0	4	?
2	?	8	5	7	?
1	?	?	?	7	?
	a	b	c	d	e

You are at square c3 = ☐.

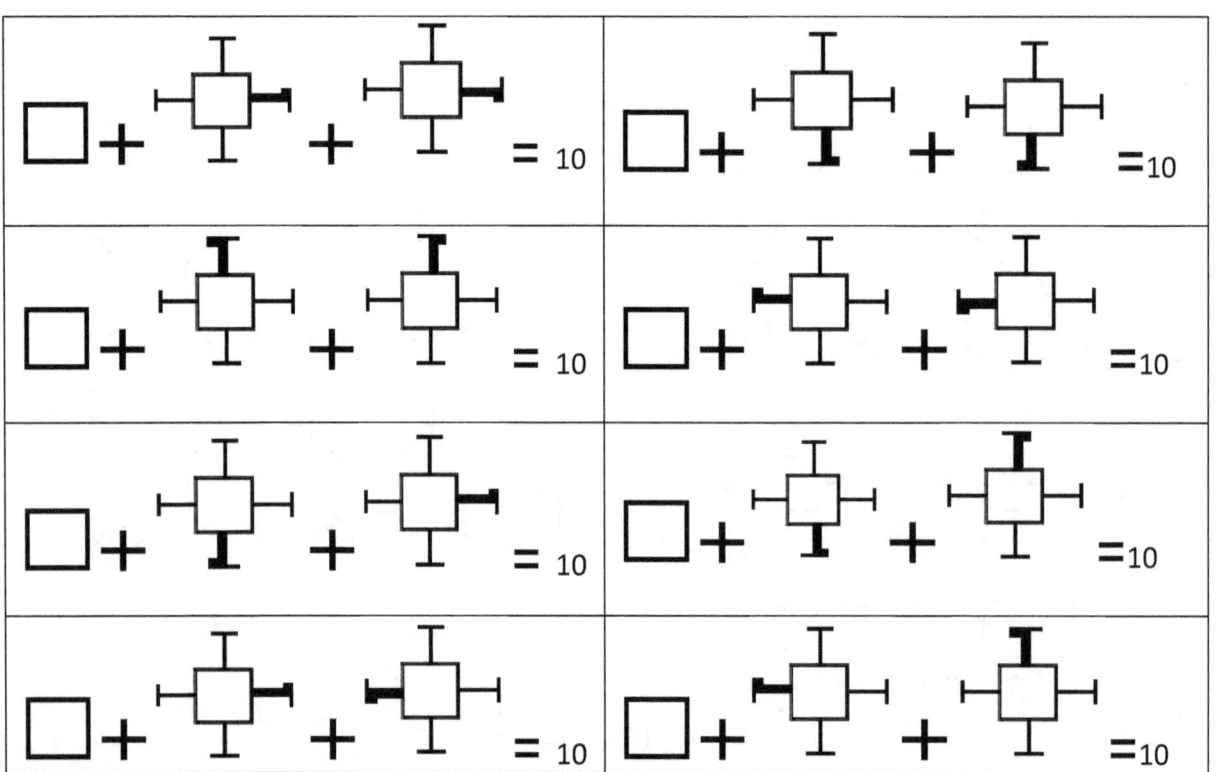

055 073
046 028
055 073
055 028

Ho Math Chess 何数棋谜　妈！我会棋谜式加法啦！
Mom! I Learn Addition Using Math-Chess-Puzzles Connection
Contents include both traditional and Math-Chess-Puzzles combined methods. Extra strength
©2008 – 2018 Frank Ho, Amanda Ho　　All rights reserved. www.homathchess.com

Student Name _____ Date _____

Making 10 or adding around 10
Only transfer each number to a correct position. Do not the transfer the 3 by 3 square.

2	2	1
2	8	1
1	3	3

You are a chess piece located at a square indicated by a shaded square.

10 10 10 10
9 9 9 9
9 9 9 9
10 10 10 10

217

Ho Math Chess 何数棋谜 妈！我会棋谜式加法啦！
Mom! I Learn Addition Using Math-Chess-Puzzles Connection
Contents include both traditional and Math-Chess-Puzzles combined methods. Extra strength
©2008 — 2018 Frank Ho, Amanda Ho   All rights reserved. www.homathchess.com

Student Name _____ Date _____

Making 10 or adding around 10
Only transfer each number to a correct position. Do not the transfer the 3 by 3 square.

You are a chess piece located at a square indicated by a shaded square.

○ = +

10 10 10 10
10 10 10 9
10 10 10 11
9 9 9 10

218

Ho Math Chess 何数棋谜　妈！我会棋谜式加法啦！

Mom! I Learn Addition Using Math-Chess-Puzzles Connection

Contents include both traditional and Math-Chess-Puzzles combined methods. Extra strength

©2008 – 2018 Frank Ho, Amanda Ho　All rights reserved. www.homathchess.com

Student Name _____ Date _____

Making 10 or adding around 10

Only transfer each number to a correct position. Do not the transfer the 3 by 3 square.

You are a chess piece located at a square indicated by a shaded square.

11 11 11 9
10 10 10 10
10 10 10 9
9 9 9 11

219

Ho Math Chess 何数棋谜   妈！我会棋谜式加法啦！
Mom! I Learn Addition Using Math-Chess-Puzzles Connection
Contents include both traditional and Math-Chess-Puzzles combined methods. Extra strength
©2008 – 2018 Frank Ho, Amanda Ho   All rights reserved. www.homathchess.com

Student Name _____ Date _____

Making 10 or adding around 10
Only transfer each number to a correct position. Do not the transfer the 3 by 3 square.

You are a chess piece located at a square indicated by a shaded square.

○ = +

9 9 9 11
11 11 11 9
10 10 10 10
10 10 10 9

220

Ho Math Chess 何数棋谜　妈！我会棋谜式加法啦！
Mom! I Learn Addition Using Math-Chess-Puzzles Connection

Contents include both traditional and Math-Chess-Puzzles combined methods. Extra strength

©2008 − 2018 Frank Ho, Amanda Ho　　All rights reserved. www.homathchess.com

Student Name _____ Date _____

Making 10 or adding around 10

Only transfer each number to a correct position. Do not the transfer the 3 by 3 square.

You are a chess piece located at a square indicated by a shaded square.

○ = +

10 10 10 9
9 9 9 11
11 11 11 10
10 10 10 11

Ho Math Chess 何数棋谜　妈！我会棋谜式加法啦！
Mom! I Learn Addition Using Math-Chess-Puzzles Connection
Contents include both traditional and Math-Chess-Puzzles combined methods. Extra strength
©2008 – 2018 Frank Ho, Amanda Ho　　All rights reserved. www.homathchess.com

Student Name_____ Date_____

Making 10 or adding around 10
Only transfer each number to a correct position. Do not the transfer the 3 by 3 square.

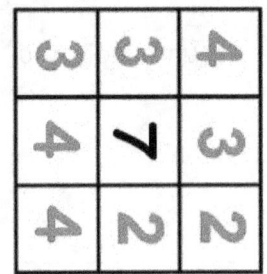

You are a chess piece located at a square indicated by a shaded square.

10 10 10 11
10 10 10 9
9 9 9 11
11 11 11 10

222

Ho Math Chess 何数棋谜   妈！我会棋谜式加法啦！
Mom! I Learn Addition Using Math-Chess-Puzzles Connection

Contents include both traditional and Math-Chess-Puzzles combined methods. Extra strength

©2008 － 2018 Frank Ho, Amanda Ho   All rights reserved. www.homathchess.com

Student Name _____   Date _____

Making 10 or adding around 10

Only transfer each number to a correct position. Do not the transfer the 3 by 3 square.

You are a chess piece located at a square indicated by a shaded square.

11 11 11 10
10 10 10 11
10 10 10 9
11 11 9 10

223

Ho Math Chess 何数棋谜 妈！我会棋谜式加法啦！
Mom! I Learn Addition Using Math-Chess-Puzzles Connection
Contents include both traditional and Math-Chess-Puzzles combined methods. Extra strength
©2008 – 2018 Frank Ho, Amanda Ho   All rights reserved. www.homathchess.com

Student Name_____ Date_____

Making 10 or adding around 10
Only transfer each number to a correct position. Do not the transfer the 3 by 3 square.

You are a chess piece located at a square indicated by a shaded square.

○ = +

9 9 9 11
11 11 11 10
10 10 10 11
10 10 10 9

224

Ho Math Chess 何数棋谜   妈！我会棋谜式加法啦！
Mom! I Learn Addition Using Math-Chess-Puzzles Connection

Contents include both traditional and Math-Chess-Puzzles combined methods. Extra strength

©2008 − 2018 Frank Ho, Amanda Ho   All rights reserved. www.homathchess.com

Student Name _____ Date _____

Making 10 or adding around 10 or doubling

Only transfer each number to a correct position. Do not the transfer the 3 by 3 square.

You are a chess piece located at a square indicated by a shaded square.

10 10 10 9
9 9 9 10
11 11 11 9
10 10 10 9

225

Ho Math Chess 何数棋谜   妈！我会棋谜式加法啦！
Mom! I Learn Addition Using Math-Chess-Puzzles Connection
Contents include both traditional and Math-Chess-Puzzles combined methods. Extra strength
©2008 – 2018 Frank Ho, Amanda Ho   All rights reserved. www.homathchess.com

Student Name _____ Date _____

Making 10 or adding around 10 or doubling
Only transfer each number to a correct position. Do not the transfer the 3 by 3 square.

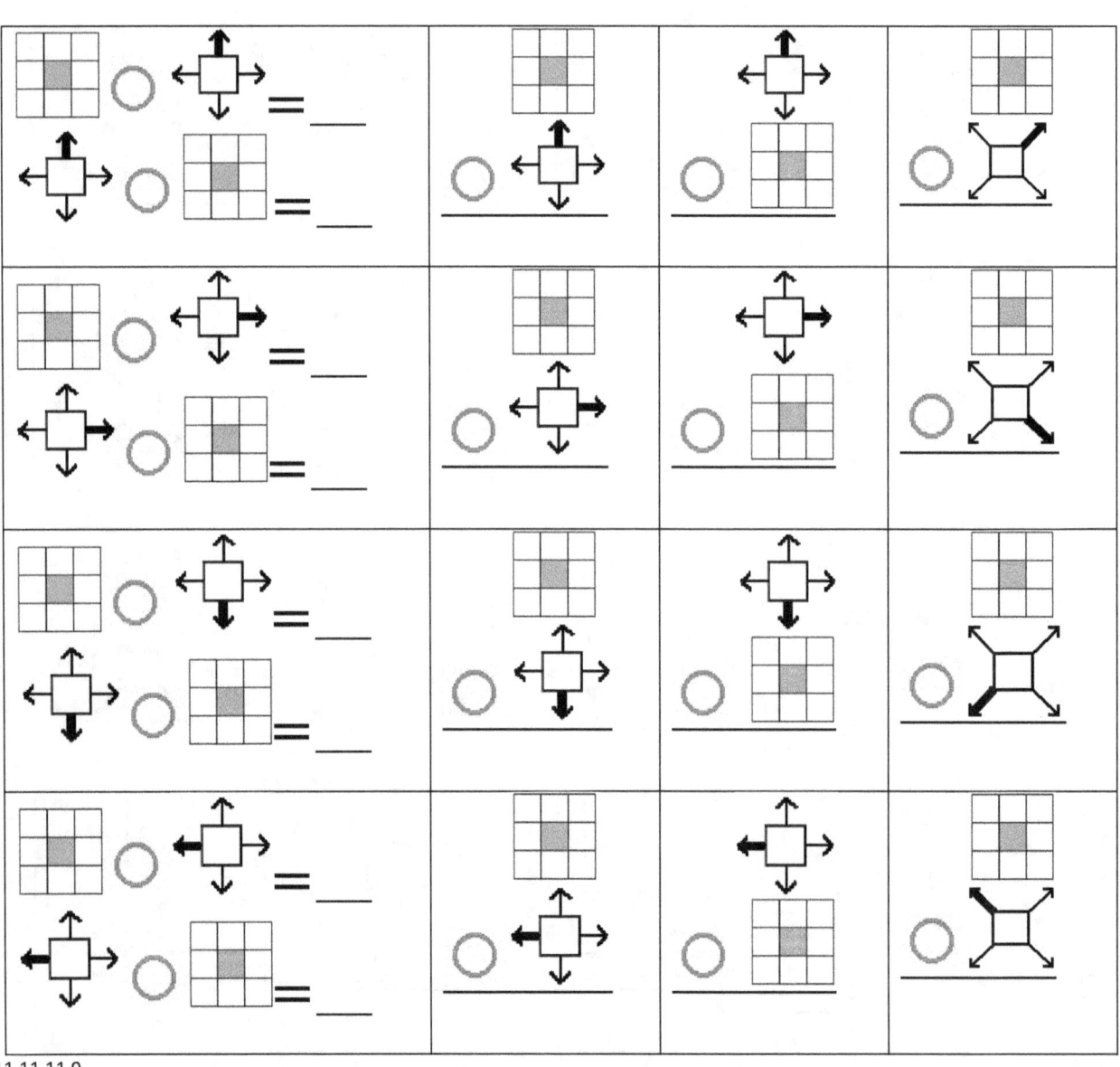

11 11 11 9
10 10 10 9
10 10 10 9
9 9 9 10

Mom! I Learn Addition Using Math-Chess-Puzzles Connection

Contents include both traditional and Math-Chess-Puzzles combined methods. Extra strength

©2008 − 2018 Frank Ho, Amanda Ho   All rights reserved. www.homathchess.com

Student Name _____ Date _____

Making 10 or adding around 10 or doubling

Only transfer each number to a correct position. Do not the transfer the 3 by 3 square.

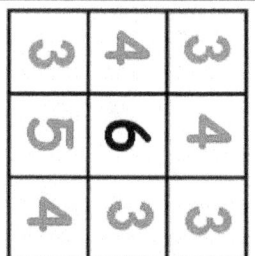

You are a chess piece located at a square indicated by a shaded square.

10 10 10 9
10 10 10 9
9 9 9 10
11 11 11 9

227

Ho Math Chess 何数棋谜 妈！我会棋谜式加法啦！
Mom! I Learn Addition Using Math-Chess-Puzzles Connection
Contents include both traditional and Math-Chess-Puzzles combined methods. Extra strength
©2008 – 2018 Frank Ho, Amanda Ho   All rights reserved. www.homathchess.com

Student Name _____ Date _____

Making 10 or adding around 10 or doubling
Only transfer each number to a correct position. Do not the transfer the 3 by 3 square.

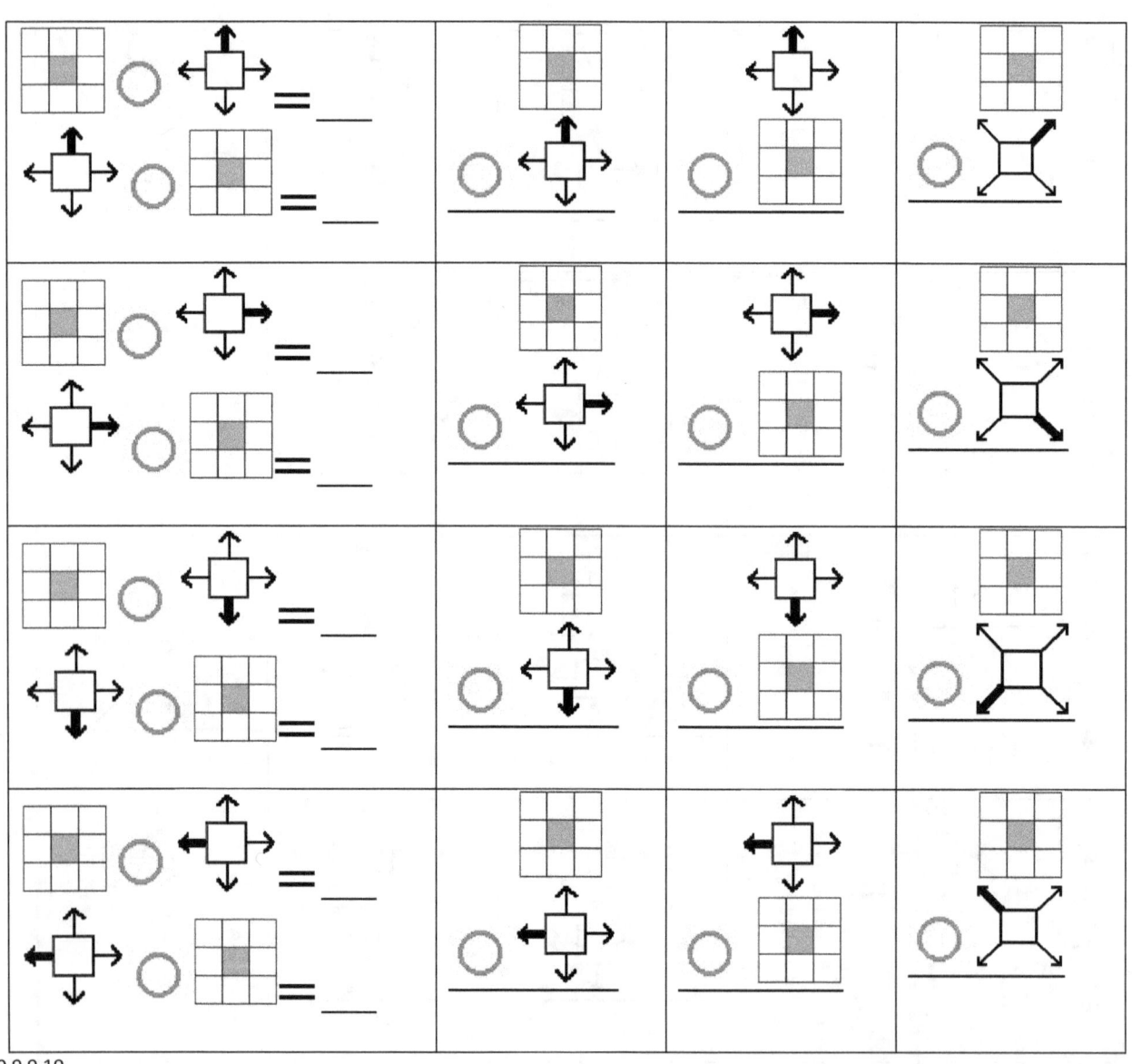

You are a chess piece located at a square indicated by a shaded square.

○ = +

9 9 9 10
11 11 11 9
10 10 10 9
10 10 10 9

Ho Math Chess 何数棋谜　妈！我会棋谜式加法啦！
Mom! I Learn Addition Using Math-Chess-Puzzles Connection

Contents include both traditional and Math-Chess-Puzzles combined methods. Extra strength

©2008 — 2018 Frank Ho, Amanda Ho   All rights reserved. www.homathchess.com

Student Name _____ Date _____

Making 10 or adding around 10 or doubling

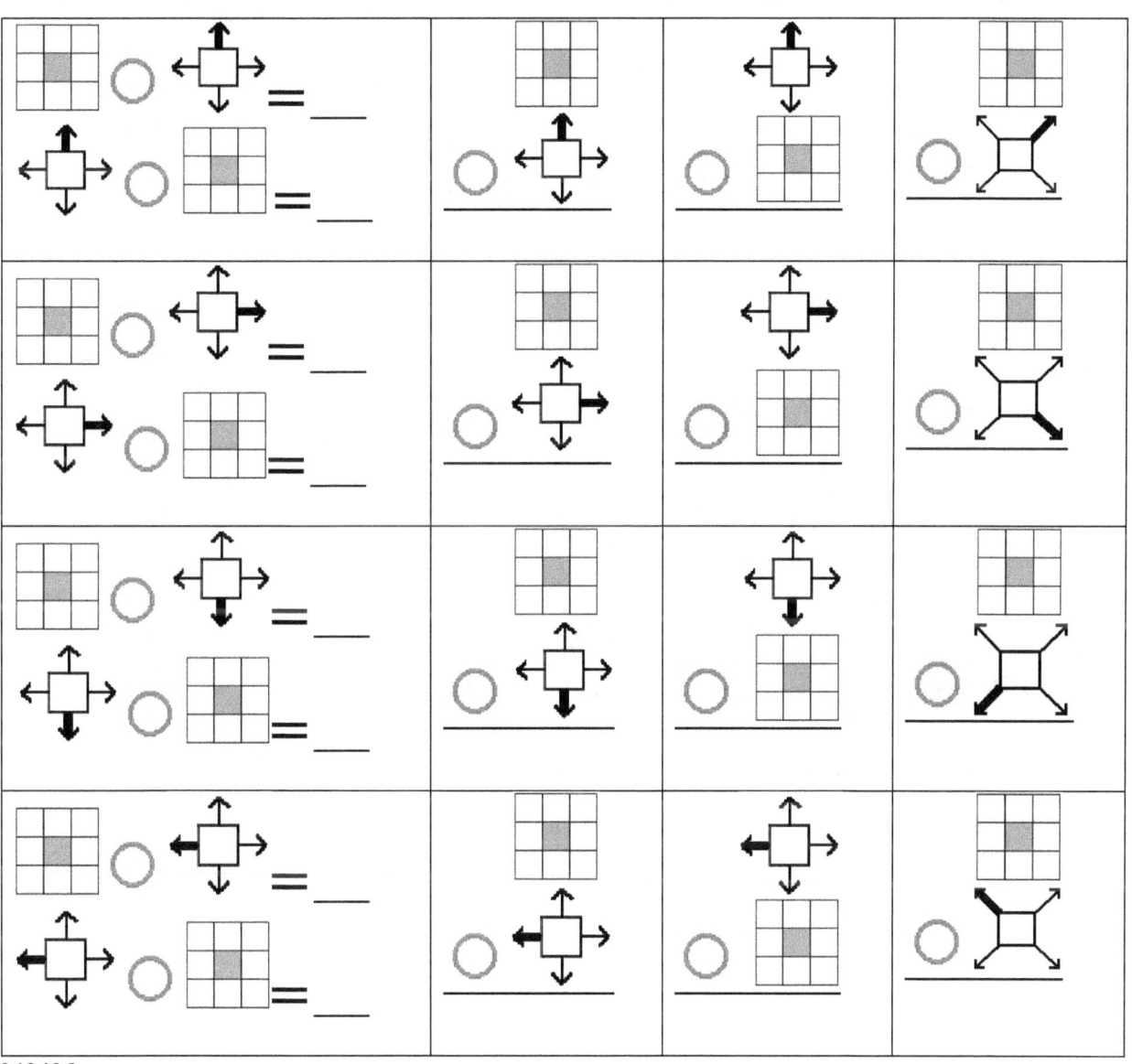

10 10 10 9
9 9 9 10
11 11 11 11
10 10 10 9

Ho Math Chess 何数棋谜 妈！我会棋谜式加法啦！
Mom! I Learn Addition Using Math-Chess-Puzzles Connection
Contents include both traditional and Math-Chess-Puzzles combined methods. Extra strength
©2008 – 2018 Frank Ho, Amanda Ho   All rights reserved. www.homathchess.com

Student Name_____ Date_____

Making 10 or adding around 10 or doubling

Only transfer each number to a correct position. Do not the transfer the 3 by 3 square.

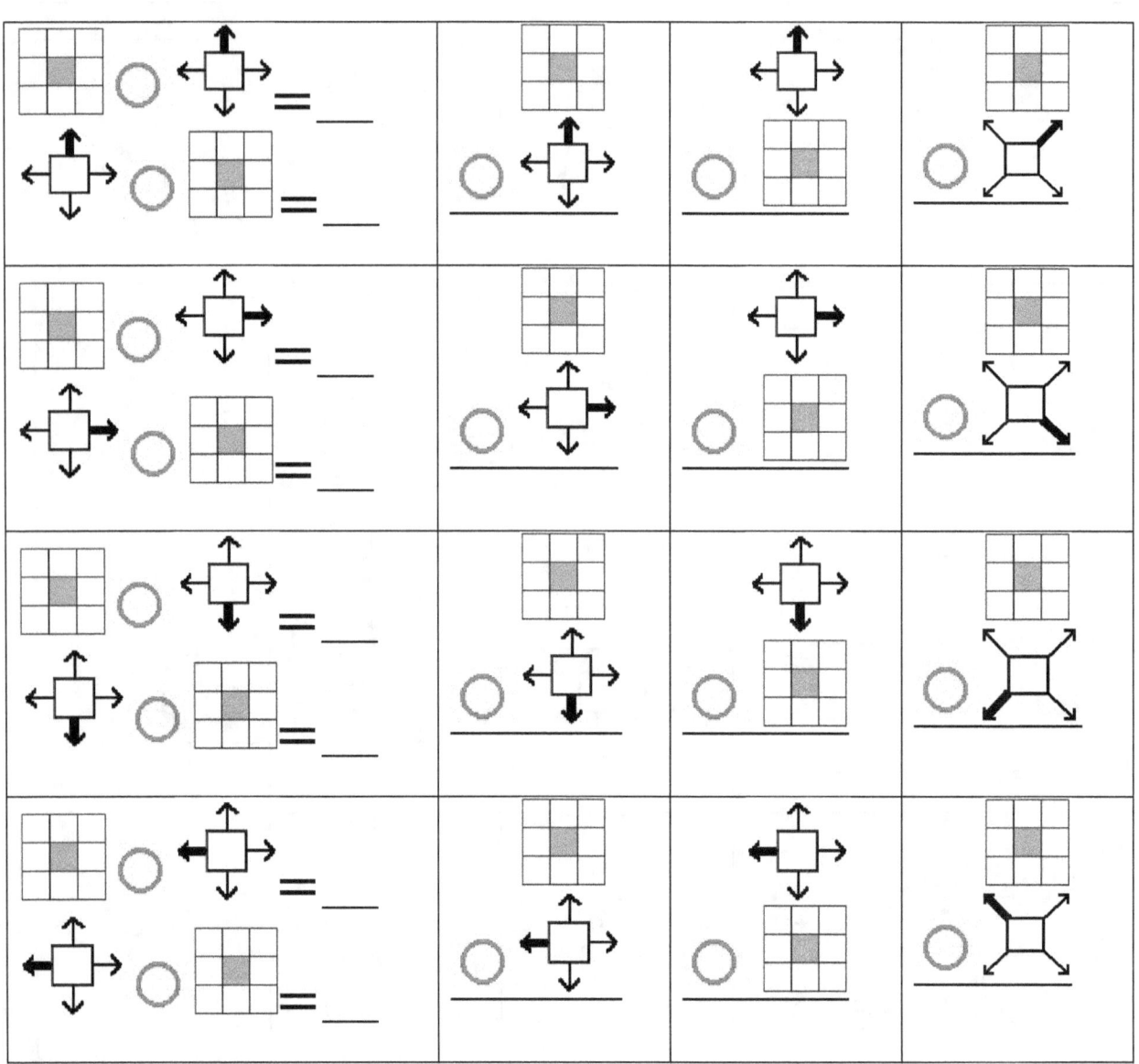

10 10 10 9
10 10 10 9
9 9 9 10
11 11 11 11

230

Ho Math Chess 何数棋谜　妈！我会棋谜式加法啦！

Mom! I Learn Addition Using Math-Chess-Puzzles Connection

Contents include both traditional and Math-Chess-Puzzles combined methods. Extra strength

©2008 – 2018 Frank Ho, Amanda Ho　All rights reserved. www.homathchess.com

Student Name _____ Date _____

Making 10 or adding around 10 or doubling

Only transfer each number to a correct position. Do not the transfer the 3 by 3 square.

You are a chess piece located at a square indicated by a shaded square.

◯ = +

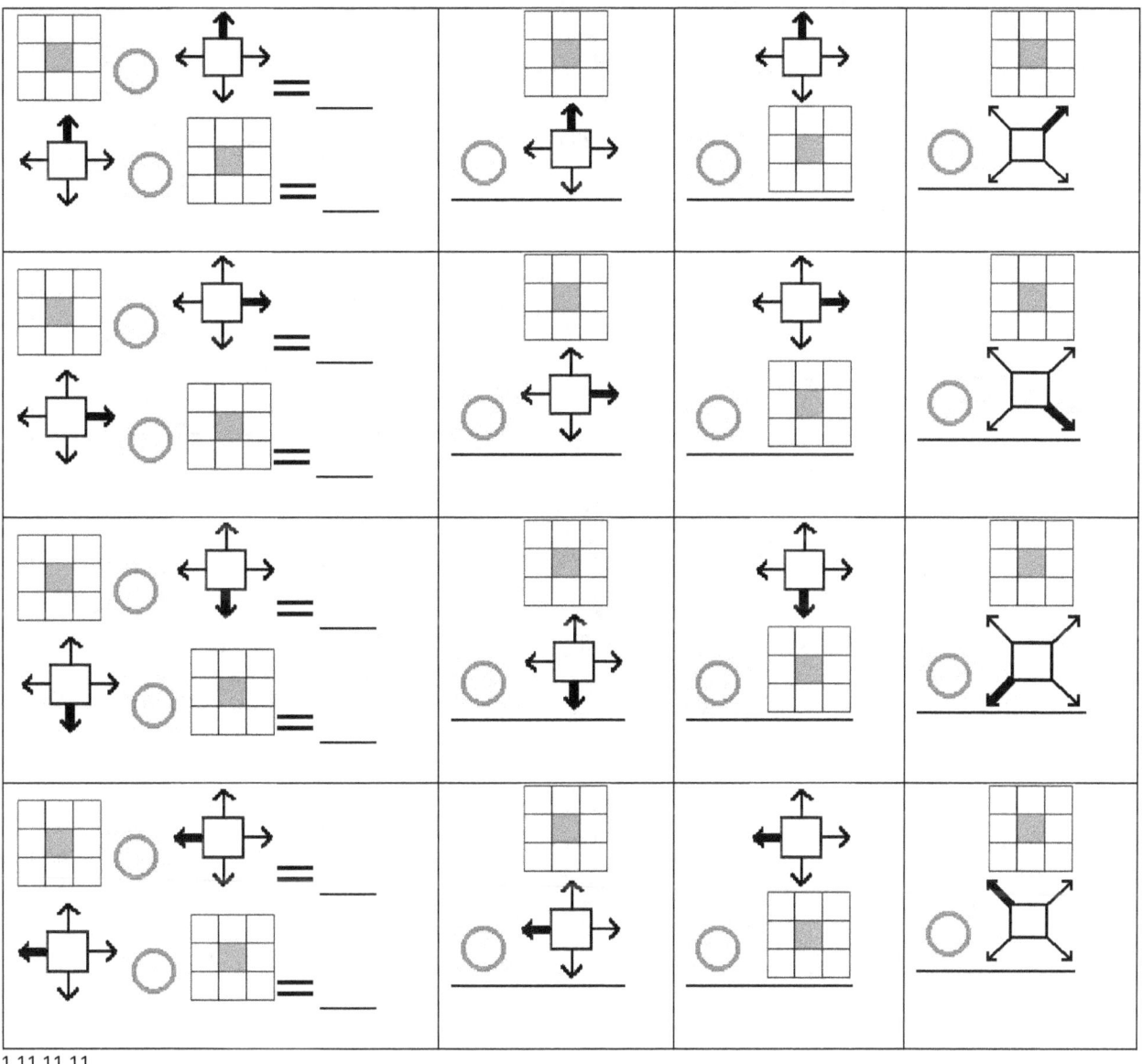

11 11 11 11
10 10 10 9
10 10 10 9
9 9 9 10

231

Ho Math Chess 何数棋谜   妈！我会棋谜式加法啦！
Mom! I Learn Addition Using Math-Chess-Puzzles Connection
Contents include both traditional and Math-Chess-Puzzles combined methods. Extra strength
©2008 − 2018 Frank Ho, Amanda Ho   All rights reserved. www.homathchess.com

Student Name _____ Date _____

Making 10 or adding around 10 or doubling
Only transfer each number to a correct position. Do not the transfer the 3 by 3 square.

You are a chess piece located at a square indicated by a shaded square.

○ = +

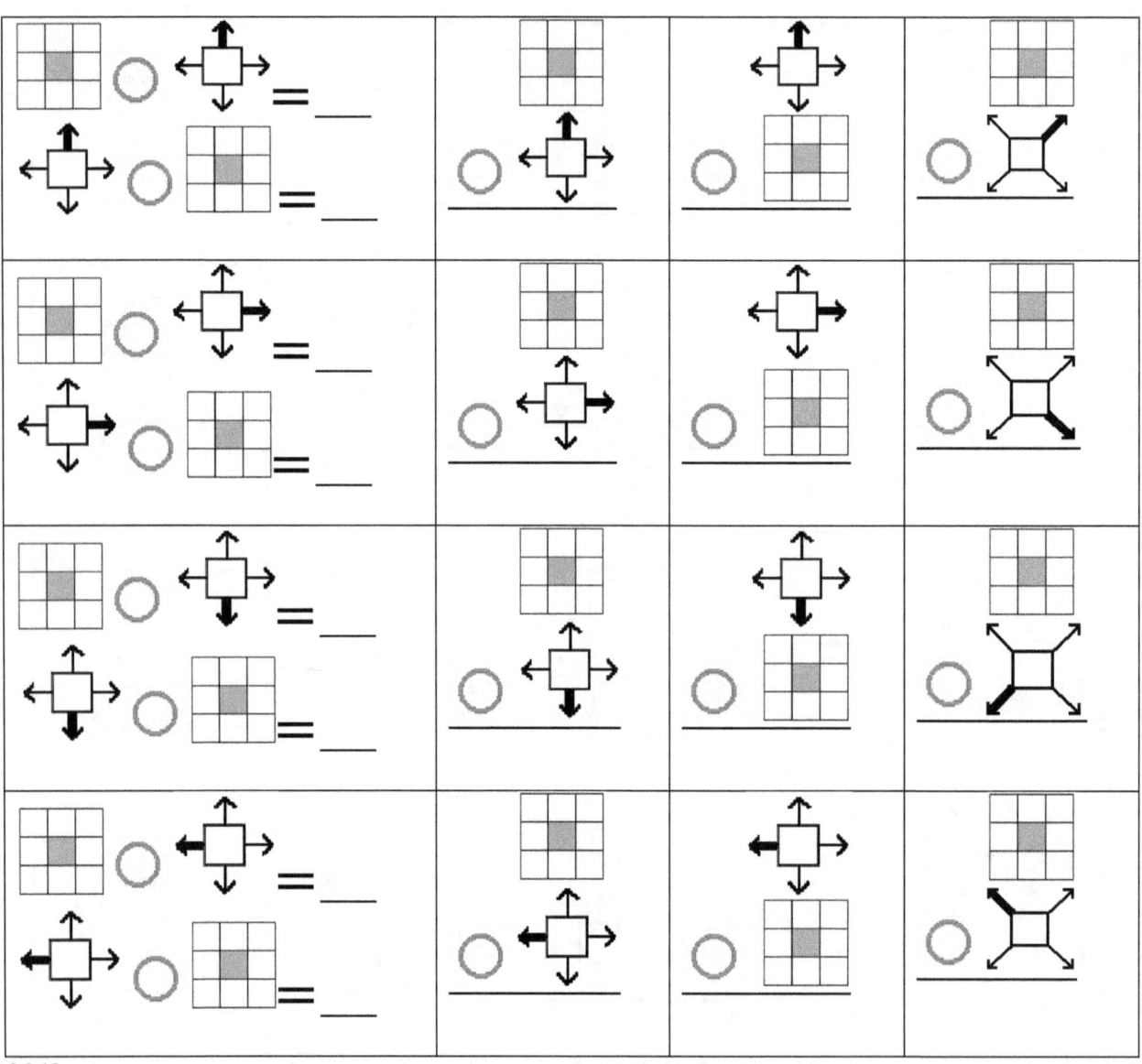

9 9 9 10
9 9 9 11
10 10 10 9
10 10 10 9

232

Mom! I Learn Addition Using Math-Chess-Puzzles Connection

Contents include both traditional and Math-Chess-Puzzles combined methods. Extra strength

©2008 − 2018 Frank Ho, Amanda Ho   All rights reserved. www.homathchess.com

Student Name _____ Date _____

Making 10 or adding around 10 or doubling

Only transfer each number to a correct position. Do not the transfer the 3 by 3 square.

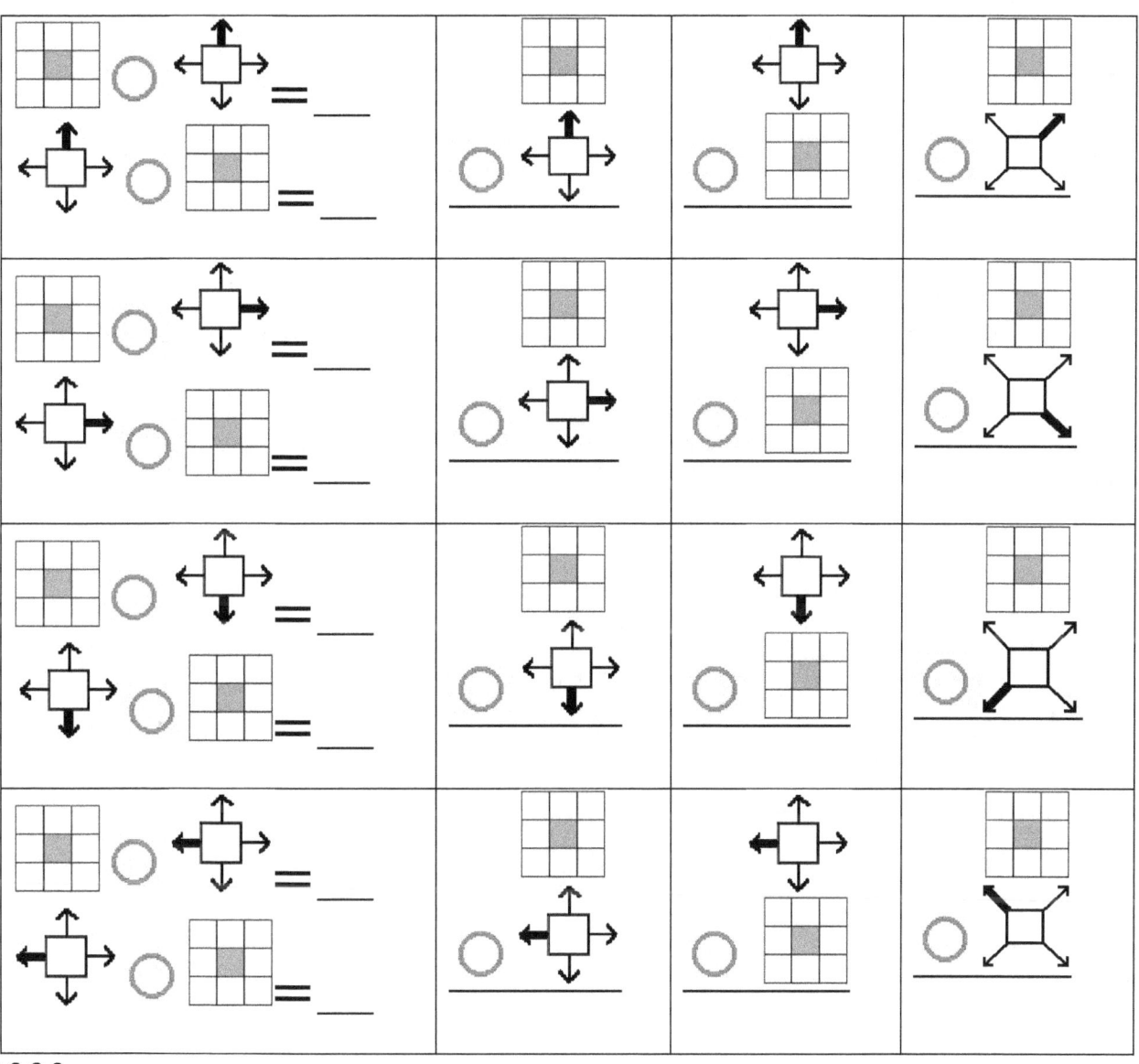

9 9 9 9
10 10 10 9
10 10 10 11
11 11 11 11

233

Ho Math Chess 何数棋谜　妈！我会棋谜式加法啦！
Mom! I Learn Addition Using Math-Chess-Puzzles Connection

Contents include both traditional and Math-Chess-Puzzles combined methods. Extra strength
©2008 – 2018 Frank Ho, Amanda Ho   All rights reserved. www.homathchess.com

Student Name _____   Date _____

Making 10 or adding around 10 or doubling
Only transfer each number to a correct position. Do not the transfer the 3 by 3 square.

You are a chess piece located at a square indicated by a shaded square.

 = +

11 11 11 10
9 9 9 9
10 10 10 9
10 10 10 11

234

Mom! I Learn Addition Using Math-Chess-Puzzles Connection

Contents include both traditional and Math-Chess-Puzzles combined methods. Extra strength

©2008 – 2018 Frank Ho, Amanda Ho   All rights reserved. www.homathchess.com

Student Name _____ Date _____

Making 10 or adding around 10 or doubling

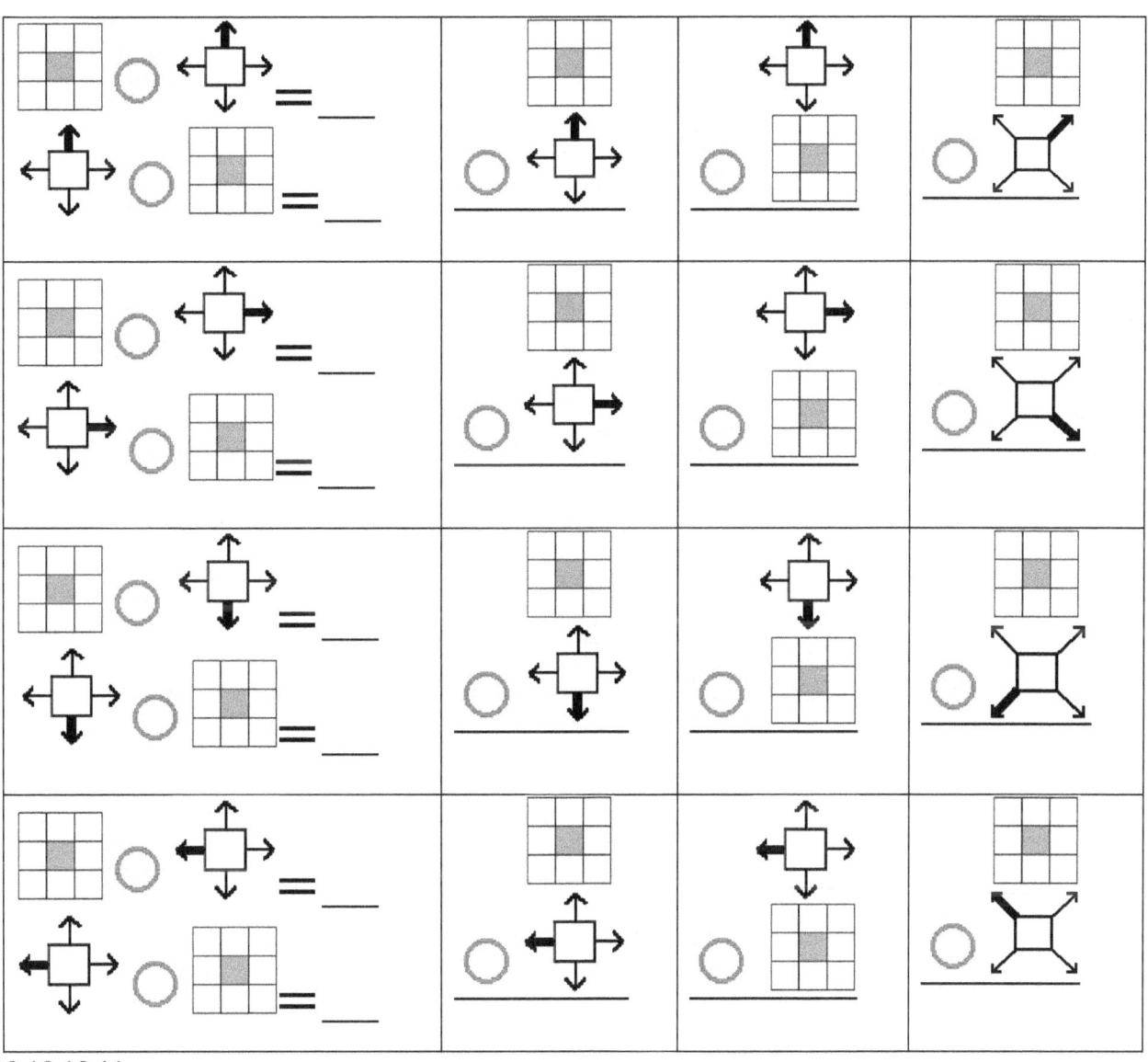

10 10 10 11
11 11 11 9
11 11 11 10
10 10 10 11

235

Ho Math Chess 何数棋谜   妈！我会棋谜式加法啦！
Mom! I Learn Addition Using Math-Chess-Puzzles Connection
Contents include both traditional and Math-Chess-Puzzles combined methods. Extra strength
©2008 – 2018 Frank Ho, Amanda Ho   All rights reserved. www.homathchess.com

Student Name_____   Date_____

Making 10 or adding around 10 or doubling
Only transfer each number to a correct position. Do not the transfer the 3 by 3 square.

You are a chess piece located at a square indicated by a shaded square.

○ = +

10 10 10 11
10 10 10 11
9 9 9 9
11 11 11 10

Ho Math Chess 何数棋谜　妈！我会棋谜式加法啦！
Mom! I Learn Addition Using Math-Chess-Puzzles Connection

Contents include both traditional and Math-Chess-Puzzles combined methods. Extra strength

©2008 — 2018 Frank Ho, Amanda Ho　All rights reserved. www.homathchess.com

Student Name _____ Date _____

Making 10 or adding around 10 or doubling

Only transfer each number to a correct position. Do not the transfer the 3 by 3 square.

You are a chess piece located at a square indicated by a shaded square.

11 11 11 10
10 10 10 11
10 10 10 11
9 9 9 9

237

Ho Math Chess 何数棋谜　妈！我会棋谜式加法啦！
Mom! I Learn Addition Using Math-Chess-Puzzles Connection

Contents include both traditional and Math-Chess-Puzzles combined methods. Extra strength
©2008 − 2018 Frank Ho, Amanda Ho　　All rights reserved. www.homathchess.com

Student Name _____ Date _____

Making 10 or adding around 10 or doubling
Only transfer each number to a correct position. Do not the transfer the 3 by 3 square.

You are a chess piece located at a square indicated by a shaded square.

9 9 9 9
11 11 11 10
10 10 10 11
10 10 10 11

Ho Math Chess 何数棋谜　妈！我会棋谜式加法啦！
Mom! I Learn Addition Using Math-Chess-Puzzles Connection

Contents include both traditional and Math-Chess-Puzzles combined methods. Extra strength

©2008 – 2018 Frank Ho, Amanda Ho   All rights reserved. www.homathchess.com

Student Name_____ Date_____

Making 10 or adding around 10

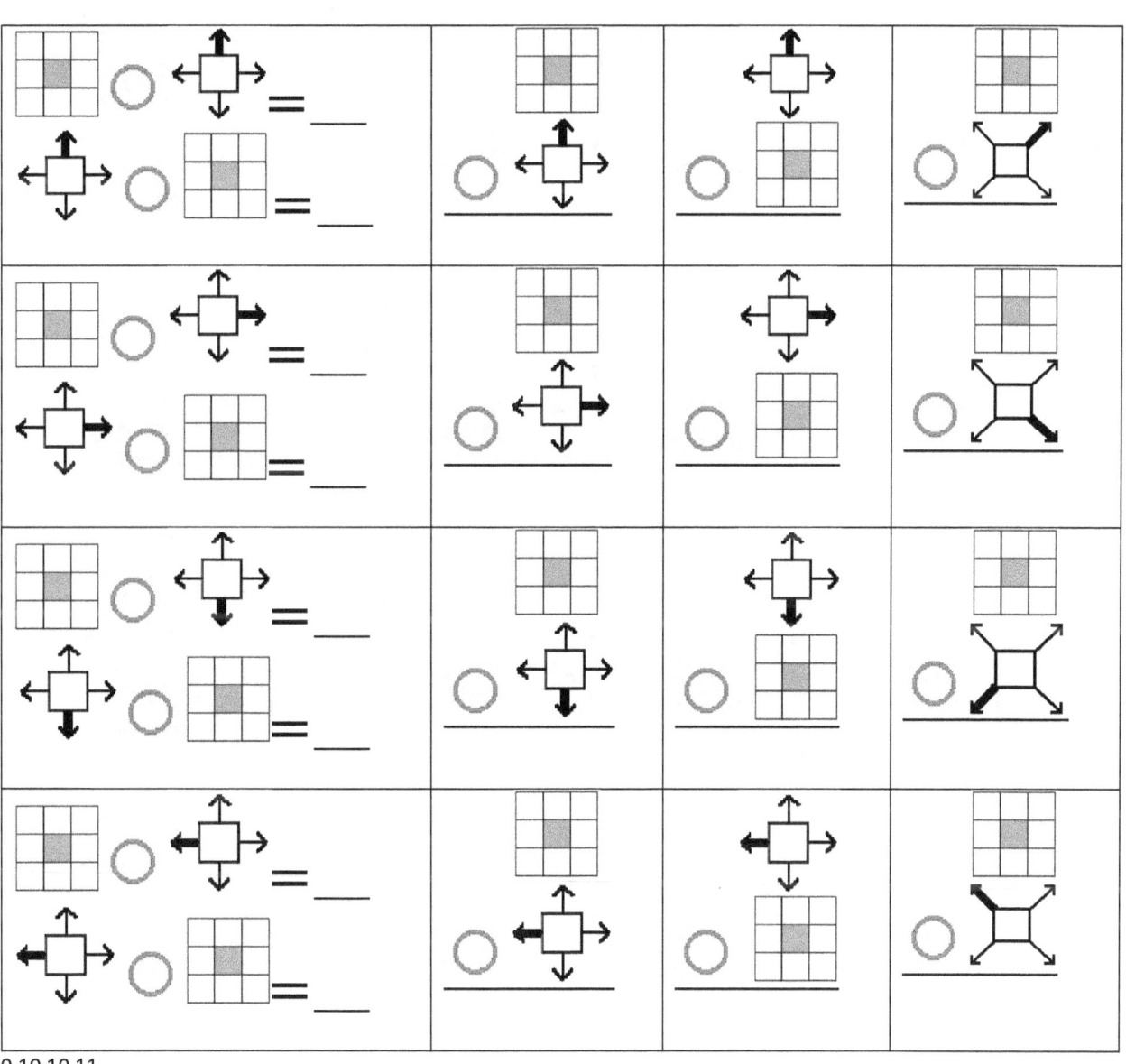

10 10 10 11
9 9 9 10
11 11 11 9
9 9 9 10

239

Ho Math Chess 何数棋谜　妈！我会棋谜式加法啦！
Mom! I Learn Addition Using Math-Chess-Puzzles Connection
Contents include both traditional and Math-Chess-Puzzles combined methods. Extra strength
©2008 – 2018 Frank Ho, Amanda Ho　　All rights reserved. www.homathchess.com

Student Name _____ Date _____

Making 10 or adding around 10
Only transfer each number to a correct position. Do not the transfer the 3 by 3 square.

You are a chess piece located at a square indicated by a shaded square.

○ = +

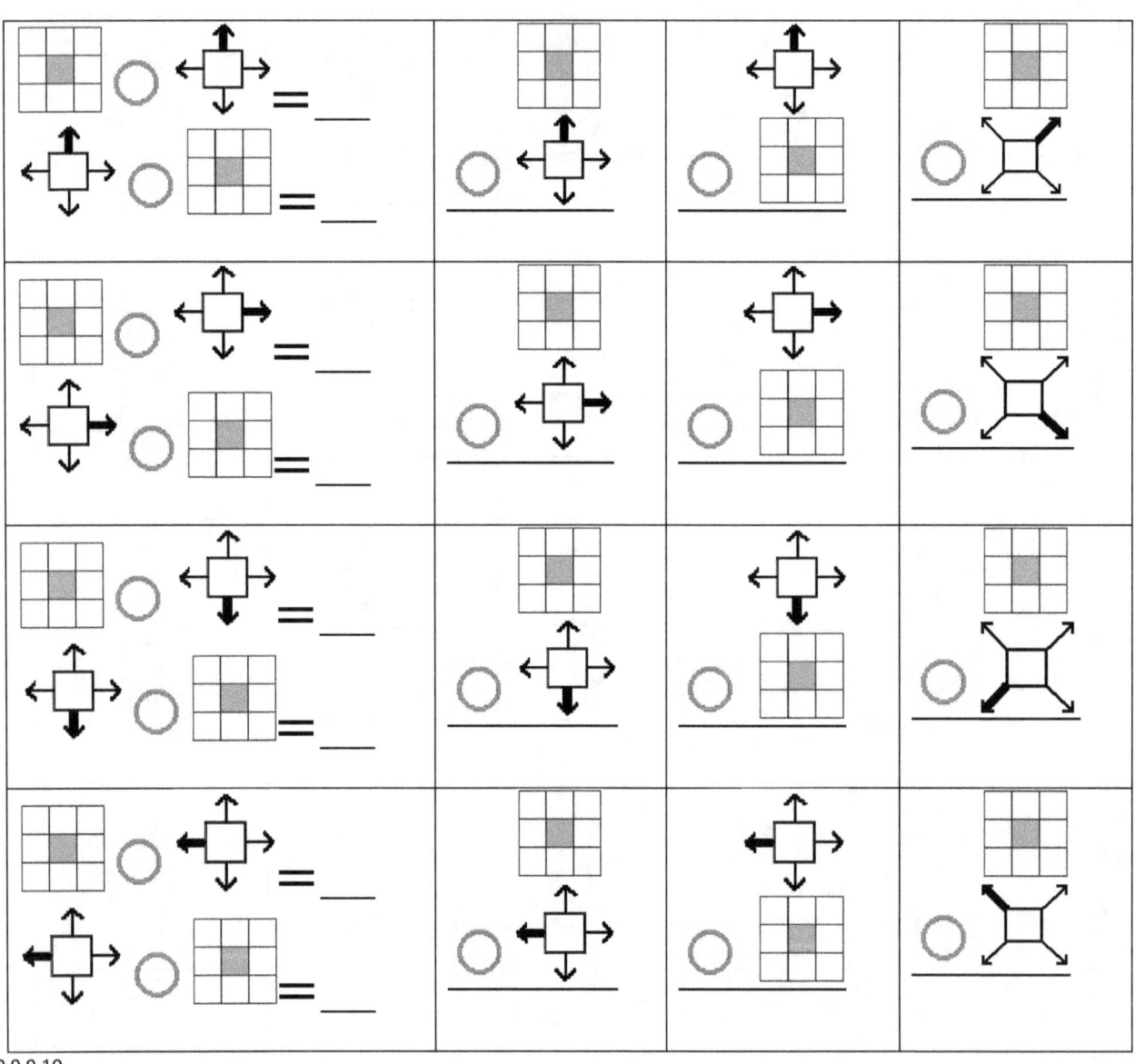

9 9 9 10
10 10 10 11
9 9 9 10
11 11 11 9

240

Making 10 or adding around 10

Only transfer each number to a correct position. Do not the transfer the 3 by 3 square.

You are a chess piece located at a square indicated by a shaded square.

11 11 11 9
9 9 9 10
10 10 10 11
9 9 9 10

Ho Math Chess 何数棋谜　妈！我会棋谜式加法啦！
Mom! I Learn Addition Using Math-Chess-Puzzles Connection
Contents include both traditional and Math-Chess-Puzzles combined methods. Extra strength
©2008 – 2018 Frank Ho, Amanda Ho　All rights reserved. www.homathchess.com

Student Name _____ Date _____

Making 10 or adding around 10
Only transfer each number to a correct position. Do not the transfer the 3 by 3 square.

You are a chess piece located at a square indicated by a shaded square.

10 10 10 10
9 9 9 9
11 11 11 10
9 9 9 11

Mom! I Learn Addition Using Math-Chess-Puzzles Connection

Contents include both traditional and Math-Chess-Puzzles combined methods. Extra strength

©2008 – 2018 Frank Ho, Amanda Ho   All rights reserved. www.homathchess.com

Student Name _____   Date _____

Making 10 or adding around 10

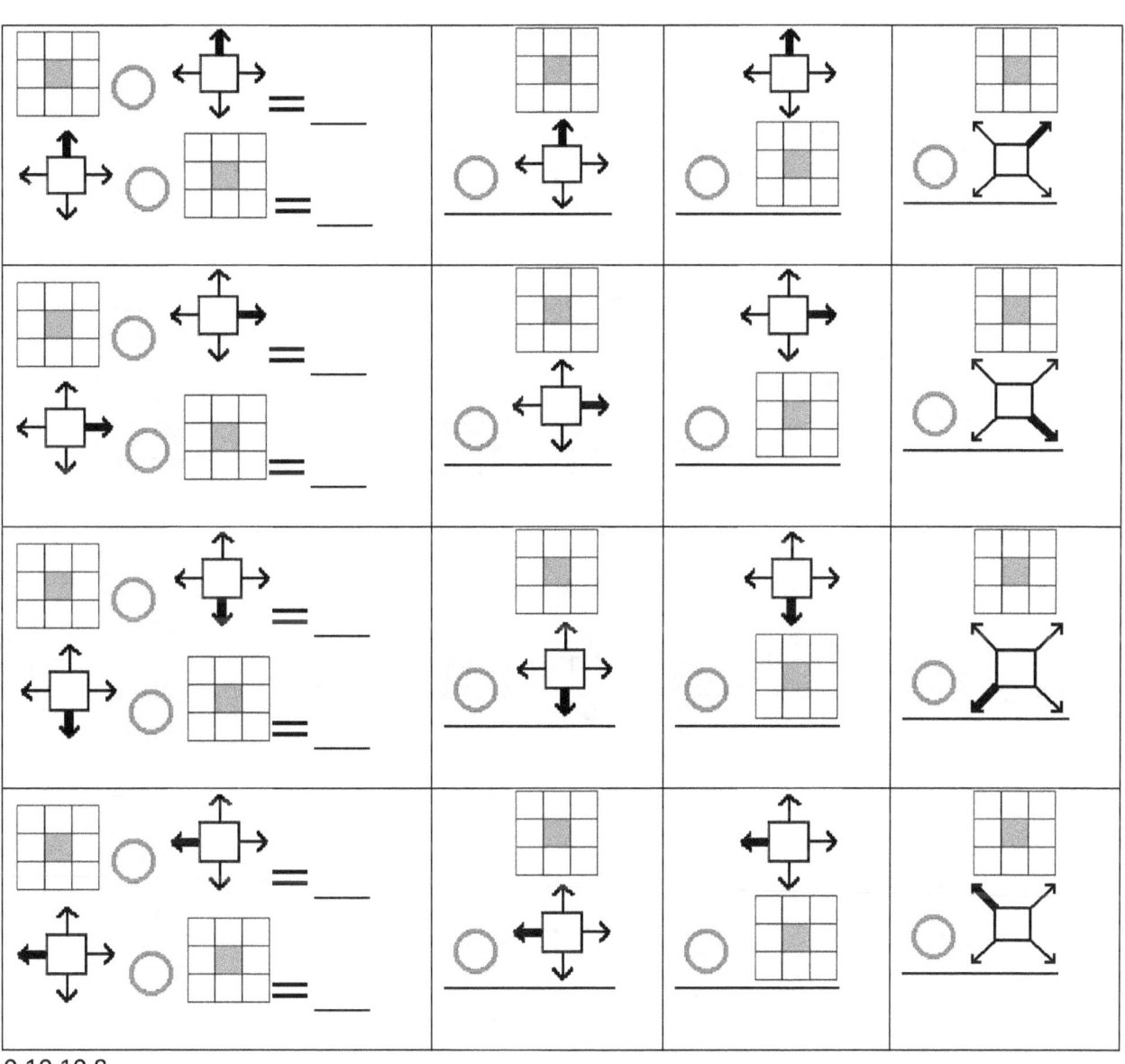

10 10 10 8
9 9 9 9
11 11 11 10
10 10 10 9

243

Making 10 or adding around 10
Only transfer each number to a correct position. Do not the transfer the 3 by 3 square.

You are a chess piece located at a square indicated by a shaded square.

10 10 10 9
10 10 10 11
9 9 9 9
11 11 11 10

244

Ho Math Chess 何数棋谜　妈！我会棋谜式加法啦！
Mom! I Learn Addition Using Math-Chess-Puzzles Connection
Contents include both traditional and Math-Chess-Puzzles combined methods. Extra strength
©2008 – 2018 Frank Ho, Amanda Ho   All rights reserved. www.homathchess.com

Student Name _____ Date _____

Making 10 or adding around 10
Only transfer each number to a correct position. Do not the transfer the 3 by 3 square.

You are a chess piece located at a square indicated by a shaded square.

○ = +

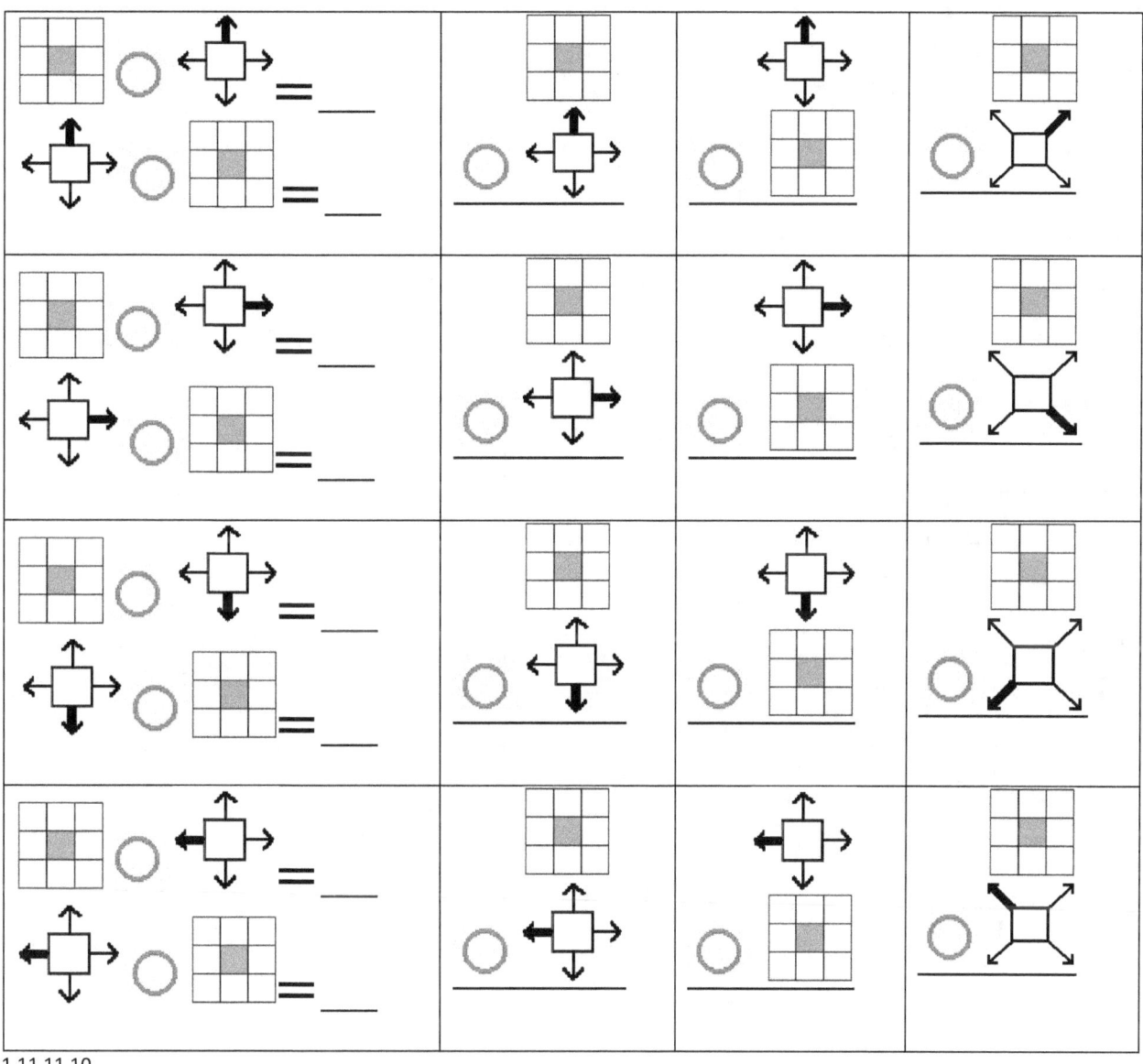

11 11 11 10
10 10 10 9
10 10 10 9
9 9 9 9

245

Ho Math Chess 何数棋谜 妈！我会棋谜式加法啦！
Mom! I Learn Addition Using Math-Chess-Puzzles Connection

Contents include both traditional and Math-Chess-Puzzles combined methods. Extra strength
©2008 – 2018 Frank Ho, Amanda Ho   All rights reserved. www.homathchess.com

Student Name _____ Date _____

Making 10 or adding around 10

Only transfer each number to a correct position. Do not the transfer the 3 by 3 square.

You are a chess piece located at a square indicated by a shaded square.

 = +

9 9 9 9
11 11 11 10
10 10 10 9
10 10 10 11

246

Ho Math Chess 何数棋谜　妈！我会棋谜式加法啦！

Mom! I Learn Addition Using Math-Chess-Puzzles Connection

Contents include both traditional and Math-Chess-Puzzles combined methods. Extra strength

©2008 – 2018 Frank Ho, Amanda Ho　　All rights reserved. www.homathchess.com

Student Name _____ Date _____

Making 10 or adding around 10

You are a chess piece located at a square indicated by a shaded square.

○ = +

10 10 10 9
9 9 9 8
10 10 10 9
9 9 9 8

247

Ho Math Chess 何数棋谜   妈！我会棋谜式加法啦！
Mom! I Learn Addition Using Math-Chess-Puzzles Connection
Contents include both traditional and Math-Chess-Puzzles combined methods. Extra strength
©2008 – 2018 Frank Ho, Amanda Ho   All rights reserved. www.homathchess.com

Student Name _____   Date _____

Making 10 or adding around 10

Only transfer each number to a correct position. Do not the transfer the 3 by 3 square.

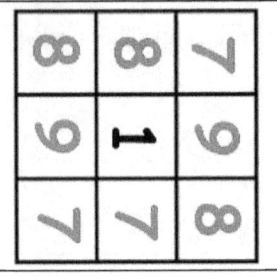

You are a chess piece located at a square indicated by a shaded square.

  = +

9 9 9 8
10 10 10 9
8 8 8 8
10 10 10 9

248

Ho Math Chess 何数棋谜　妈！我会棋谜式加法啦！
Mom! I Learn Addition Using Math-Chess-Puzzles Connection
Contents include both traditional and Math-Chess-Puzzles combined methods. Extra strength
©2008 – 2018 Frank Ho, Amanda Ho   All rights reserved. www.homathchess.com

Student Name _____   Date _____

Making 10 or adding around 10

Only transfer each number to a correct position. Do not the transfer the 3 by 3 square.

You are a chess piece located at a square indicated by a shaded square.

10 10 10 9
9 9 9 8
10 10 10 9
8 8 8 8

249

Ho Math Chess 何数棋谜　妈！我会棋谜式加法啦！
Mom! I Learn Addition Using Math-Chess-Puzzles Connection
Contents include both traditional and Math-Chess-Puzzles combined methods. Extra strength
©2008 − 2018 Frank Ho, Amanda Ho　　All rights reserved. www.homathchess.com

Student Name _____ Date _____

Making 10 or adding around 10
Only transfer each number to a correct position. Do not the transfer the 3 by 3 square.

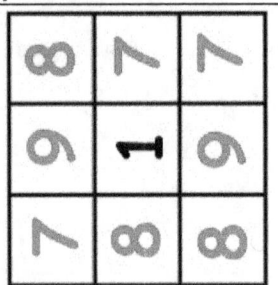

You are a chess piece located at a square indicated by a shaded square.

8 8 8 8
10 10 10 9
9 9 9 8
10 10 10 9

250

Ho Math Chess 何数棋谜　妈！我会棋谜式加法啦！
Mom! I Learn Addition Using Math-Chess-Puzzles Connection
Contents include both traditional and Math-Chess-Puzzles combined methods. Extra strength
©2008 – 2018 Frank Ho, Amanda Ho　All rights reserved. www.homathchess.com
Student Name _____ Date _____

Knight moves to make 10
For those who know how to move knights.

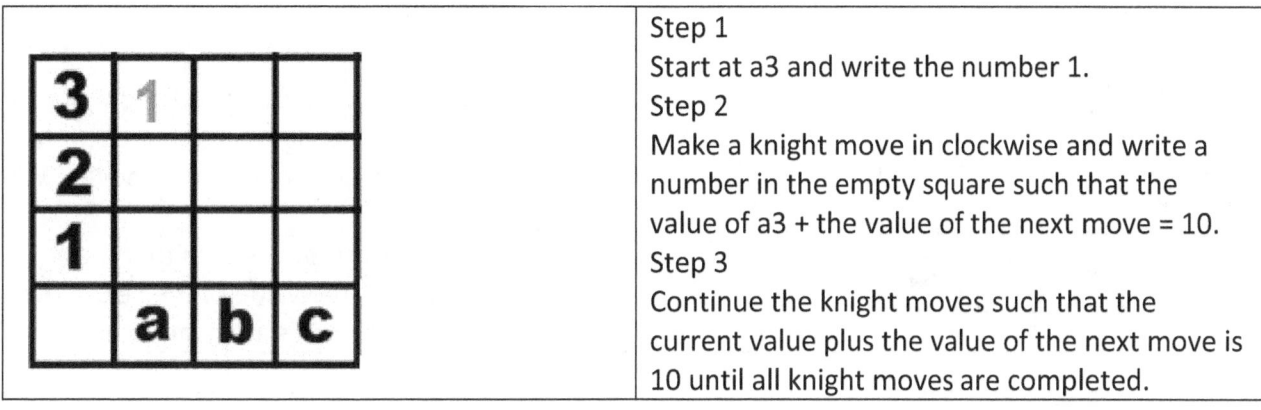

Step 1
Start at a3 and write the number 1.
Step 2
Make a knight move in clockwise and write a number in the empty square such that the value of a3 + the value of the next move = 10.
Step 3
Continue the knight moves such that the current value plus the value of the next move is 10 until all knight moves are completed.

3

To solve the problem, first you position yourself at the square of a3 which is 1 for this problem.

You move in the dirction of the knight's move ⌐. The answer is 1 + 9 = 10.

The top to down answer for L is also 1 + 9 = 10.

a

251

Knight moves to make 10

Step 1
Start at a2 and write the number 2.
Step 2
Make a knight move in clockwise and write a number in the empty square such that the value of a2 + the value of the next move = 10.
Step 3
Continue the knight moves such that the current value plus the value of the next move is 10 until all knight moves are completed.

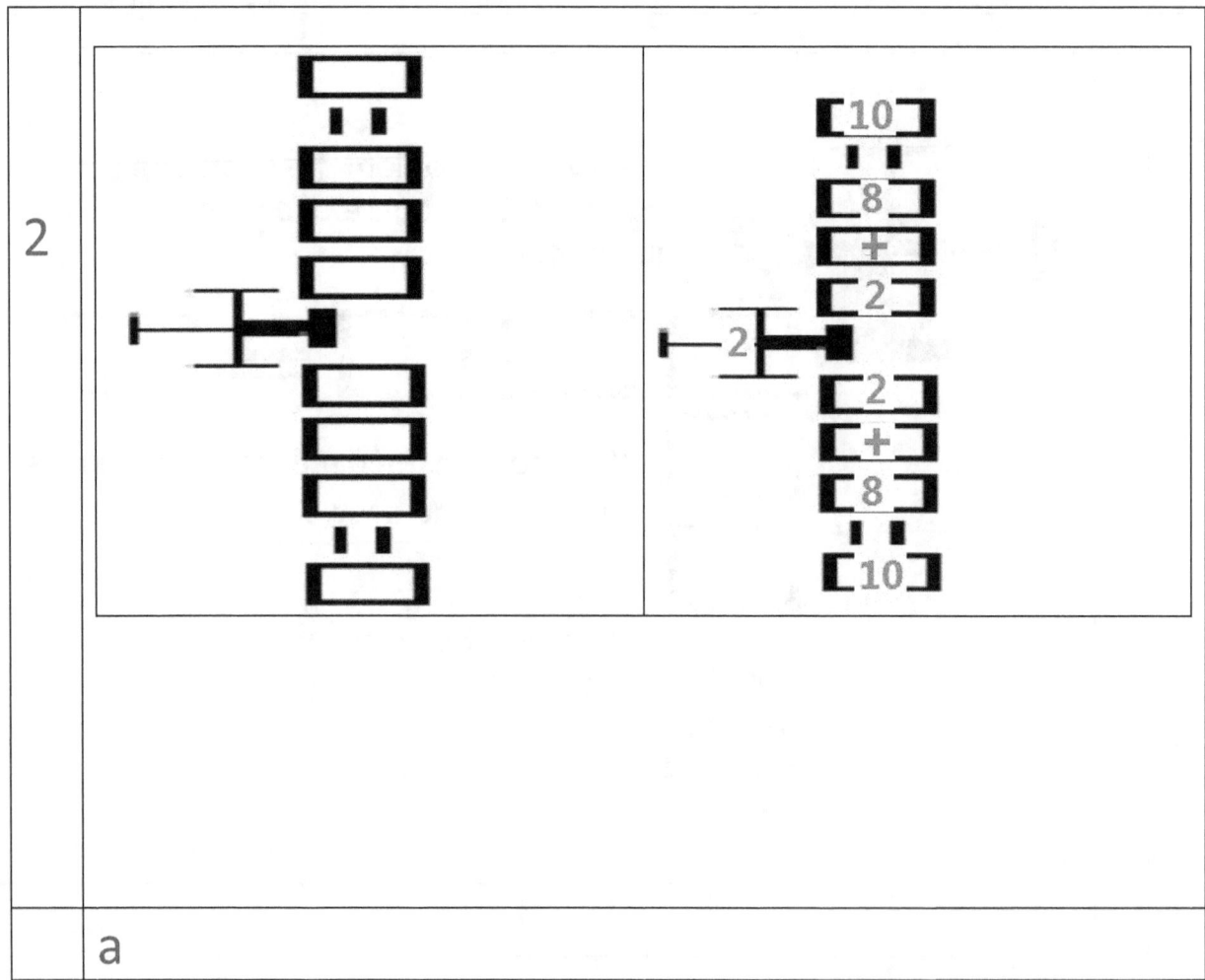

a

Knight moves to make 10

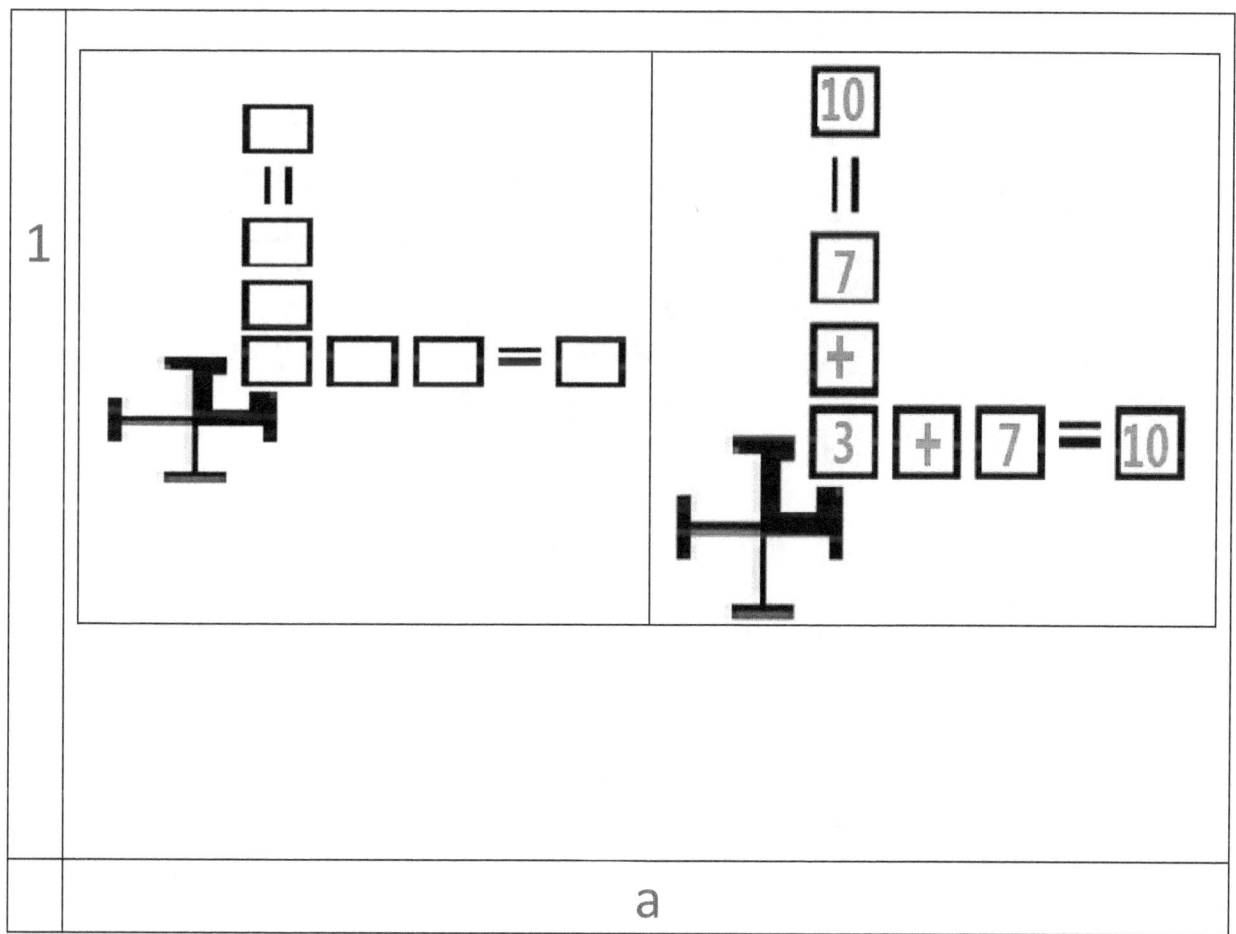

Knight moves to make 10

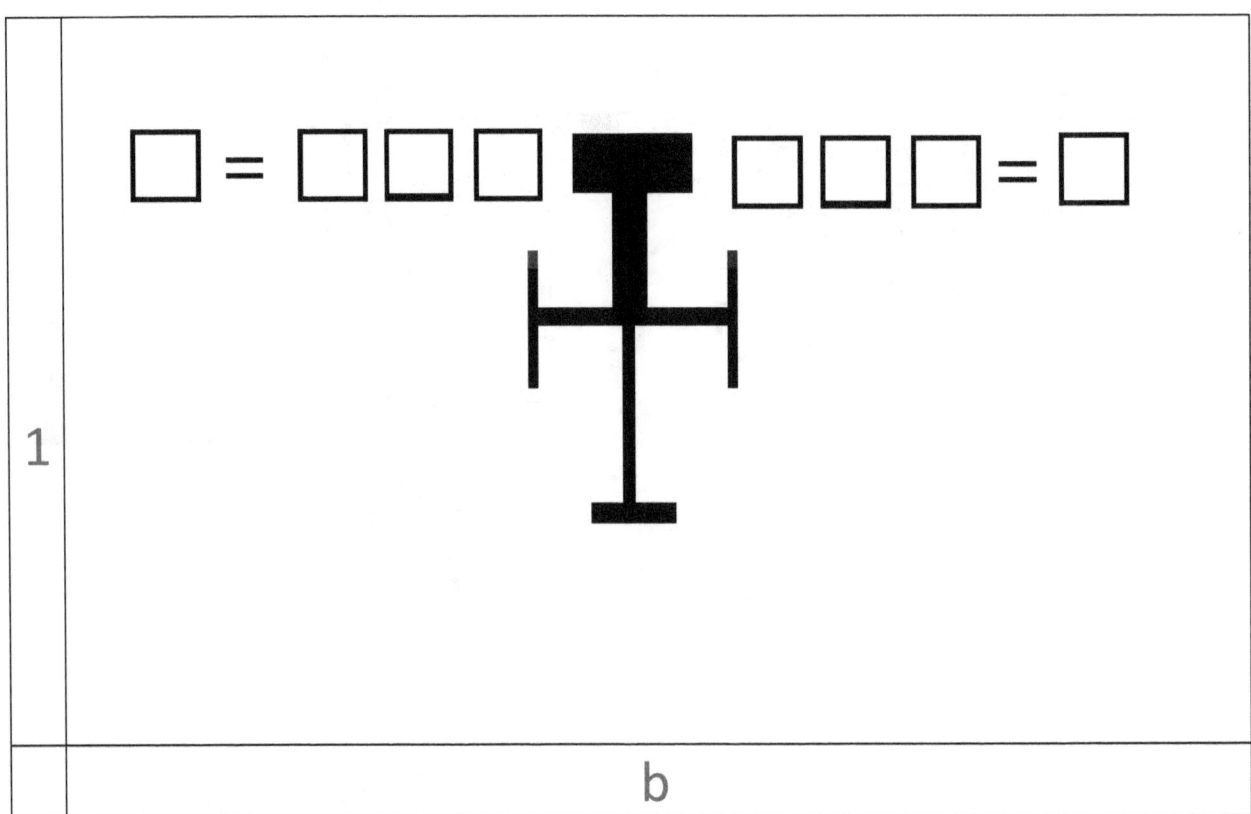

Step 1
Start at b1 and write the number 4.
Step 2
Make a knight move in clockwise and write a number in the empty square such that the value of b1 + the value of the next move = 10.
Step 3
Continue the knight moves such that the current value plus the value of the next move is 10 until all knight moves are completed.

Knight moves to make 10

| | Step 1 Start at b3 and write the number 5. Step 2 Make a knight move in clockwise and write a number in the empty square such that the value of b3 + the value of the next move = 10. Step 3 Continue the knight moves such that the current value plus the value of the next move is 10 until all knight moves are completed. |

3

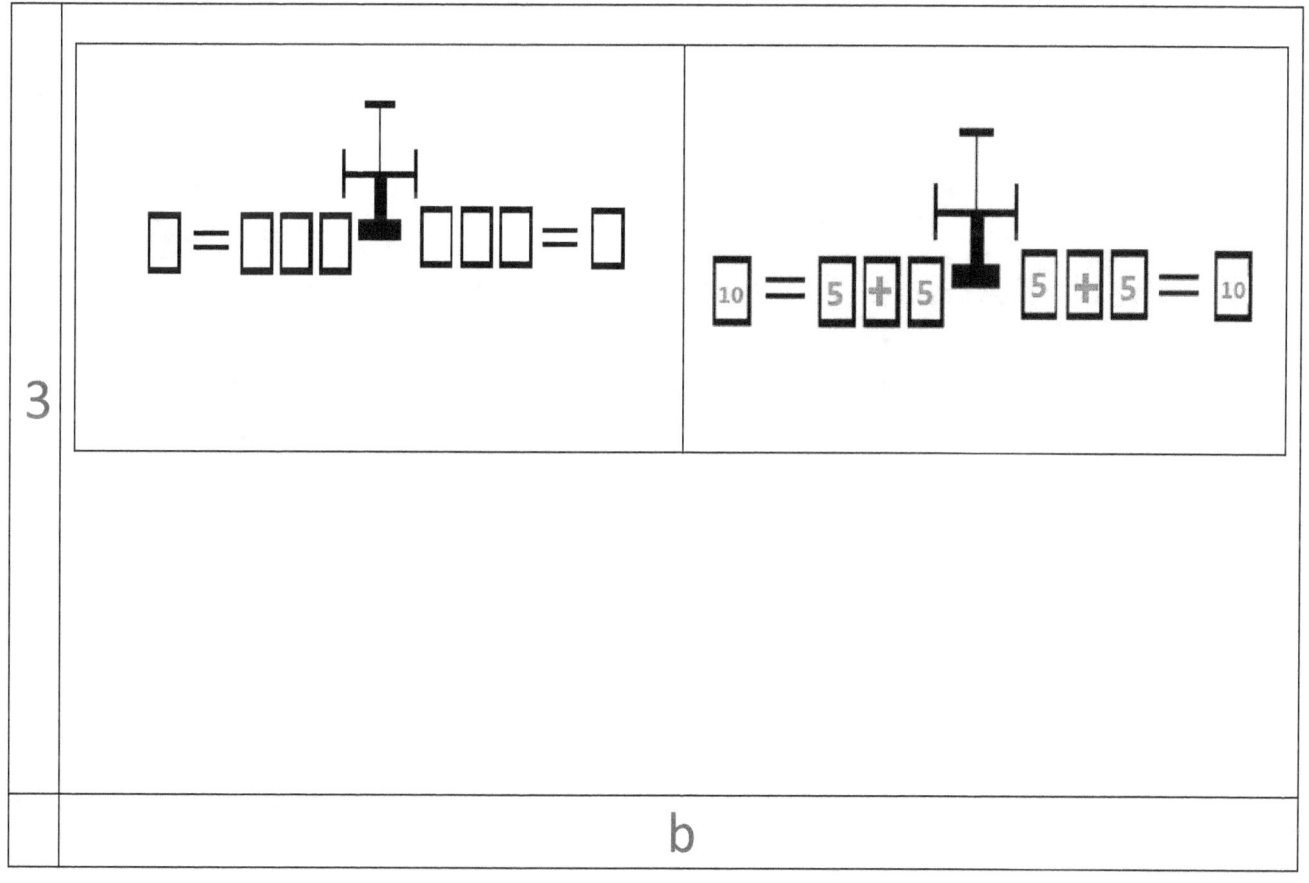

b

Ho Math Chess 何数棋谜　妈！我会棋谜式加法啦！
Mom! I Learn Addition Using Math-Chess-Puzzles Connection
Contents include both traditional and Math-Chess-Puzzles combined methods. Extra strength
©2008 − 2018 Frank Ho, Amanda Ho   All rights reserved. www.homathchess.com

Student Name _____   Date _____

Knight moves to make 10

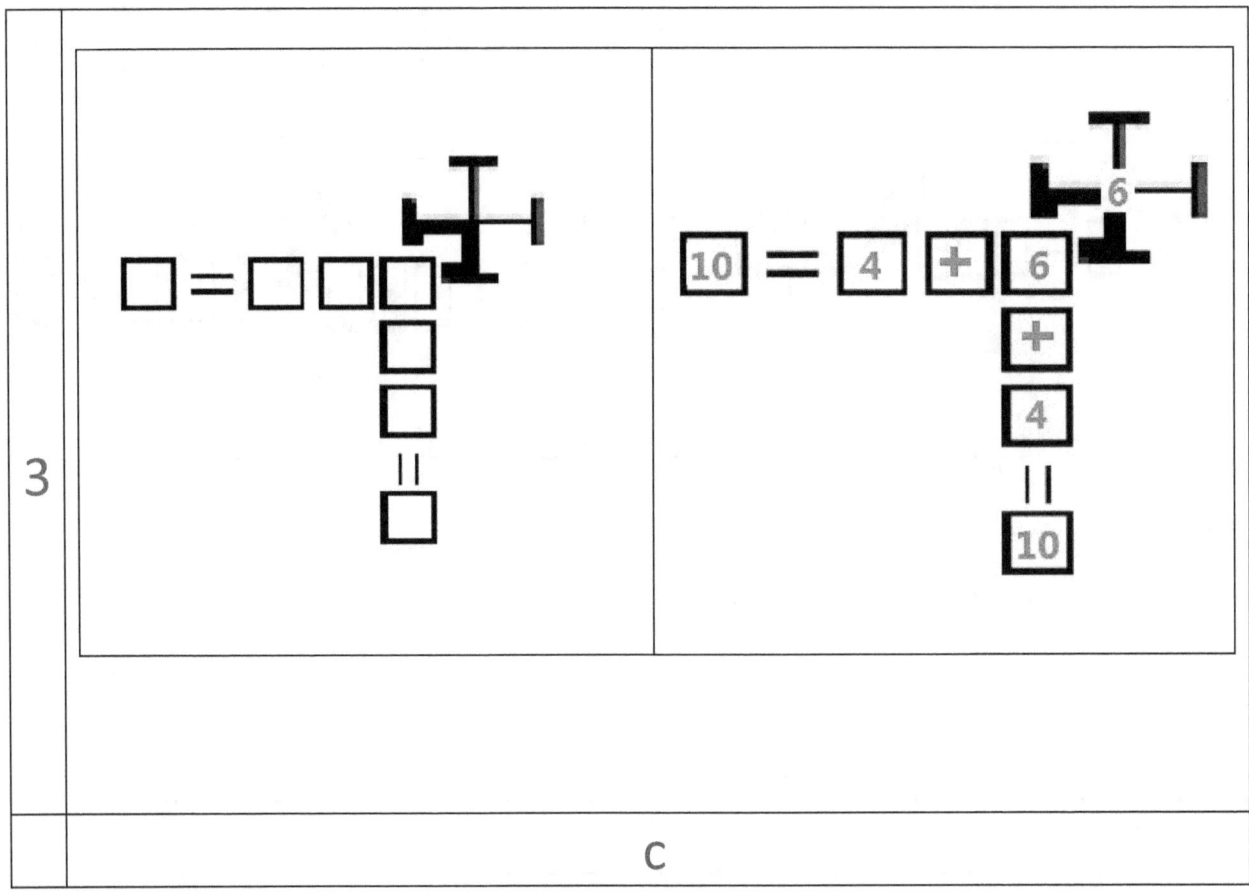

Step 1
Start at c3 and write the number 6.
Step 2
Make a knight move in clockwise and write a number in the empty square such that the value of c3 + the value of the next move = 10.
Step 3
Continue the knight moves such that the current value plus the value of the next move is 10 until all knight moves are completed.

256

Mom! I Learn Addition Using Math-Chess-Puzzles Connection

Contents include both traditional and Math-Chess-Puzzles combined methods. Extra strength
©2008 – 2018 Frank Ho, Amanda Ho   All rights reserved. www.homathchess.com

Student Name _____   Date _____

Doubling, adding 10, or 9 + d

Students need to write the entire question before writing the result 10 to reinforce the idea what 2 numbers added together to make 10.

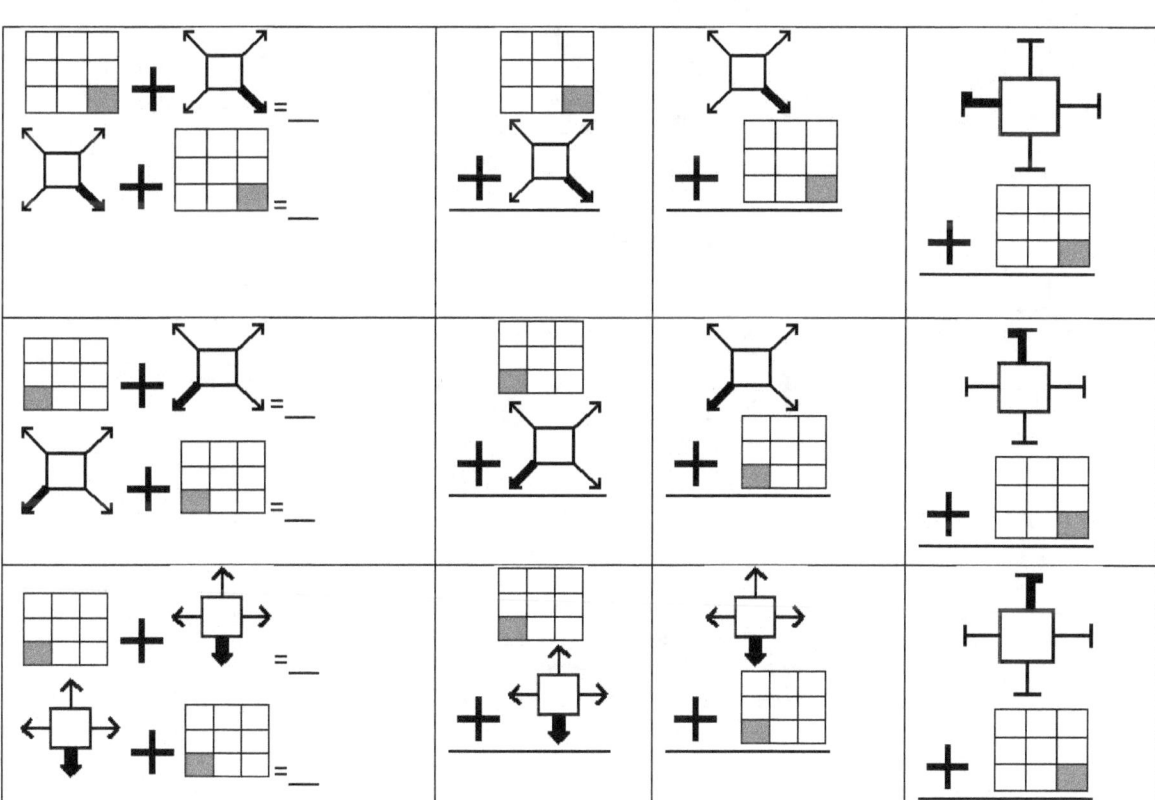

10 10 10 10
10 10 10 10
10 10 10 10

Making 10 or adding around 10 or doubling

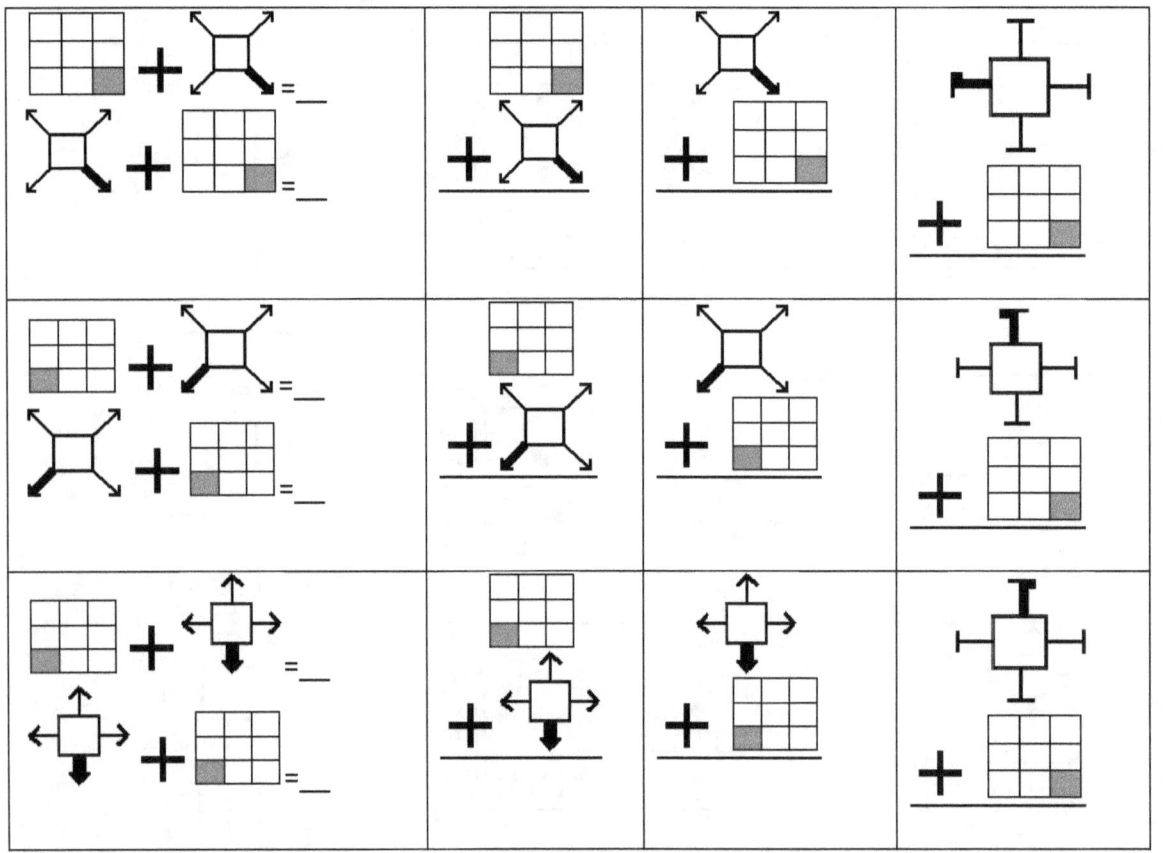

9 9 9 9
10 10 10 9
10 10 10 9

Ho Math Chess 何数棋谜　妈！我会棋谜式加法啦！
Mom! I Learn Addition Using Math-Chess-Puzzles Connection
Contents include both traditional and Math-Chess-Puzzles combined methods. Extra strength
©2008 – 2018 Frank Ho, Amanda Ho　All rights reserved. www.homathchess.com

Student Name _____ Date _____

Making 10 or adding around 10 or doubling

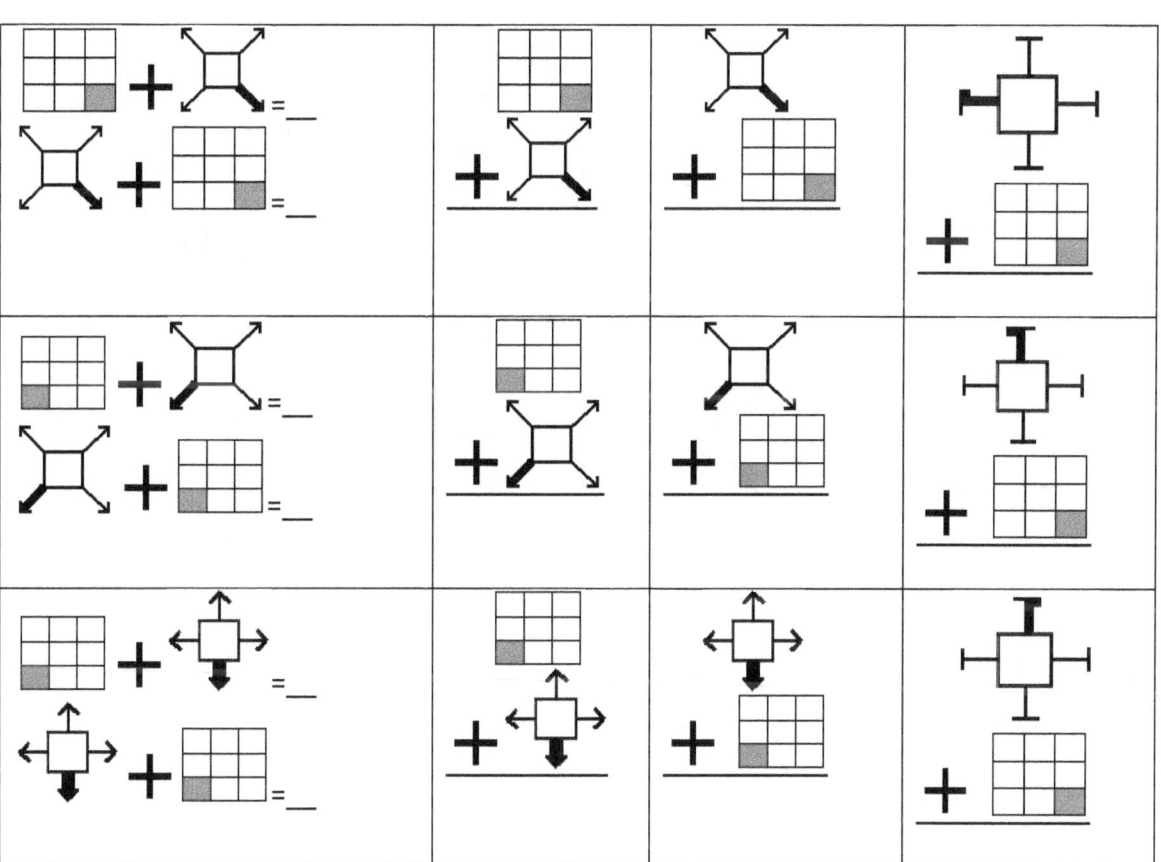

10 10 10 10
9 9 9 10
9 9 9 10

Making 10 or adding around 10 or doubling

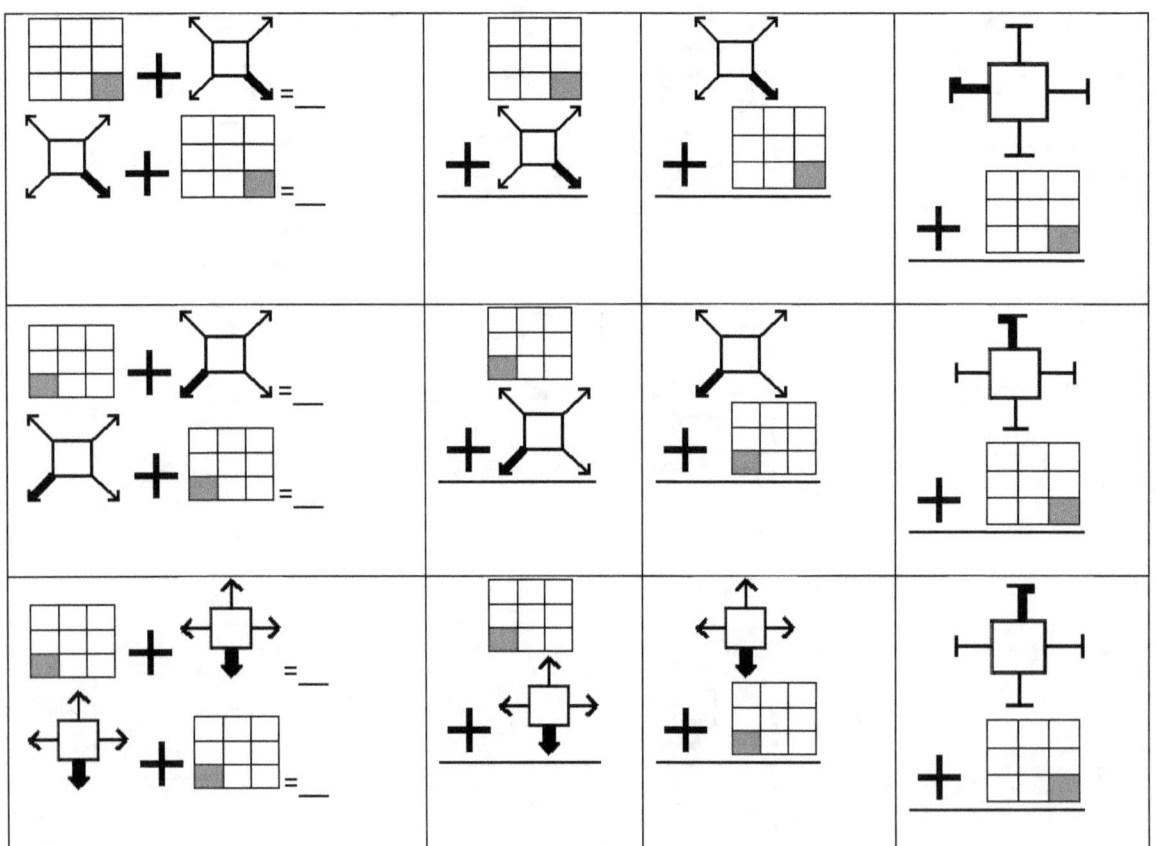

13 13 13 13
12 12 12 13
12 12 12 13

Ho Math Chess 何数棋谜   妈！我会棋谜式加法啦！
Mom! I Learn Addition Using Math-Chess-Puzzles Connection

Contents include both traditional and Math-Chess-Puzzles combined methods. Extra strength

©2008 − 2018 Frank Ho, Amanda Ho   All rights reserved. www.homathchess.com

Student Name _____ Date _____

Making 10 or adding around 10 or doubling

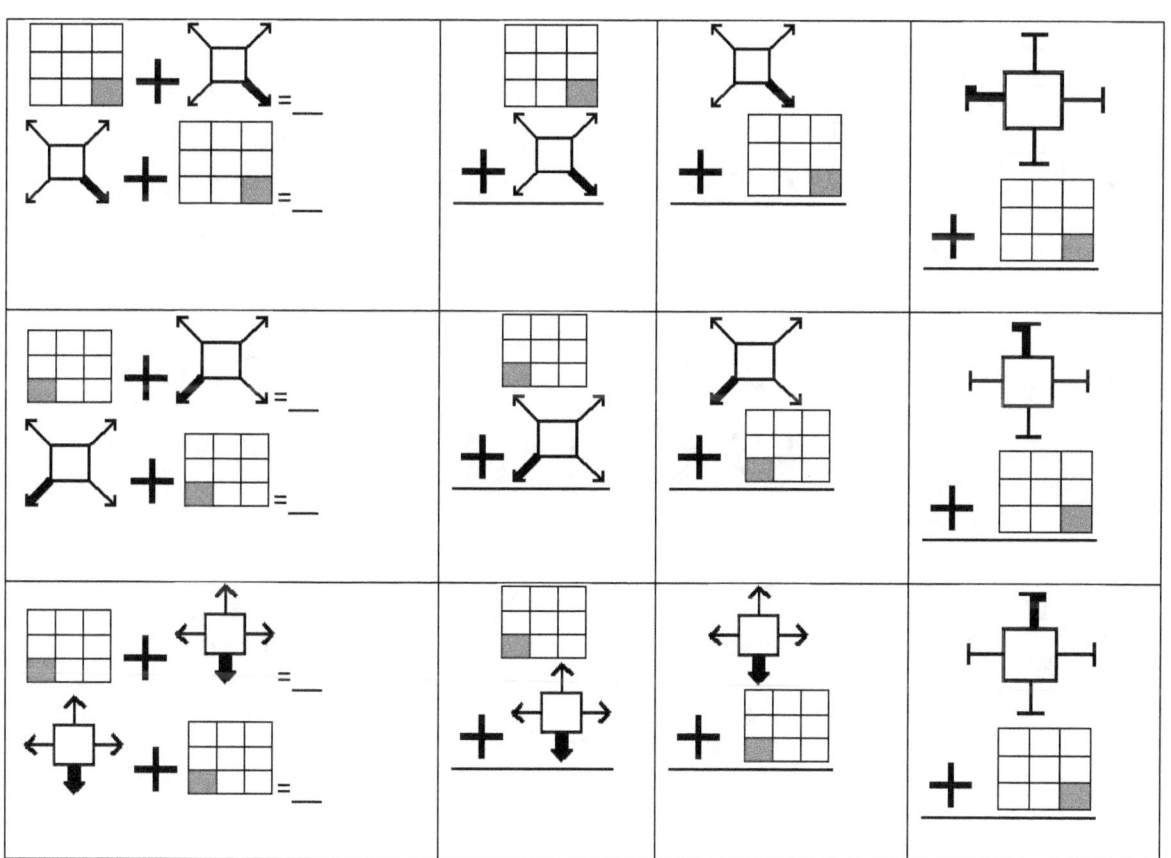

14 14 14 14
12 12 12 14
12 12 12 14

261

Ho Math Chess 何数棋谜 妈！我会棋谜式加法啦！
Mom! I Learn Addition Using Math-Chess-Puzzles Connection

Contents include both traditional and Math-Chess-Puzzles combined methods. Extra strength

©2008 － 2018 Frank Ho, Amanda Ho   All rights reserved. www.homathchess.com

Student Name _____ Date _____

Making 10 or adding around 10 or doubling

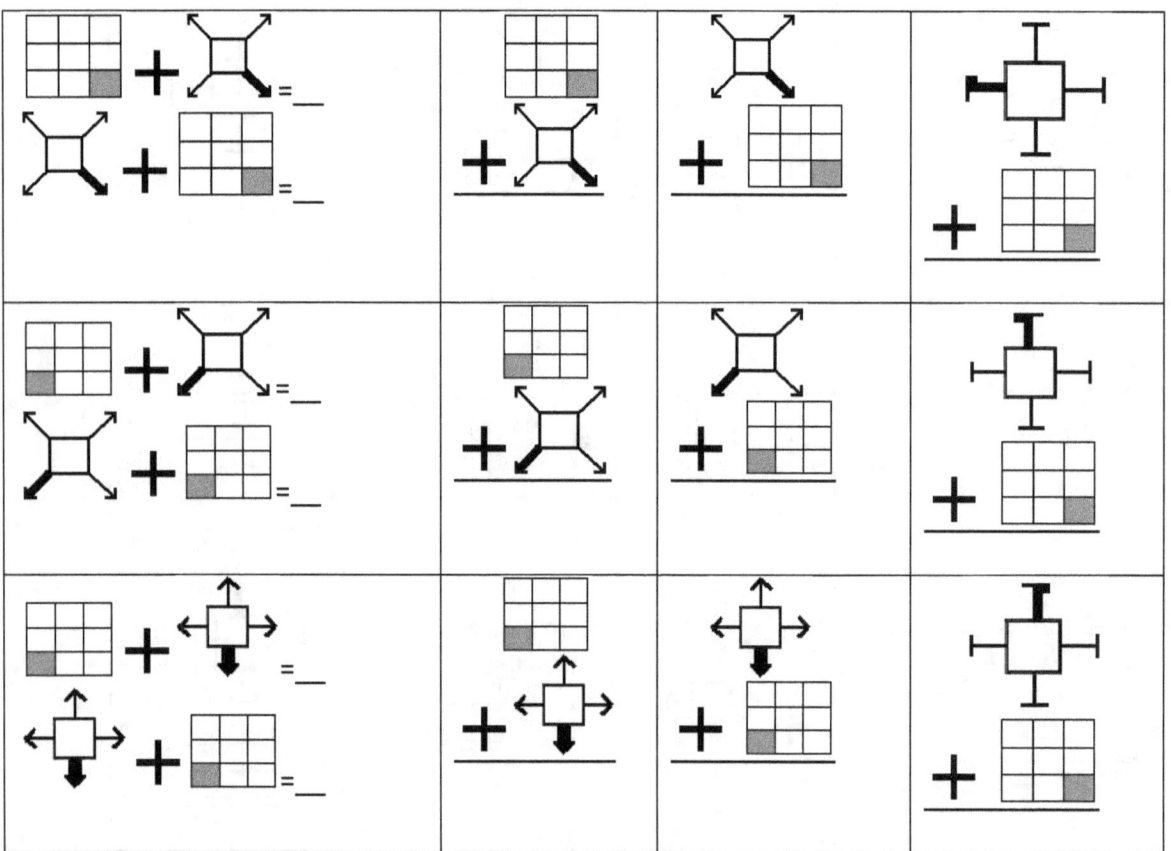

12 13 13 13
13 13 13 13
13 13 13 13

Ho Math Chess 何数棋谜　妈！我会棋谜式加法啦！
Mom! I Learn Addition Using Math-Chess-Puzzles Connection
Contents include both traditional and Math-Chess-Puzzles combined methods. Extra strength
©2008－2018 Frank Ho, Amanda Ho　All rights reserved. www.homathchess.com

Student Name _____ Date _____

Making 10 or adding around 10 or doubling

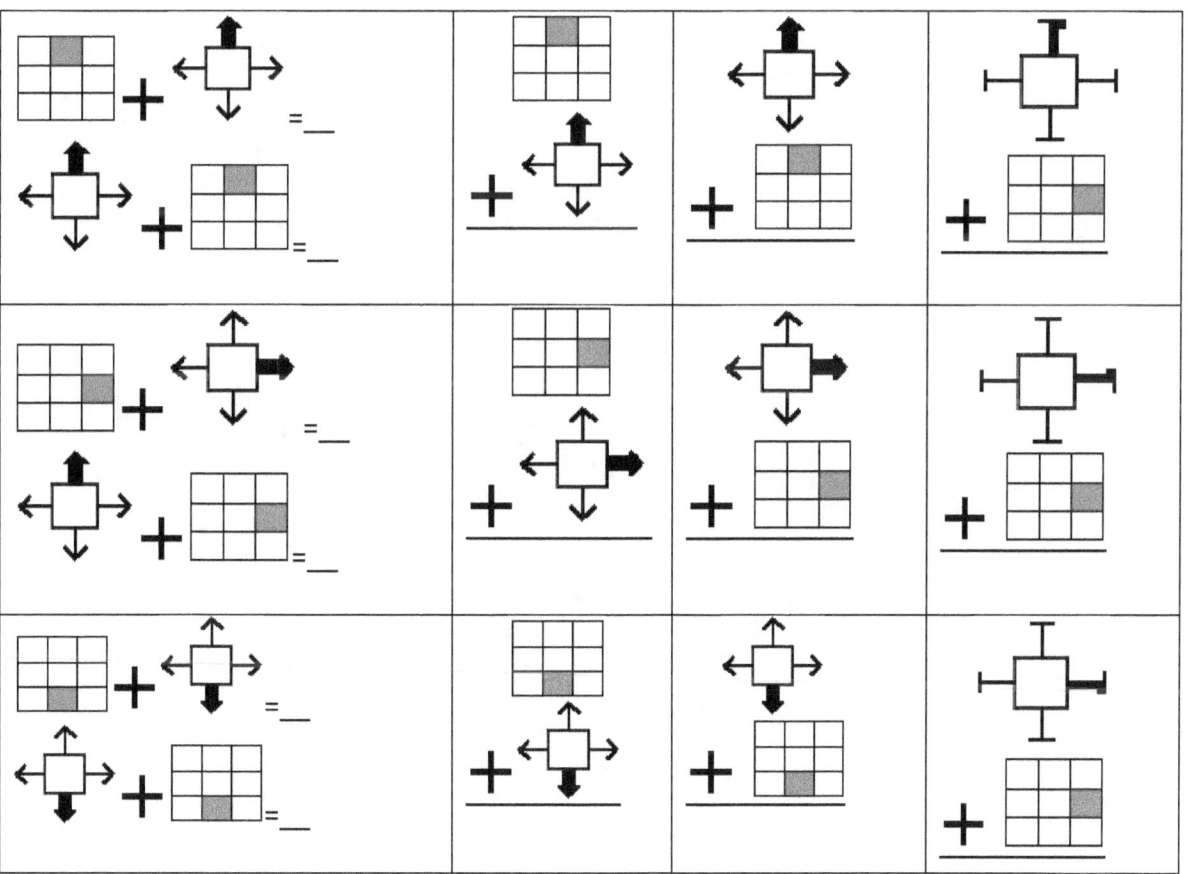

You are a chess piece located at c3.
● = 1

Only transfer each number to a correct position.
Do not the transfer the 3 by 3 square.

8889
9999
10 10 10 9

263

Ho Math Chess 何数棋谜 妈！我会棋谜式加法啦！
Mom! I Learn Addition Using Math-Chess-Puzzles Connection
Contents include both traditional and Math-Chess-Puzzles combined methods. Extra strength
©2008 – 2018 Frank Ho, Amanda Ho   All rights reserved. www.homathchess.com

Student Name _____ Date _____

Making 10 or adding around 10 or doubling

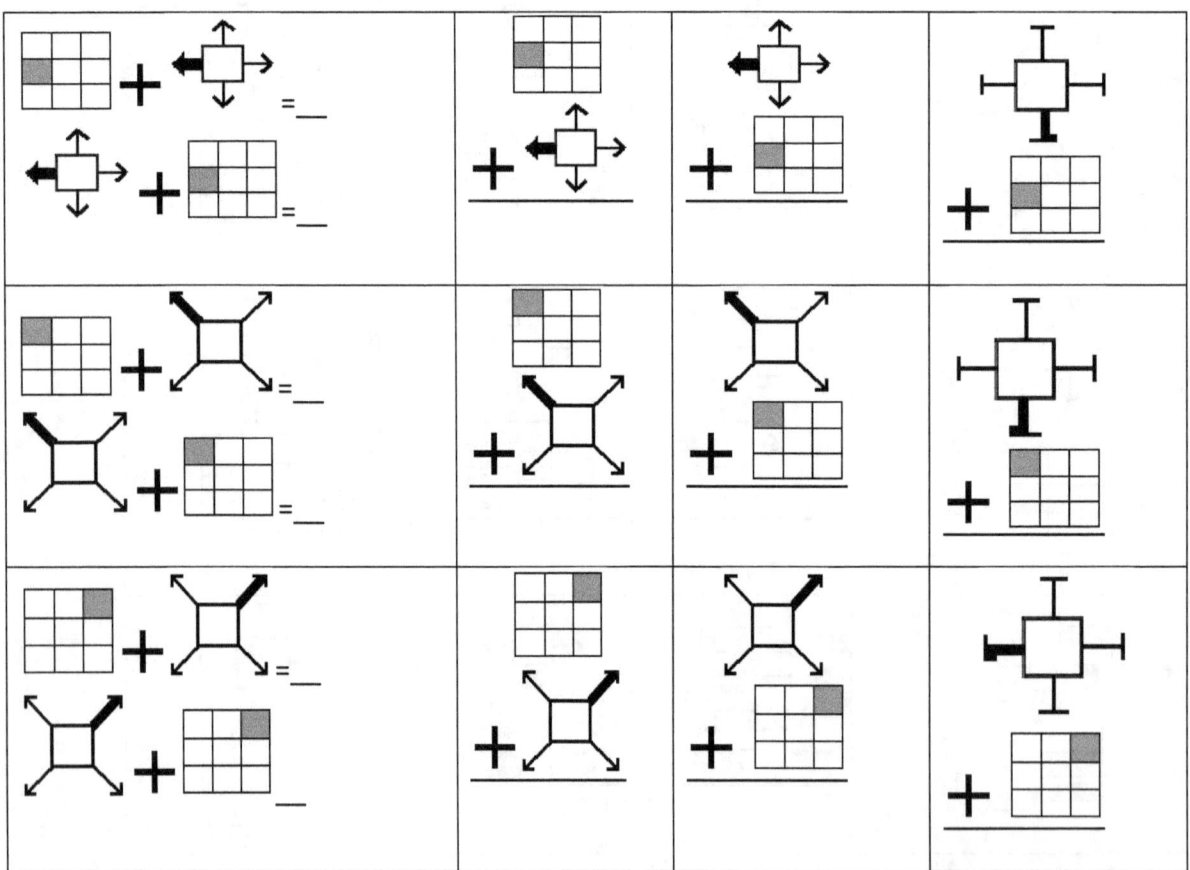

10 10 10 10
10 10 10 10
8 8 8 8

Ho Math Chess 何数棋谜　妈！我会棋谜式加法啦！
Mom! I Learn Addition Using Math-Chess-Puzzles Connection

Contents include both traditional and Math-Chess-Puzzles combined methods. Extra strength

©2008 – 2018 Frank Ho, Amanda Ho   All rights reserved. www.homathchess.com

Student Name _____ Date _____

Making 10 or adding around 10 or doubling

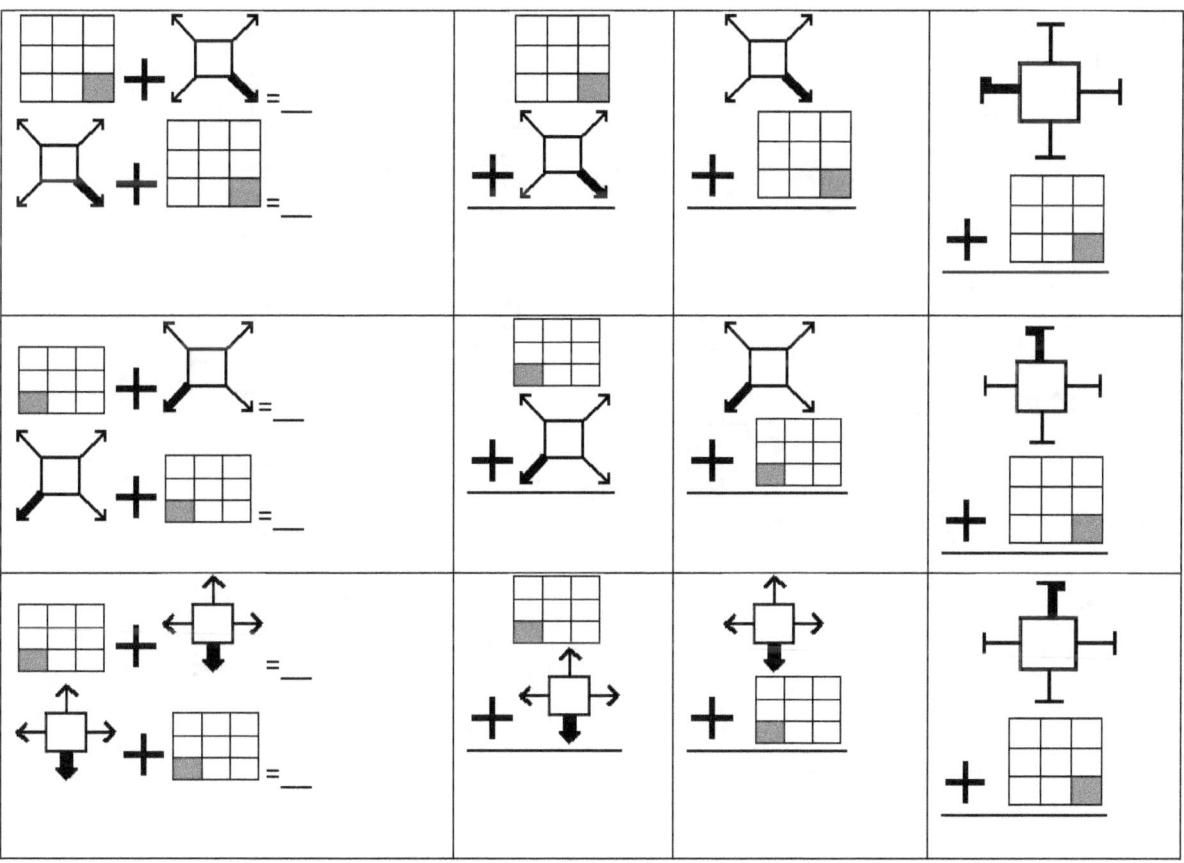

8 8 8 8
6 6 6 8
6 6 6 8

Ho Math Chess 何数棋谜　妈！我会棋谜式加法啦！
Mom! I Learn Addition Using Math-Chess-Puzzles Connection
Contents include both traditional and Math-Chess-Puzzles combined methods. Extra strength
©2008 − 2018 Frank Ho, Amanda Ho　　All rights reserved. www.homathchess.com

Student Name_____ Date_____

Making 10 or adding around 10 or doubling

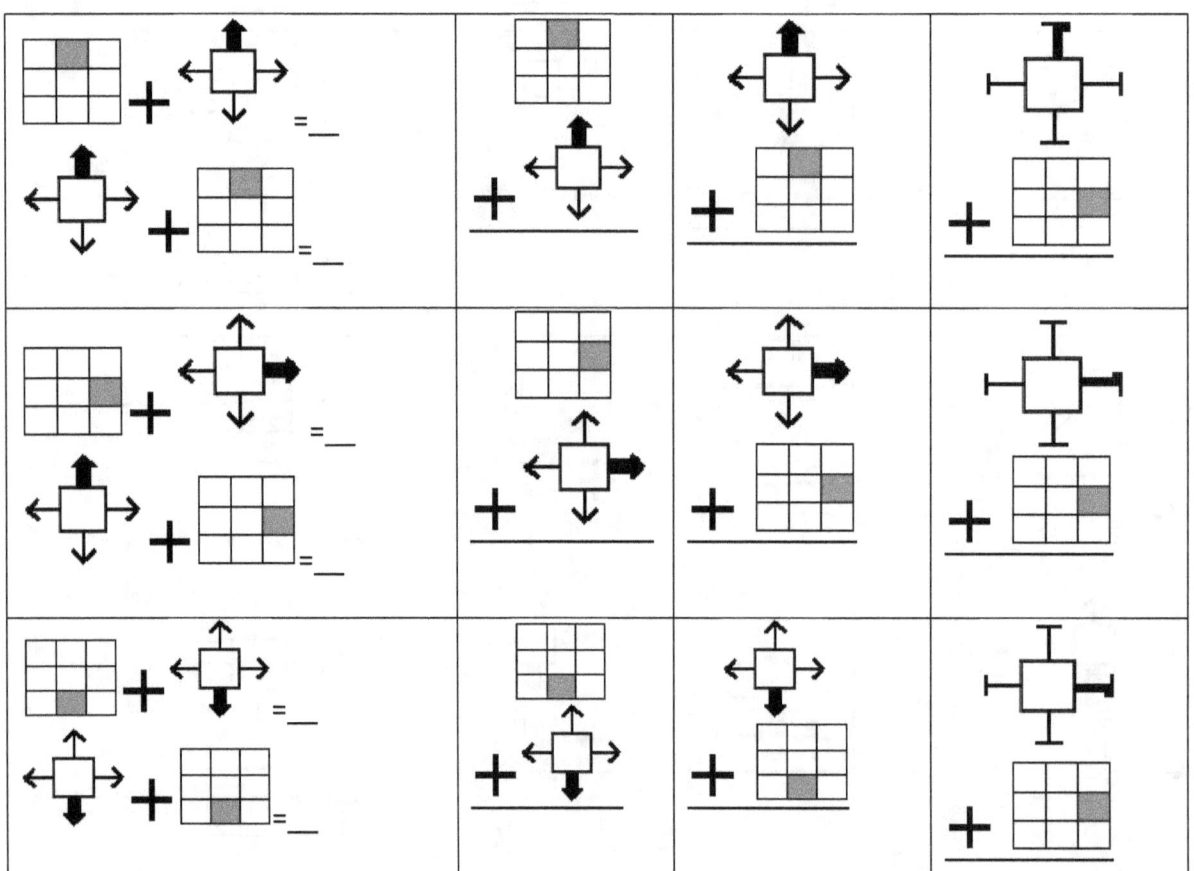

7 7 7 10
10 10 10 10
7 7 7 10

Mom! I Learn Addition Using Math-Chess-Puzzles Connection

Contents include both traditional and Math-Chess-Puzzles combined methods. Extra strength
©2008 — 2018 Frank Ho, Amanda Ho   All rights reserved. www.homathchess.com

Student Name _____ Date _____

Making 10 or adding around 10 or doubling

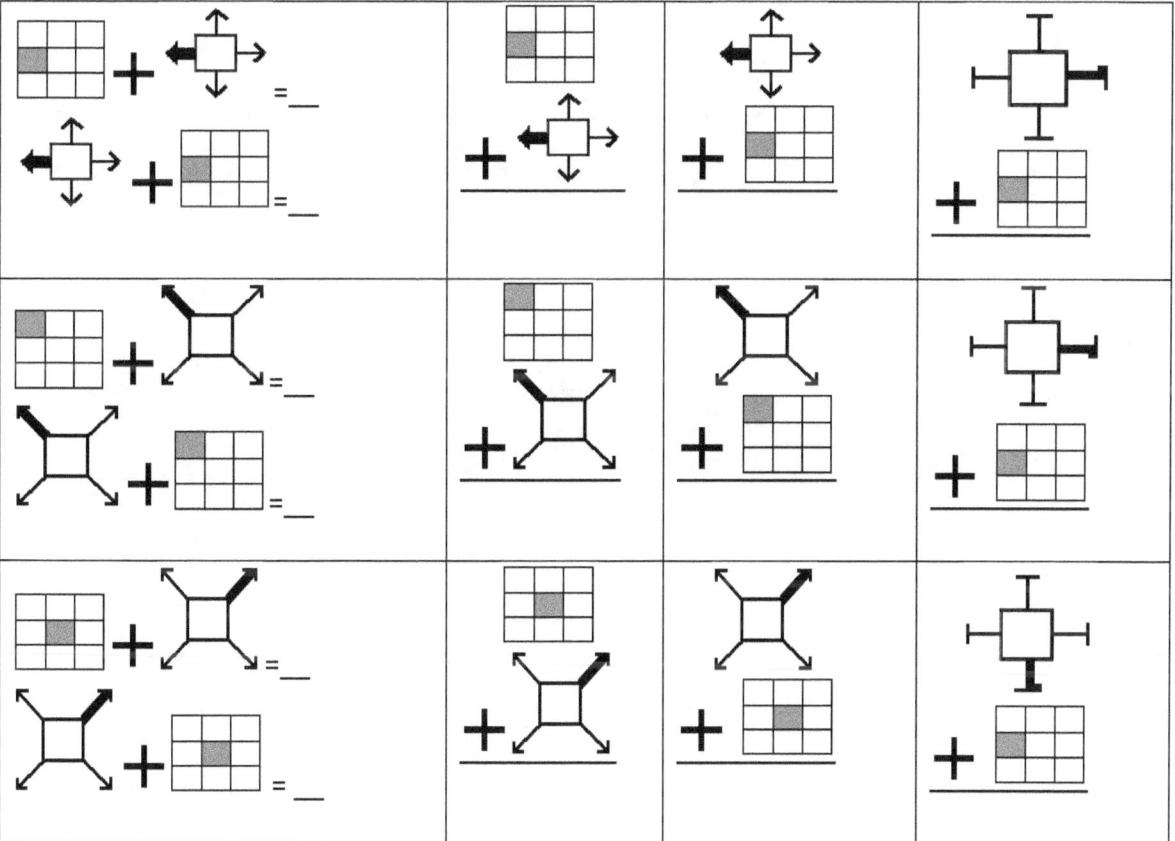

10 10 10 10
8 8 8 10
8 8 8 10

267

Ho Math Chess 何数棋谜　妈！我会棋谜式加法啦！
Mom! I Learn Addition Using Math-Chess-Puzzles Connection
Contents include both traditional and Math-Chess-Puzzles combined methods. Extra strength
©2008 — 2018 Frank Ho, Amanda Ho   All rights reserved. www.homathchess.com

Student Name _____ Date _____

Making 10 or adding around 10 or doubling

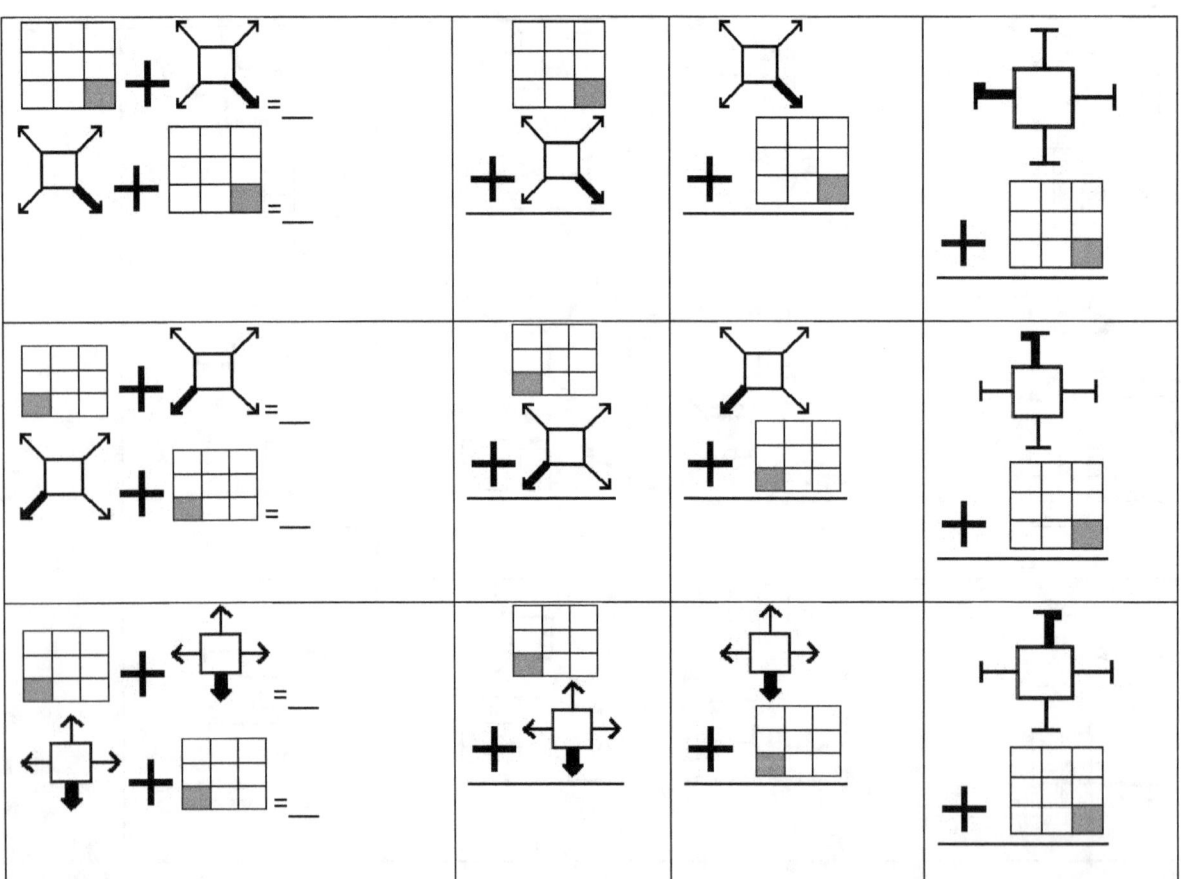

9999
6669
6669

Ho Math Chess 何数棋谜　妈！我会棋谜式加法啦！
Mom! I Learn Addition Using Math-Chess-Puzzles Connection
Contents include both traditional and Math-Chess-Puzzles combined methods. Extra strength
©2008 － 2018 Frank Ho, Amanda Ho   All rights reserved. www.homathchess.com

Student Name _____ Date _____

Making 10 or adding around 10 or doubling

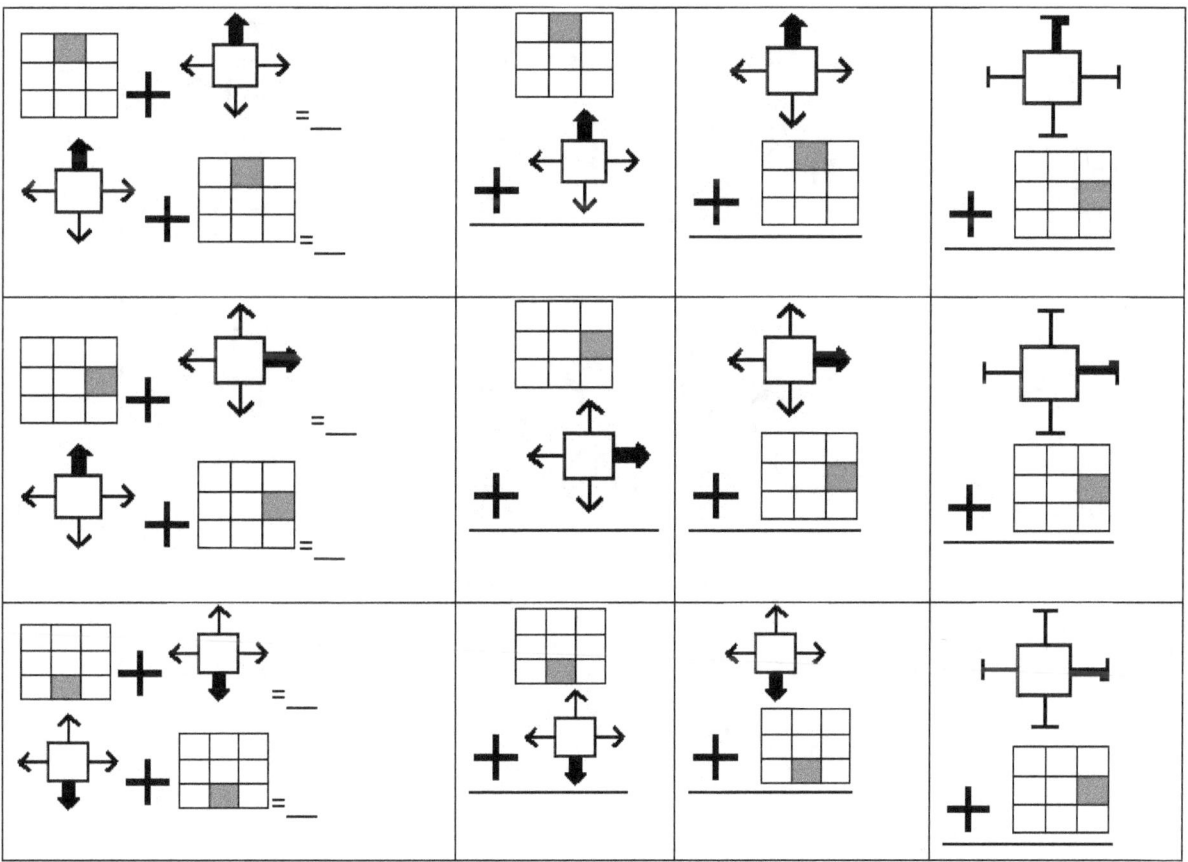

10 10 10 8
8 8 8 8
10 10 10 8

Ho Math Chess 何数棋谜 妈！我会棋谜式加法啦！
Mom! I Learn Addition Using Math-Chess-Puzzles Connection
Contents include both traditional and Math-Chess-Puzzles combined methods. Extra strength
©2008 — 2018 Frank Ho, Amanda Ho   All rights reserved. www.homathchess.com

Student Name _____ Date _____

Making 10 or adding around 10 or doubling

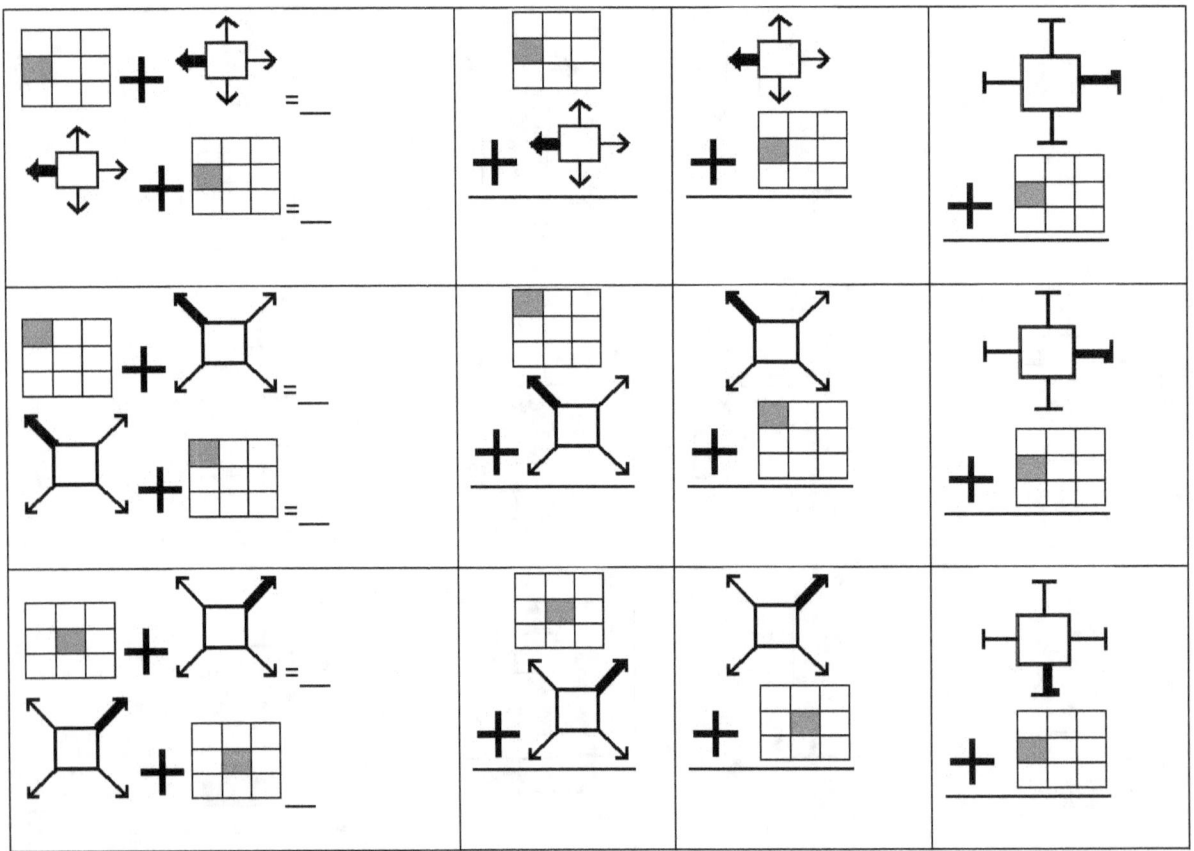

10 10 10 10
10 10 10 10
9 9 9 10

270

Ho Math Chess 何数棋谜   妈！我会棋谜式加法啦！
Mom! I Learn Addition Using Math-Chess-Puzzles Connection
Contents include both traditional and Math-Chess-Puzzles combined methods. Extra strength
©2008 — 2018 Frank Ho, Amanda Ho   All rights reserved. www.homathchess.com

Student Name_____ Date_____

Making 10 or adding around 10 or doubling

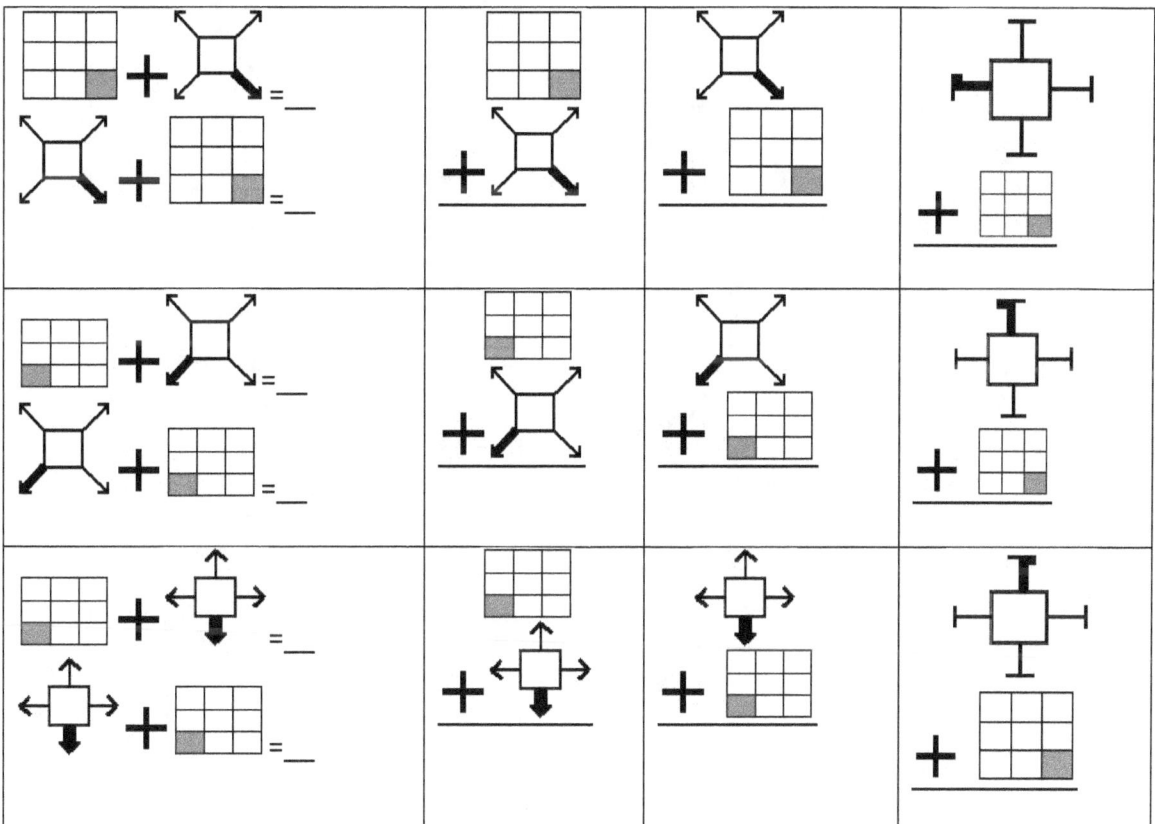

9 9 9 9
9 9 9 9
9 9 9 9

Ho Math Chess 何数棋谜   妈！我会棋谜式加法啦！
Mom! I Learn Addition Using Math-Chess-Puzzles Connection
Contents include both traditional and Math-Chess-Puzzles combined methods. Extra strength
©2008 – 2018 Frank Ho, Amanda Ho   All rights reserved. www.homathchess.com

Student Name _____   Date _____

Making 10 or adding around 10 or doubling

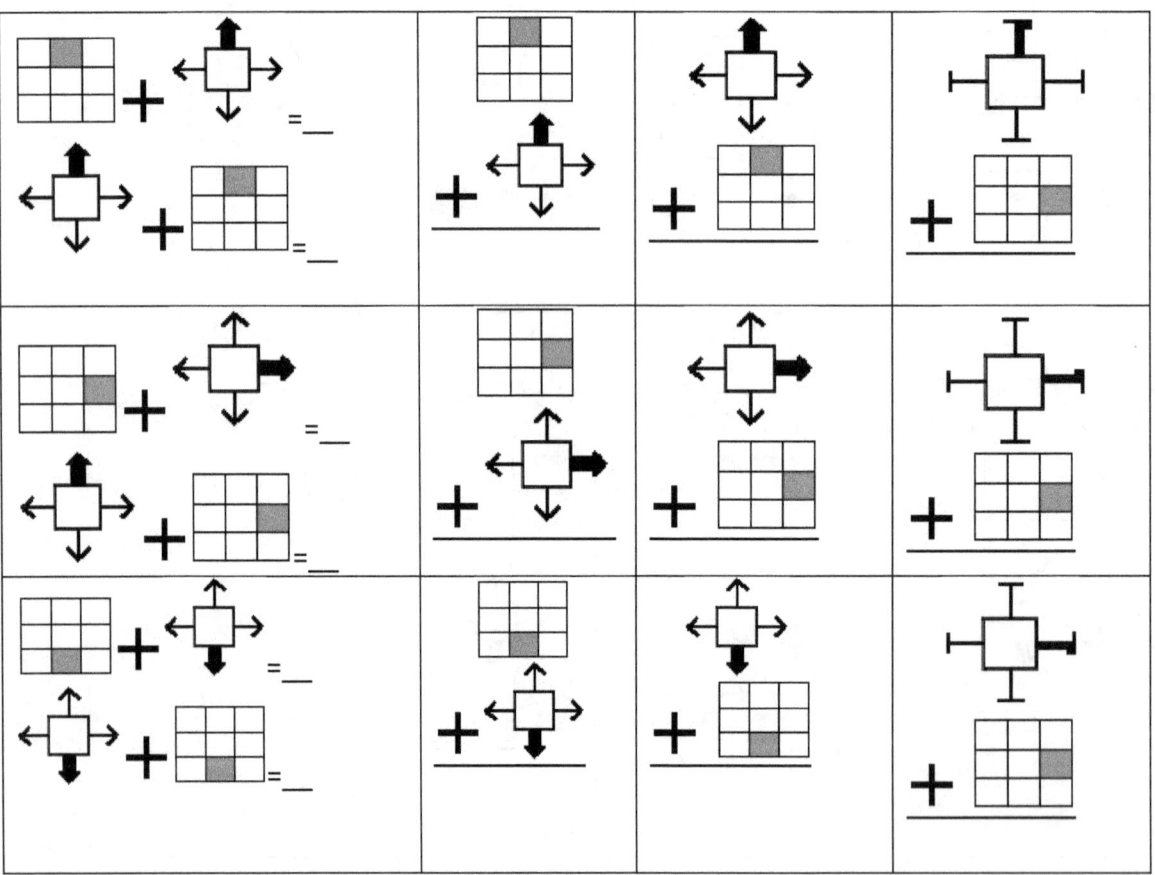

10 10 10 9
9 9 9 9
9 9 9 9

Ho Math Chess 何数棋谜　妈！我会棋谜式加法啦！

Mom! I Learn Addition Using Math-Chess-Puzzles Connection

Contents include both traditional and Math-Chess-Puzzles combined methods. Extra strength

©2008 – 2018 Frank Ho, Amanda Ho   All rights reserved. www.homathchess.com

Student Name _____ Date _____

Making 10 or adding around 10 or doubling

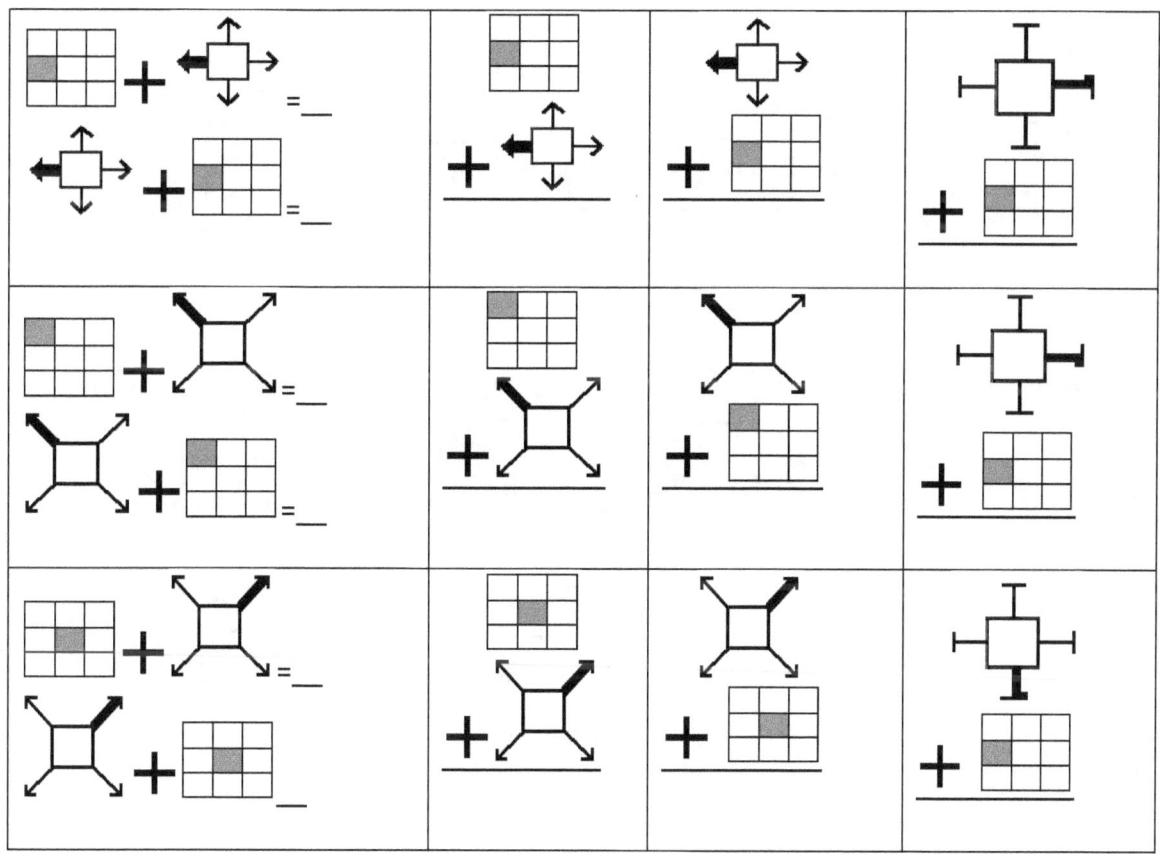

You are a chess piece located at c3.
● = 1

Only transfer each number to a correct position.
Do not the transfer the 3 by 3 square.

9 9 9 9
10 10 10 9
10 10 10 9

Ho Math Chess 何数棋谜 妈！我会棋谜式加法啦！
Mom! I Learn Addition Using Math-Chess-Puzzles Connection
Contents include both traditional and Math-Chess-Puzzles combined methods. Extra strength
©2008 — 2018 Frank Ho, Amanda Ho   All rights reserved. www.homathchess.com

Student Name _____ Date _____

Making 10 or adding around 10 or doubling

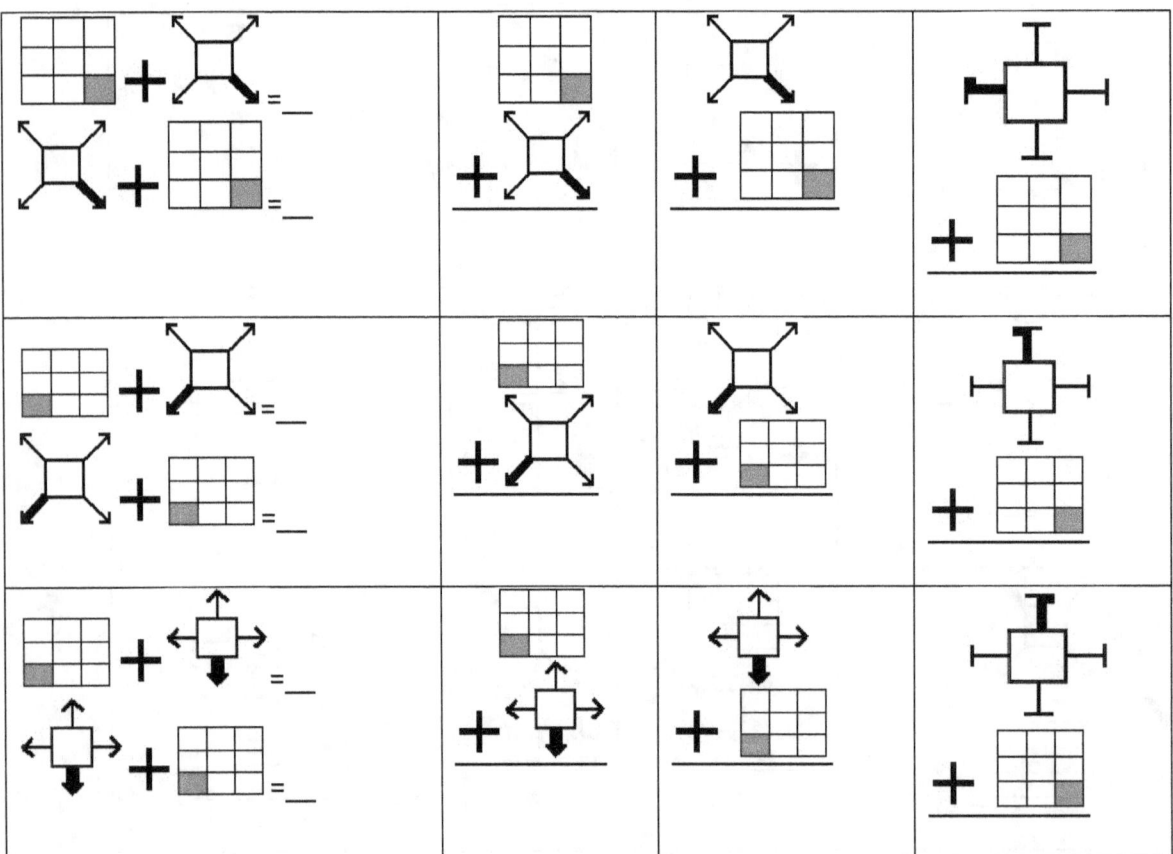

9 9 9 9
8 8 8 9
8 8 8 9

Ho Math Chess 何数棋谜   妈！我会棋谜式加法啦！
Mom! I Learn Addition Using Math-Chess-Puzzles Connection

Contents include both traditional and Math-Chess-Puzzles combined methods. Extra strength

©2008 – 2018 Frank Ho, Amanda Ho   All rights reserved. www.homathchess.com

Student Name_____ Date_____

Chess and math integrated problems

How many points altogether is the square ✱ or ☒ being attacked?

Answer _____ 16

Answer _____ 15

Answer _____ 8

Answer _____ 11

Ho Math Chess 何数棋谜　妈！我会棋谜式加法啦！
Mom! I Learn Addition Using Math-Chess-Puzzles Connection

Contents include both traditional and Math-Chess-Puzzles combined methods. Extra strength

©2008 – 2018 Frank Ho, Amanda Ho   All rights reserved. www.homathchess.com

Student Name_____   Date_____

How many points altogether is the square ✕ or ☒ being attacked?

Answer _____ 16

Answer _____ 7

Answer _____ 6

Answer _____ 20

Ho Math Chess 何数棋谜 妈！我会棋谜式加法啦！
Mom! I Learn Addition Using Math-Chess-Puzzles Connection
Contents include both traditional and Math-Chess-Puzzles combined methods. Extra strength
©2008 − 2018 Frank Ho, Amanda Ho   All rights reserved. www.homathchess.com

Student Name_____ Date_____

How many points altogether is the square ✖ or ⊠ being attacked?

Answer _____ 21

Answer _____ 15

Answer _____ 17
The crossed square is attacked by 2 rooks.

Answer _____ 17

Ho Math Chess 何数棋谜　妈！我会棋谜式加法啦！

Mom! I Learn Addition Using Math-Chess-Puzzles Connection

Contents include both traditional and Math-Chess-Puzzles combined methods. Extra strength

©2008 – 2018 Frank Ho, Amanda Ho　All rights reserved. www.homathchess.com

Student Name _____ Date _____

How many points altogether is the square ✕ or ☒ being attacked?

Answer _____ 20

Answer _____ 6

Answer _____ 8

Answer _____ 8

Ho Math Chess 何数棋谜   妈！我会棋谜式加法啦！
Mom! I Learn Addition Using Math-Chess-Puzzles Connection

Contents include both traditional and Math-Chess-Puzzles combined methods. Extra strength
©2008 – 2018 Frank Ho, Amanda Ho   All rights reserved. www.homathchess.com

Student Name_____ Date_____

8 plus

8	♙	2	8	♗
+ 1	+ 8	+ 8	+ 2	+ 8
9	9	10	10	11

8	4	8	♖	♖
+ 3	+ 8	+ 4	+ ♖	+ 8
11	12	12	10	13

8	6	6	8	7
+ 5	+ 6	+ 8	+ 6	+ 7
13	12	14	14	14

Ho Math Chess 何数棋谜 妈！我会棋谜式加法啦！
Mom! I Learn Addition Using Math-Chess-Puzzles Connection

Contents include both traditional and Math-Chess-Puzzles combined methods. Extra strength
©2008 − 2018 Frank Ho, Amanda Ho   All rights reserved. www.homathchess.com

Student Name _____ Date _____

7	7	8	♕	8
+ 8	+ 8	+ 8	+ 8	+ 9
15	15	16	17	17
6	6	8	5	8
+ 6	+ 8	+ 6	+ ♜	+ 5
12	14	14	10	13

Ho Math Chess 何数棋谜 妈！我会棋谜式加法啦！
Mom! I Learn Addition Using Math-Chess-Puzzles Connection
Contents include both traditional and Math-Chess-Puzzles combined methods. Extra strength
©2008 − 2018 Frank Ho, Amanda Ho   All rights reserved. www.homathchess.com

Student Name _____ Date _____

7 plus

7 + 1 8	♙ + 7 8	2 + 7 9	7 + 2 9	7 + 3 10
3 + 7 10	4 + 7 11	7 + 4 11	♖ + ♖ 10	5 + 7 12
7 + 5 12	6 + 6 12	6 + 7 13	7 + 6 13	7 + 7 14
7 + 8 15	8 + 7 15	7 + 7 14	10 + 7 17	7 + 9 16
♕ + 7 	8 + 8 	7 + 8 	7 + 7 	8 + 7

Mom! I Learn Addition Using Math-Chess-Puzzles Connection

6 plus

♙ + 6 = 7	6 + ♙ = 7	2 + 6 = 8	6 + 2 = 8	♗ + 6 = 9
6 + 3 = 9	4 + 6 = 10	6 + 4 = 10	♖ + 5 = 10	♖ + 6 = 11
6 + 5 = 11	6 + 6 = 12	6 + 6 = 12	7 + 6 = 13	6 + 7 = 13
6 + 6 = 12	6 + 8 = 14	8 + 6 = 14	10 + 6 = 16	♕ + 6 = 15
10 + 6 = 16	♕ + 6 = 15	6 + 6 = 12	6 + 8 = 14	8 + 6 = 14

Mom! I Learn Addition Using Math-Chess-Puzzles Connection

5 plus

1	♖	2	♖	3
+ 5	+ 1	+ 5	+ 2	+ 5
6	6	7	7	8
5	4	♖	4	♖
+ ♗	+ 4	+ 4	+ 5	+ ♖
8	8	9	9	10
♖	♖	6	7	7
+ 5	+ 6	+ 5	+ 5	+ 5
10	11	11	12	12
5	8	10	♕	♖
+ 8	+ 5	+ 5	+ 5	+ ♕
13	13	15	14	14
8	5	♖	7	♕
+ 5	+ 8	+ 7	+ 5	+ 5
13	13	12	12	14

Mom! I Learn Addition Using Math-Chess-Puzzles Connection

4 plus

4 + ♙ 5	♙ + 4 5	4 + 2 6	2 + 4 6	♗ + 4 7
4 + 3 7	4 + 4 8	4 + 5 9	♖ + 4 9	4 + 6 10
6 + 4 10	♗ + 7 10	7 + 3 10	7 + 4 11	4 + 7 11
♖ + 5 10	4 + 8 12	8 + 4 12	10 + 4 14	4 + ♕ 13
♕ + 4 13	8 + 4 12	4 + 8 12	♕ + 4 13	4 + 9 13

Additions to 11

Fill in the blank circles with numbers.

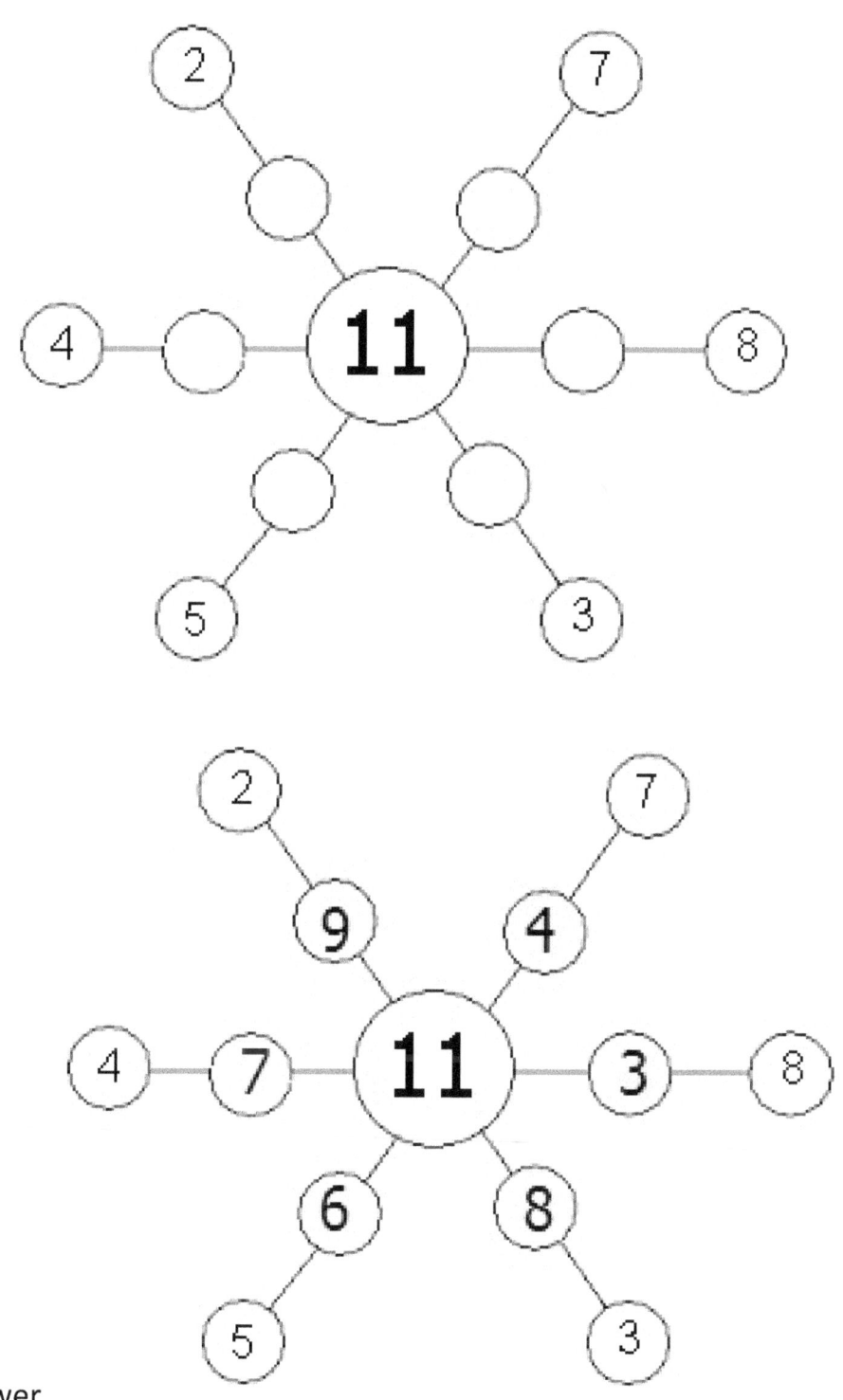

answer

Additions to 12

Fill in the blank circles with numbers.

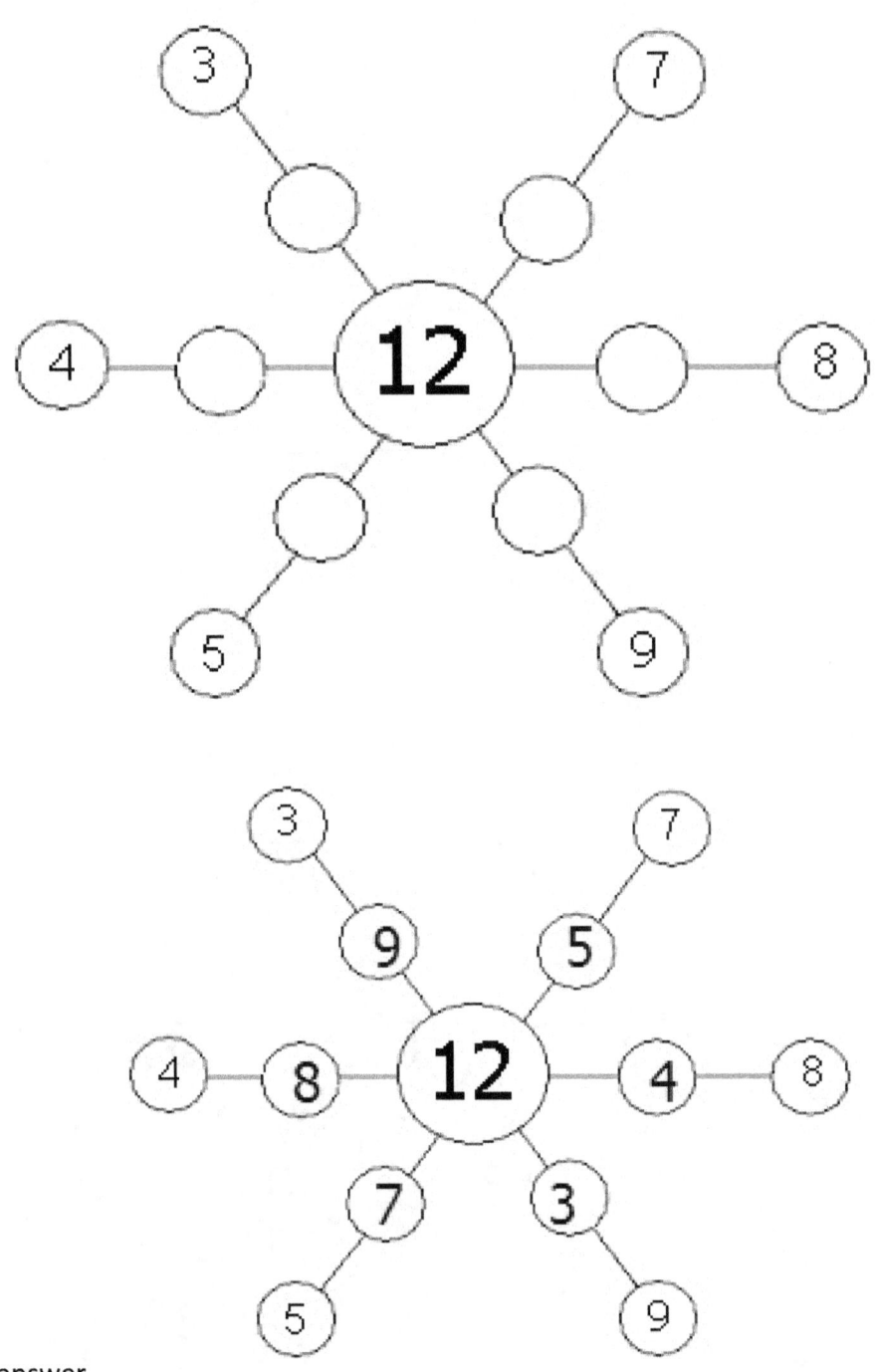

answer

Ho Math Chess 何数棋谜　妈！我会棋谜式加法啦！
Mom! I Learn Addition Using Math-Chess-Puzzles Connection
Contents include both traditional and Math-Chess-Puzzles combined methods. Extra strength
©2008 — 2018 Frank Ho, Amanda Ho　　All rights reserved. www.homathchess.com

Student Name _____ Date _____

Additions to 13

Fill in the blank circles with numbers.

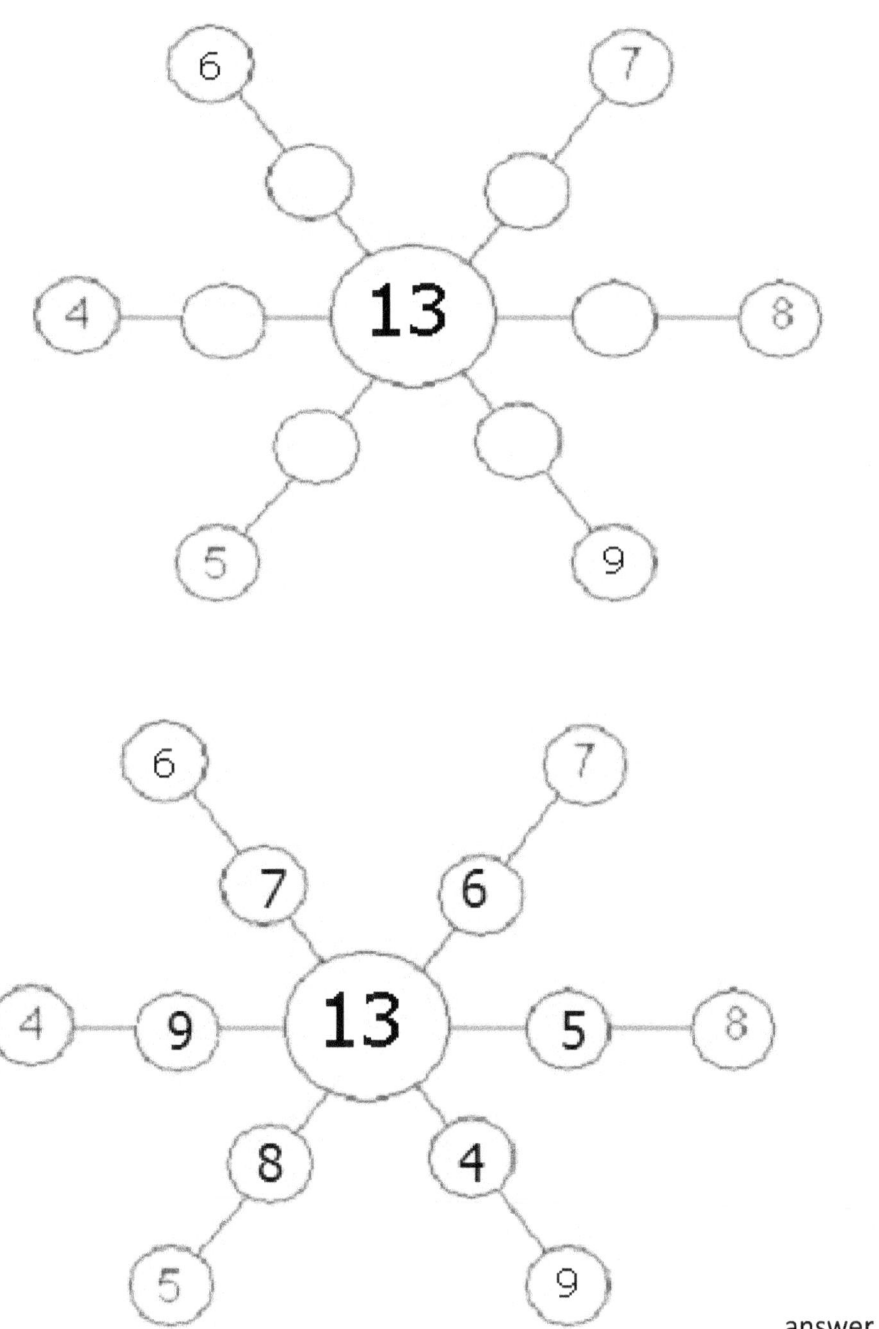

answer

Additions to 14

Fill in the blank circles with numbers.

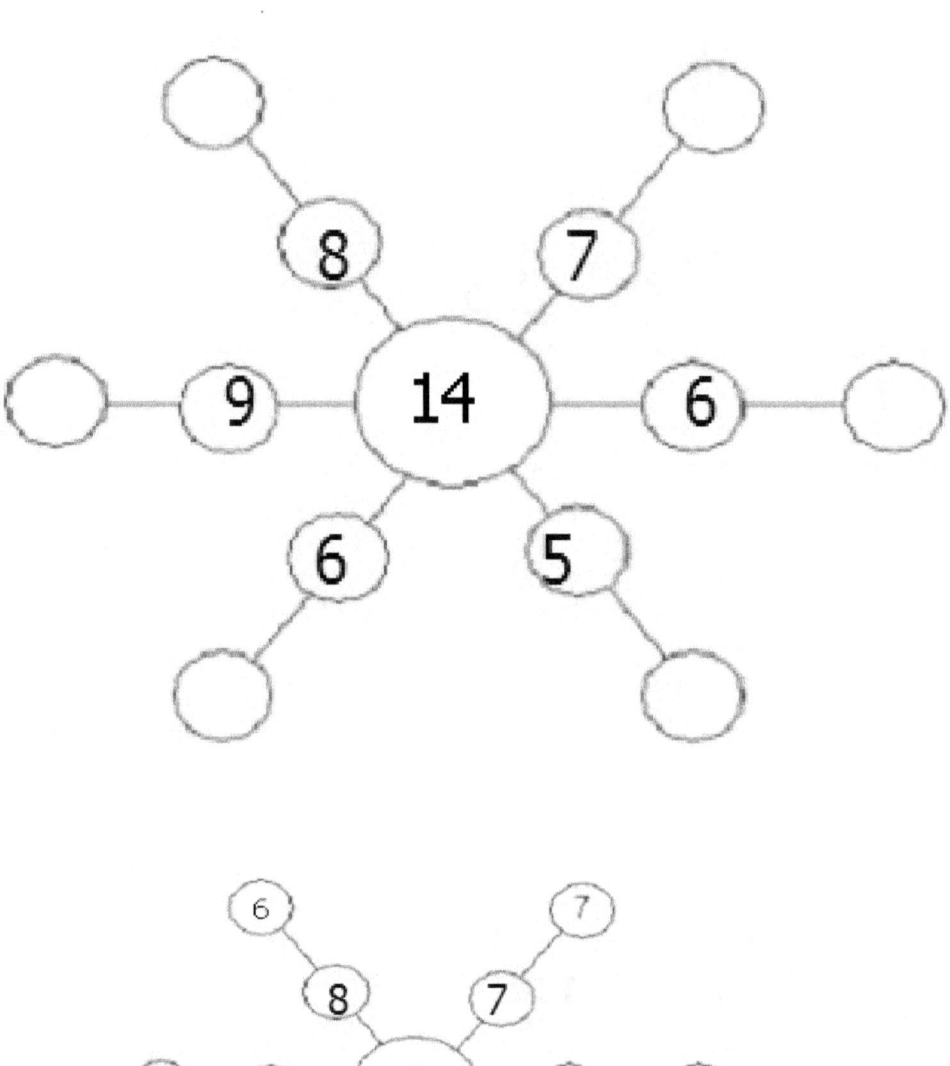

answer

Additions to 15

Fill in the blank circles with numbers.

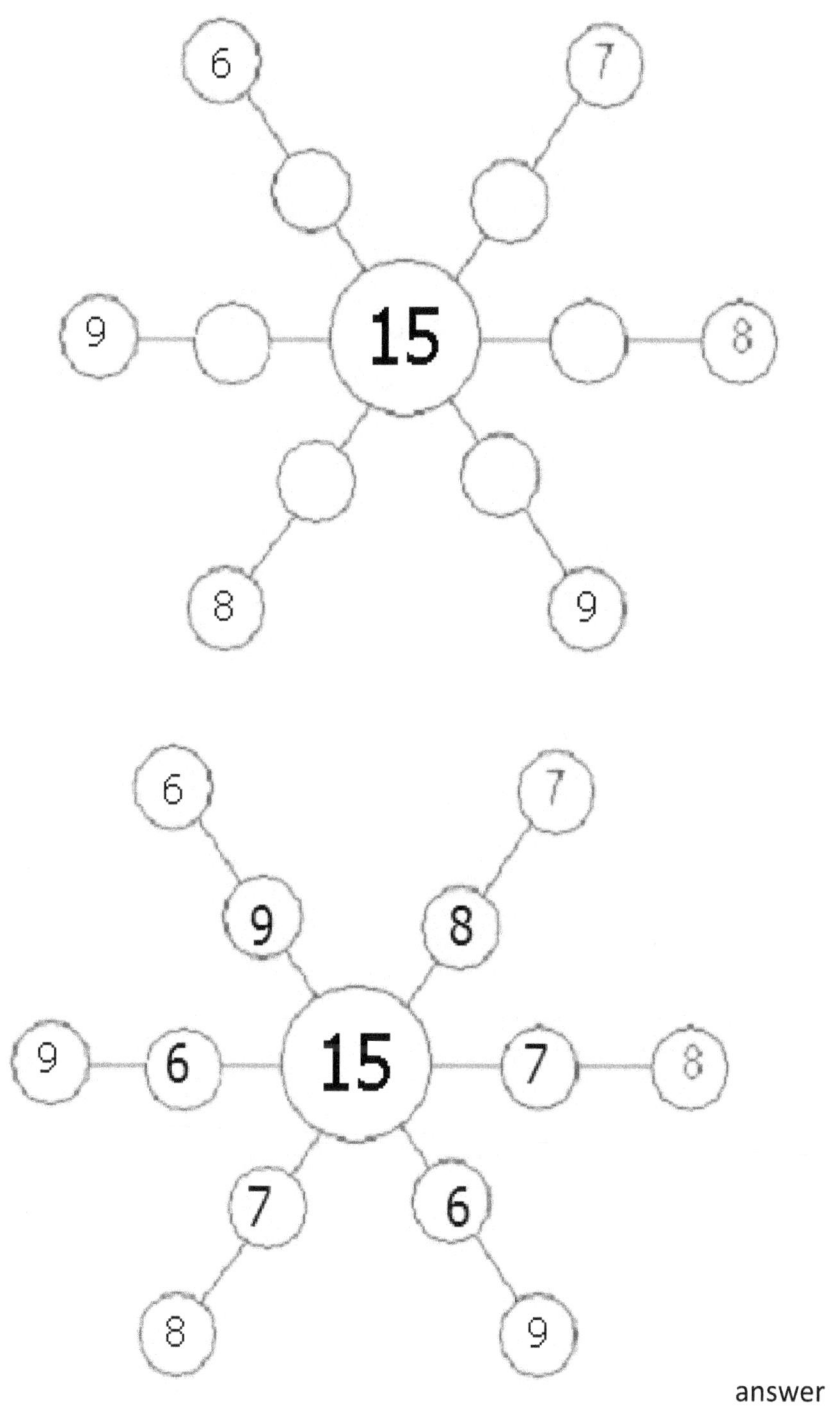

answer

Additions to 16

Fill in the blank circles with numbers.

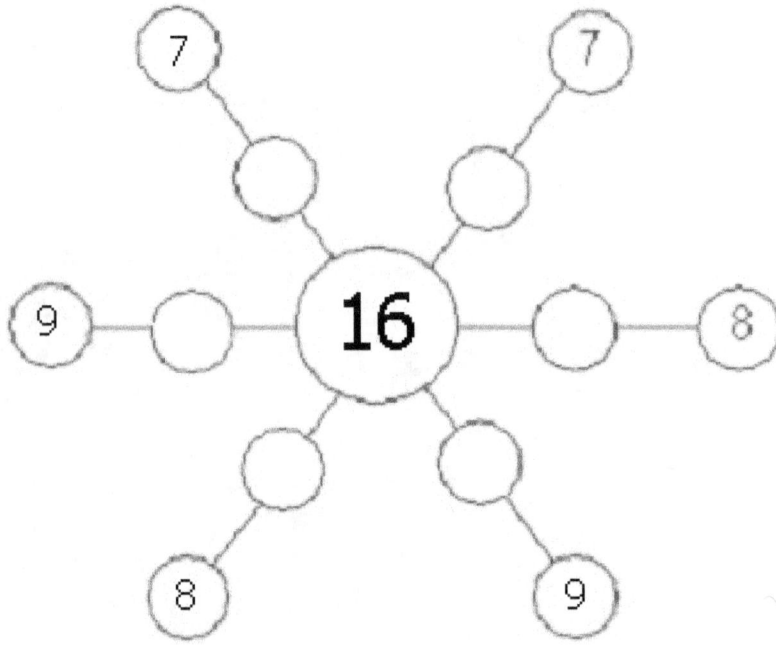

Ho Math Chess 何数棋谜　妈！我会棋谜式加法啦！
Mom! I Learn Addition Using Math-Chess-Puzzles Connection
Contents include both traditional and Math-Chess-Puzzles combined methods. Extra strength
©2008 – 2018 Frank Ho, Amanda Ho　All rights reserved. www.homathchess.com

Student Name _____ Date _____

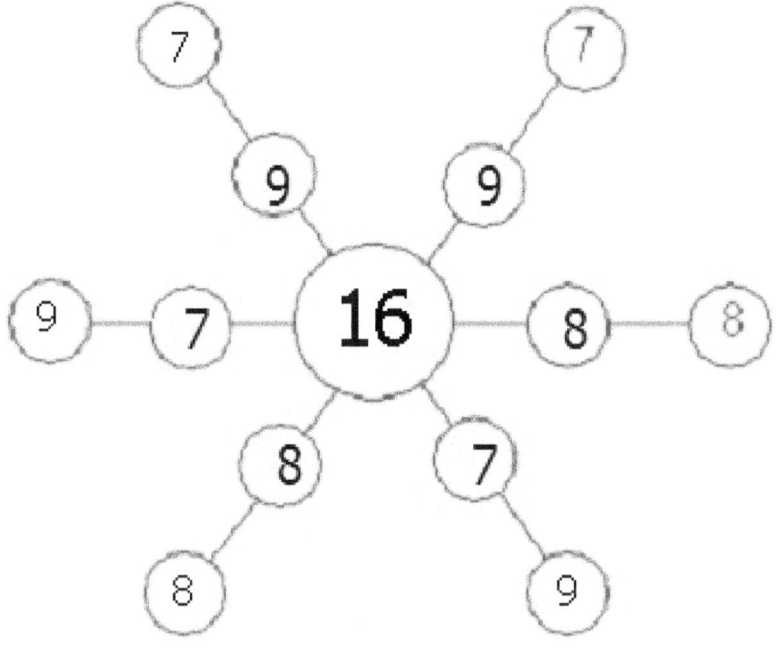

answer

Using 8 to add

Fill in the blank circles with numbers.

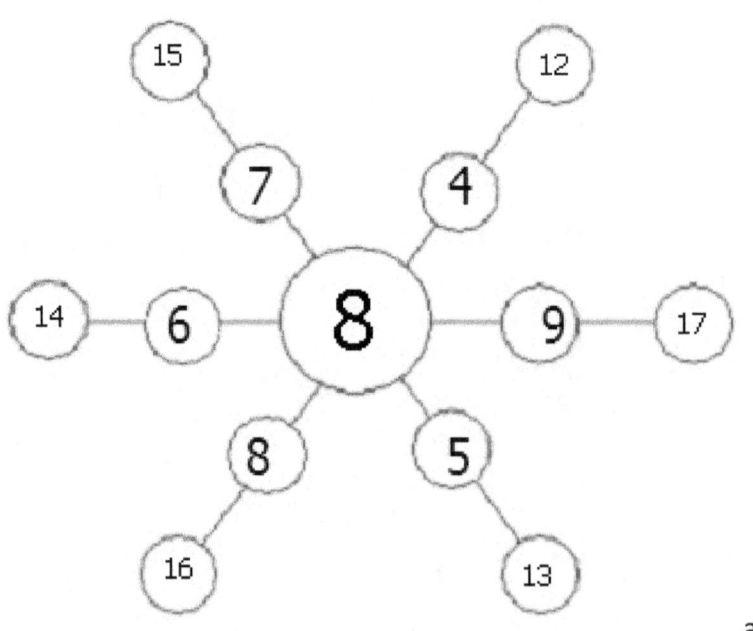

answer

Using 9 to add

Fill in the blank circles with numbers.

answer

Number relationships

1 4 5, 7 8 15, 9 1 10, 3 5 8, 8 2 10, 4 6 10, 1 9 10, 6 7 13, 1 3 4, 6 6 7 19, 4 1 1 6

Number relationships

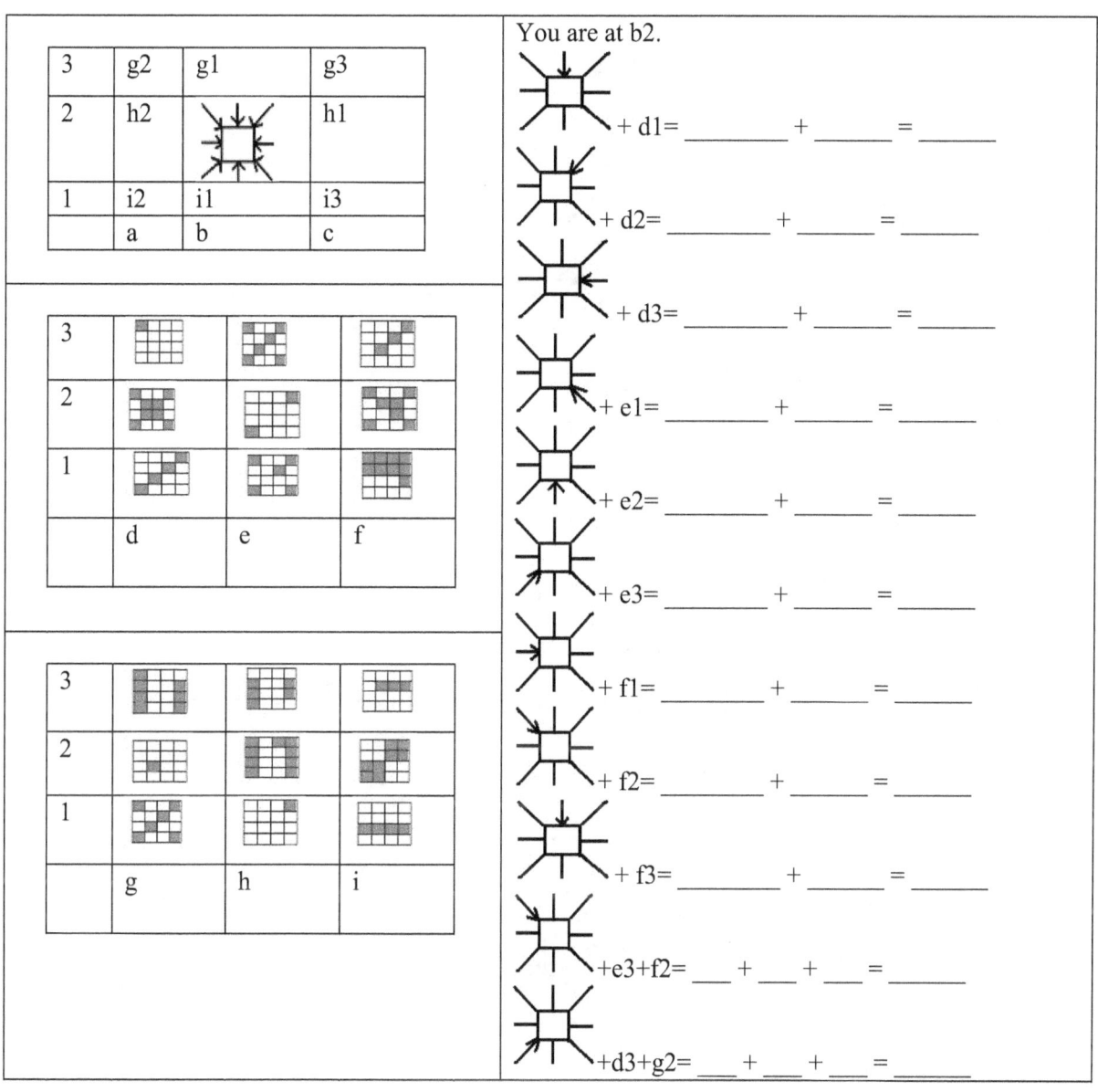

6 4 10, 7 8 15, 1 1 2, 3 5 8, 4 2 6, 8 6 14, 9 9 18, 1 7 8, 6 3 9, 1 6 7 14, 8 1 1 10

Ho Math Chess 何数棋谜 妈！我会棋谜式加法啦！
Mom! I Learn Addition Using Math-Chess-Puzzles Connection
Contents include both traditional and Math-Chess-Puzzles combined methods. Extra strength
©2008 − 2018 Frank Ho, Amanda Ho   All rights reserved. www.homathchess.com

Student Name _____   Date _____

Number relationships

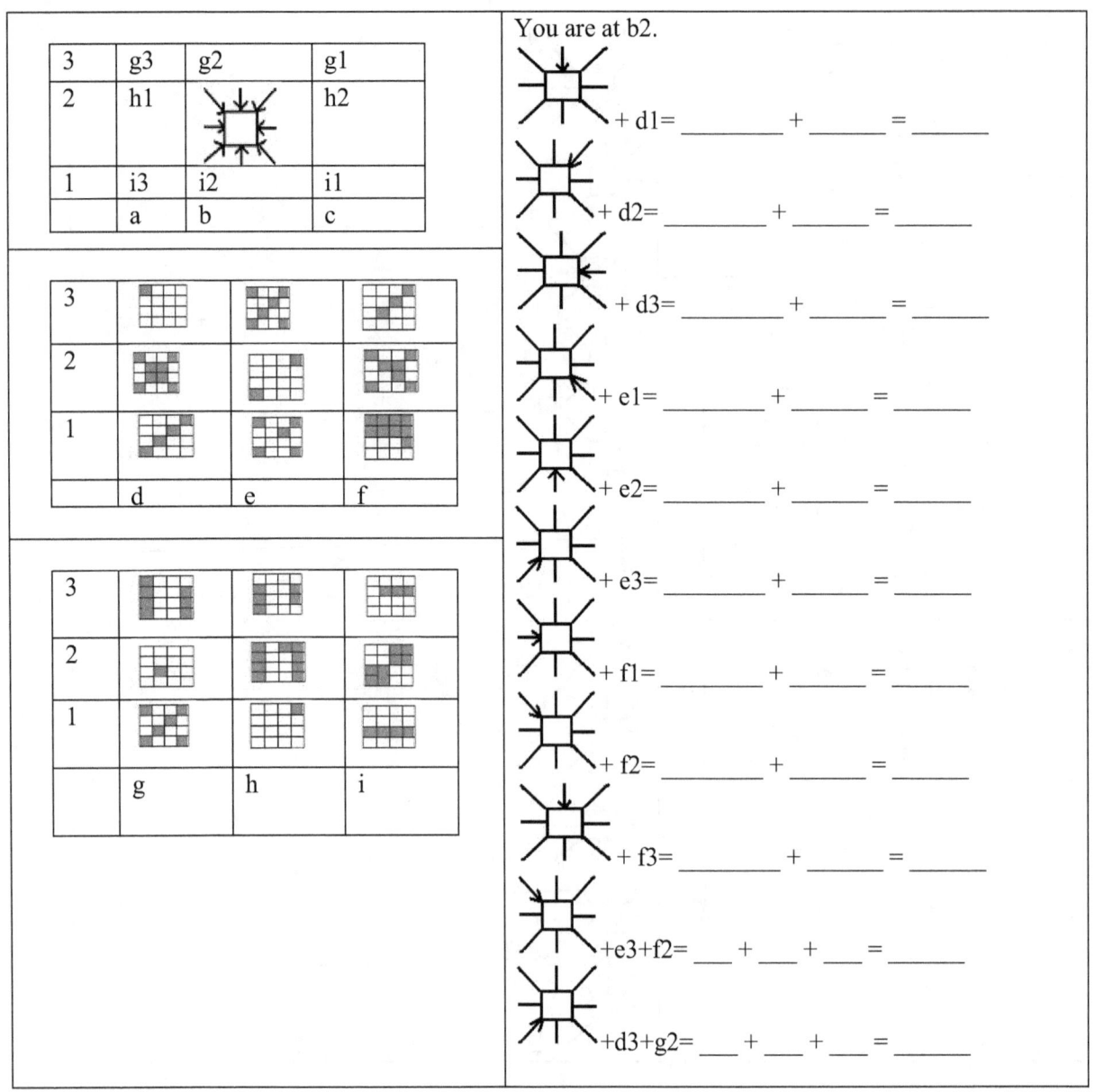

1 4 5, 6 8 14, 9 4 13, 4 5 9, 8 2 10, 3 6 9, 1 9 10, 7 7 14, 1 3 4, 7 6 7 20, 3 1 1 5

Number relationships

8 4 12, 3 8 11, 1 1 2, 5 5 10, 9 2 11, 1 6 7, 6 9 15, 4 7 11, 8 3 11, 4 6 7 17, 1 1 1 3

Number relationships

4 4 8, 3 8 11, 6 1 7, 5 5 10, 1 2 3, 9 6 15, 1 9 10, 8 7 15, 4 3 7, 8 6 7 21, 9 1 1 11

Ho Math Chess 何数棋谜　妈！我会棋谜式加法啦！
Mom! I Learn Addition Using Math-Chess-Puzzles Connection

Contents include both traditional and Math-Chess-Puzzles combined methods. Extra strength
©2008－2018 Frank Ho, Amanda Ho　　All rights reserved. www.homathchess.com

Student Name _____ Date _____

Number relationships

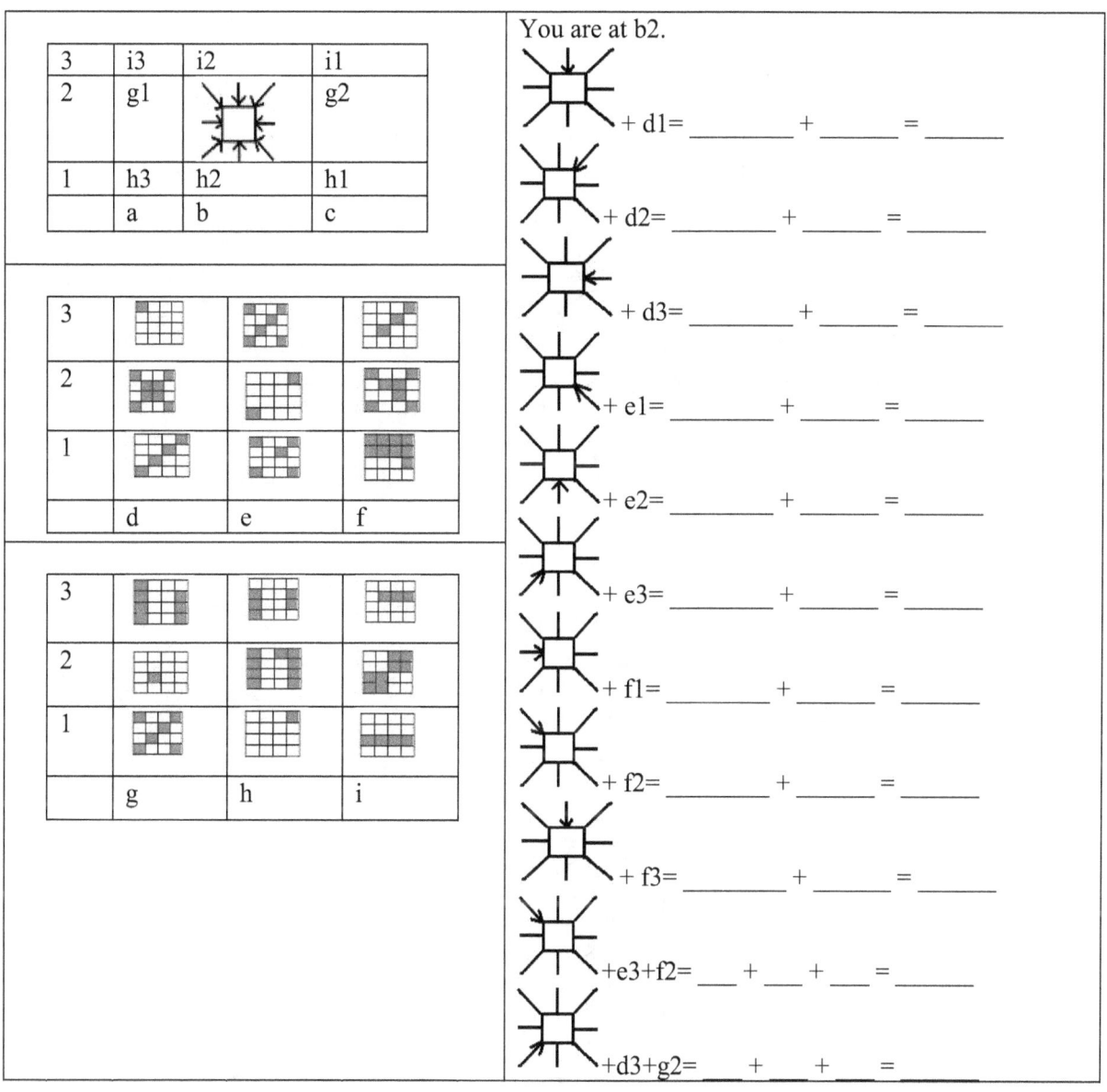

8 4 12, 1 8 9, 1 1 2, 1 5 6, 9 2 11, 5 6 11, 6 9 15, 3 7 10, 8 3 11, 3 6 7 16, 5 1 1 7

Number relationships

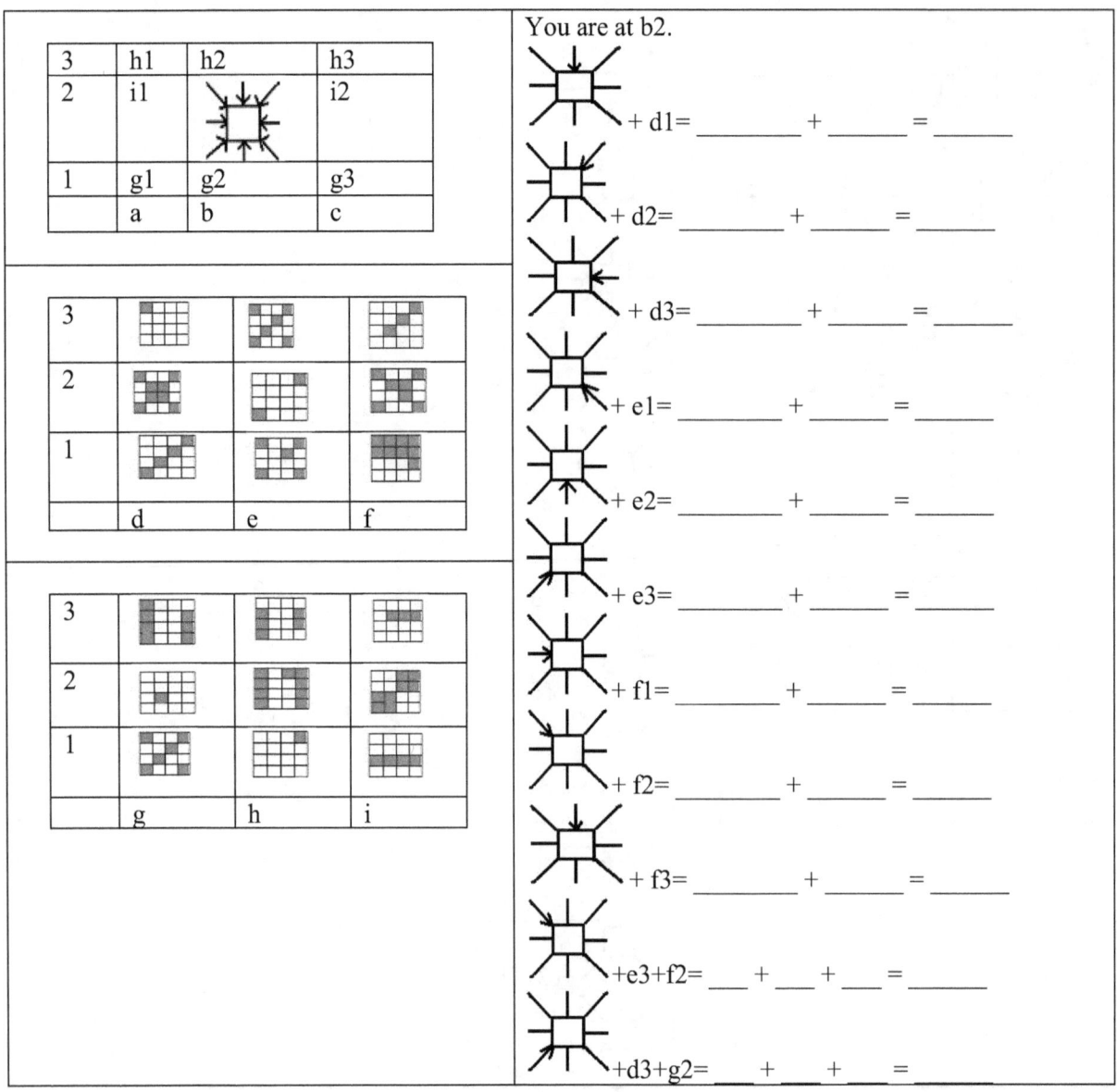

9 4 13, 5 8 13, 8 1 9, 7 5 12, 1 2 3, 6 6 12, 4 9 13, 1 9 10, 9 3 12, 1 6 7 14, 6 1 1 8

Number relationships

1 4 5, 4 8 12, 4 1 5, 7 5 12, 6 2 8, 1 6 7, 8 9 17, 9 7 16, 1 3 4, 9 6 7 22, 1 1 1 3

Ho Math Chess 何数棋谜　妈！我会棋谜式加法啦！
Mom! I Learn Addition Using Math-Chess-Puzzles Connection
Contents include both traditional and Math-Chess-Puzzles combined methods. Extra strength
©2008 − 2018 Frank Ho, Amanda Ho　　All rights reserved. www.homathchess.com

Student Name _____ Date _____

Number relationships

9 4 13, 1 8 9, 8 1 9, 6 5 11, 1 2 3, 7 6 13, 4 9 13, 5 7 12, 9 3 12, 5 6 7 18, 7 1 1 9

Number relationships

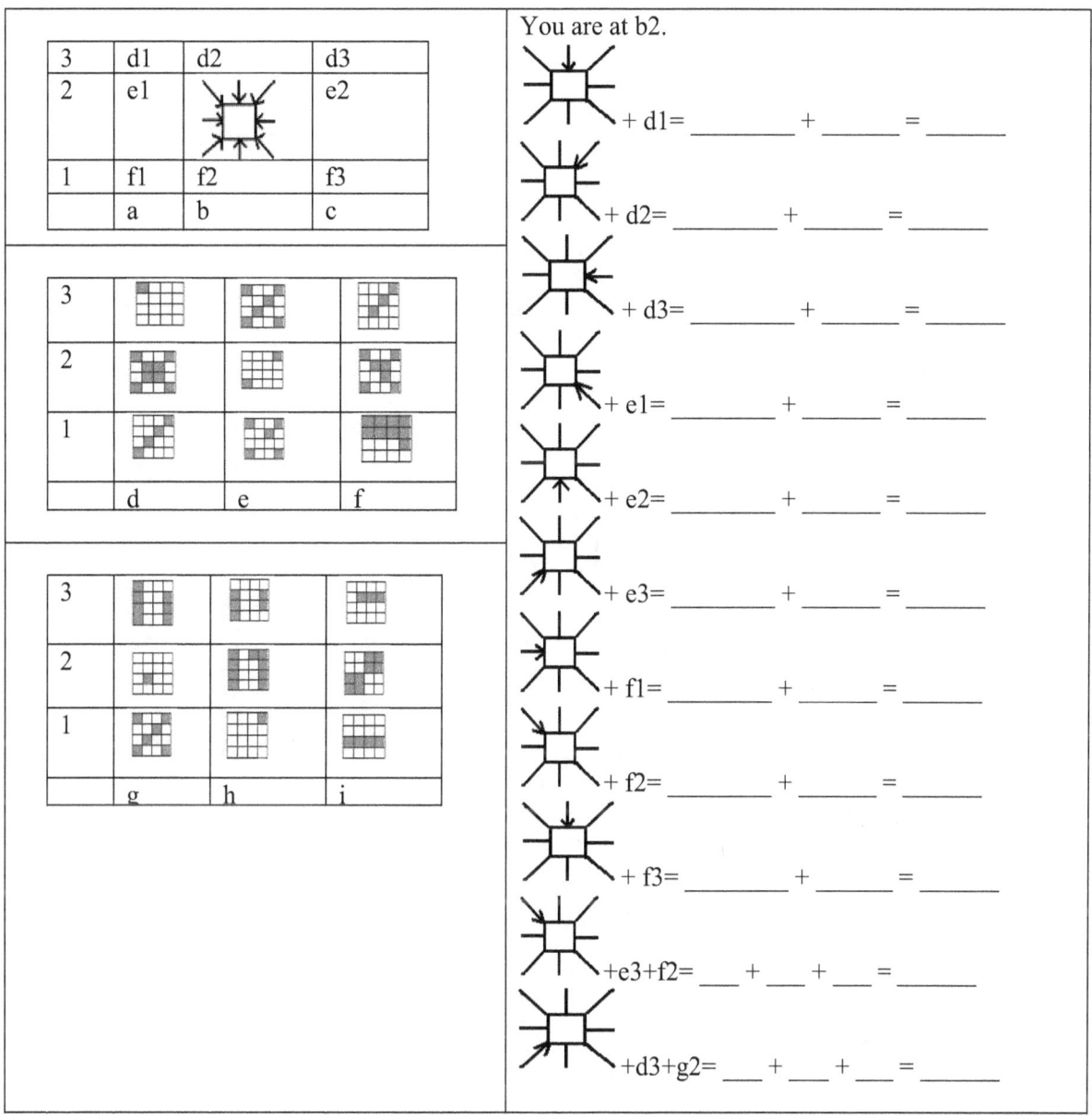

8 4 12, 1 8 9, 2 1 3, 3 5 8, 7 2 9, 9 6 15, 5 9 14, 4 7 11, 8 3 11, 4 6 7 17, 9 1 1 11

Number relationships

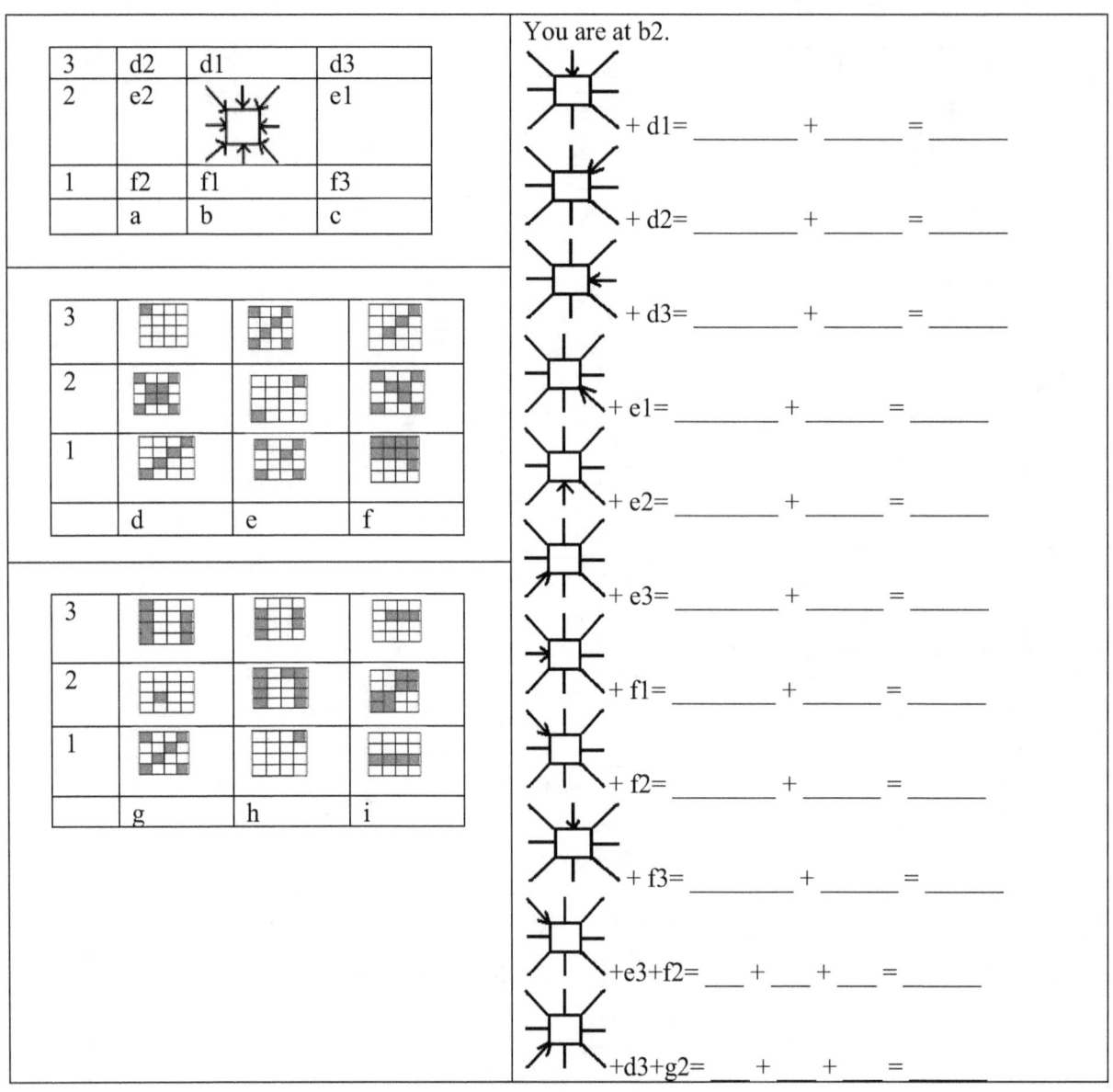

4 4 8, 1 8 9, 5 1 6, 3 5 8, 9 2 11, 7 6 13, 2 9 11, 8 7 15, 4 3 7, 8 6 7 21, 7 1 1 9

Number relationships

8 4 12, 4 8 12, 2 1 3, 9 5 14, 7 2 9, 3 6 9, 5 9 14, 1 7 8, 8 3 11, 1 6 7 14, 3 1 1 5

Number relationships

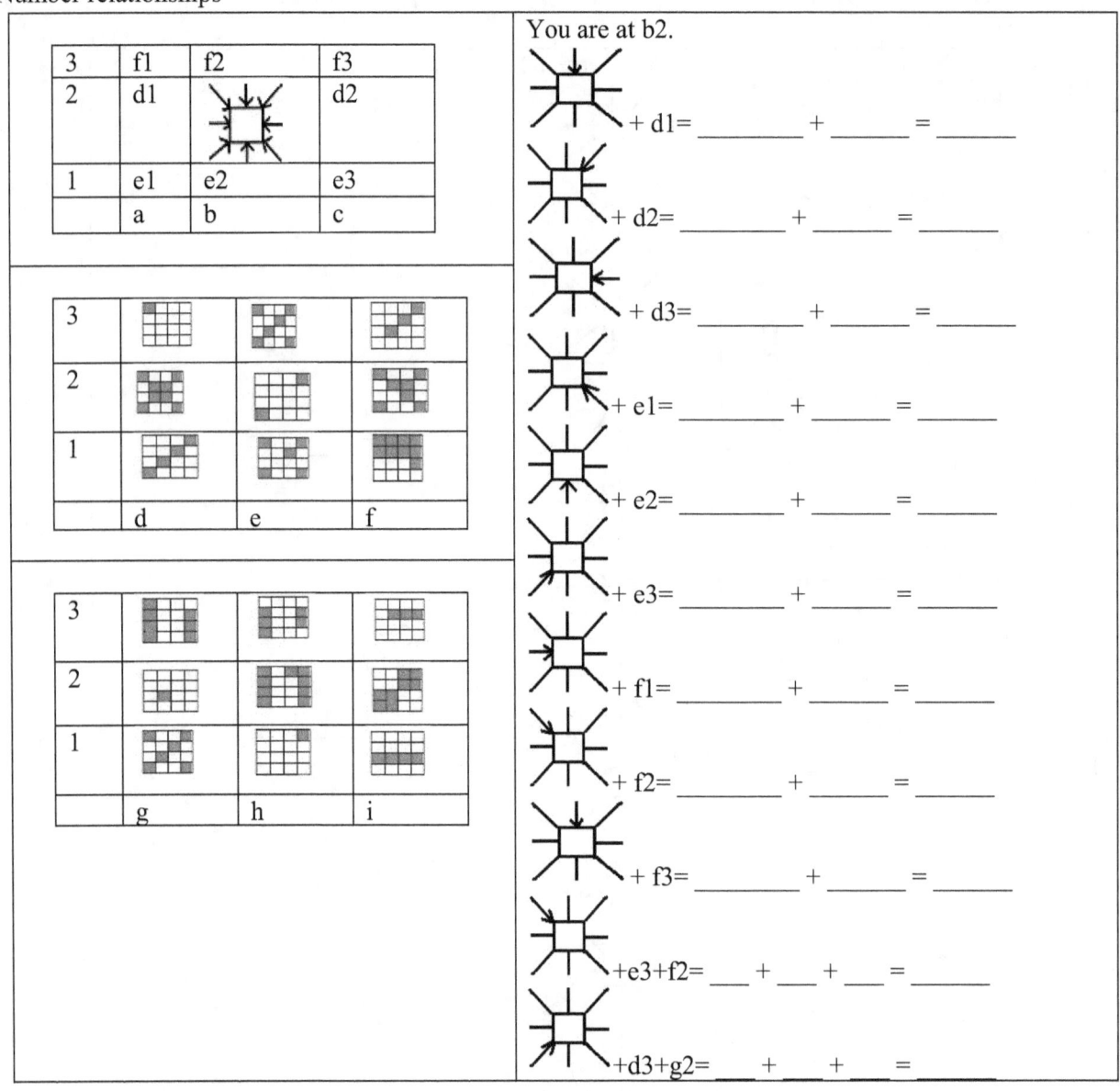

7 4 11, 3 8 11, 8 1 9, 6 5 11, 2 2 4, 5 6 11, 4 9 13, 9 7 16, 7 3 10, 9 6 7 22, 5 1 1 7

Number relationships

9 4 13, 4 8 12, 4 1 5, 6 5 11, 5 2 7, 2 6 8, 8 9 17, 7 7 14, 9 3 12, 7 6 7 20, 2 1 1 4

Number relationships

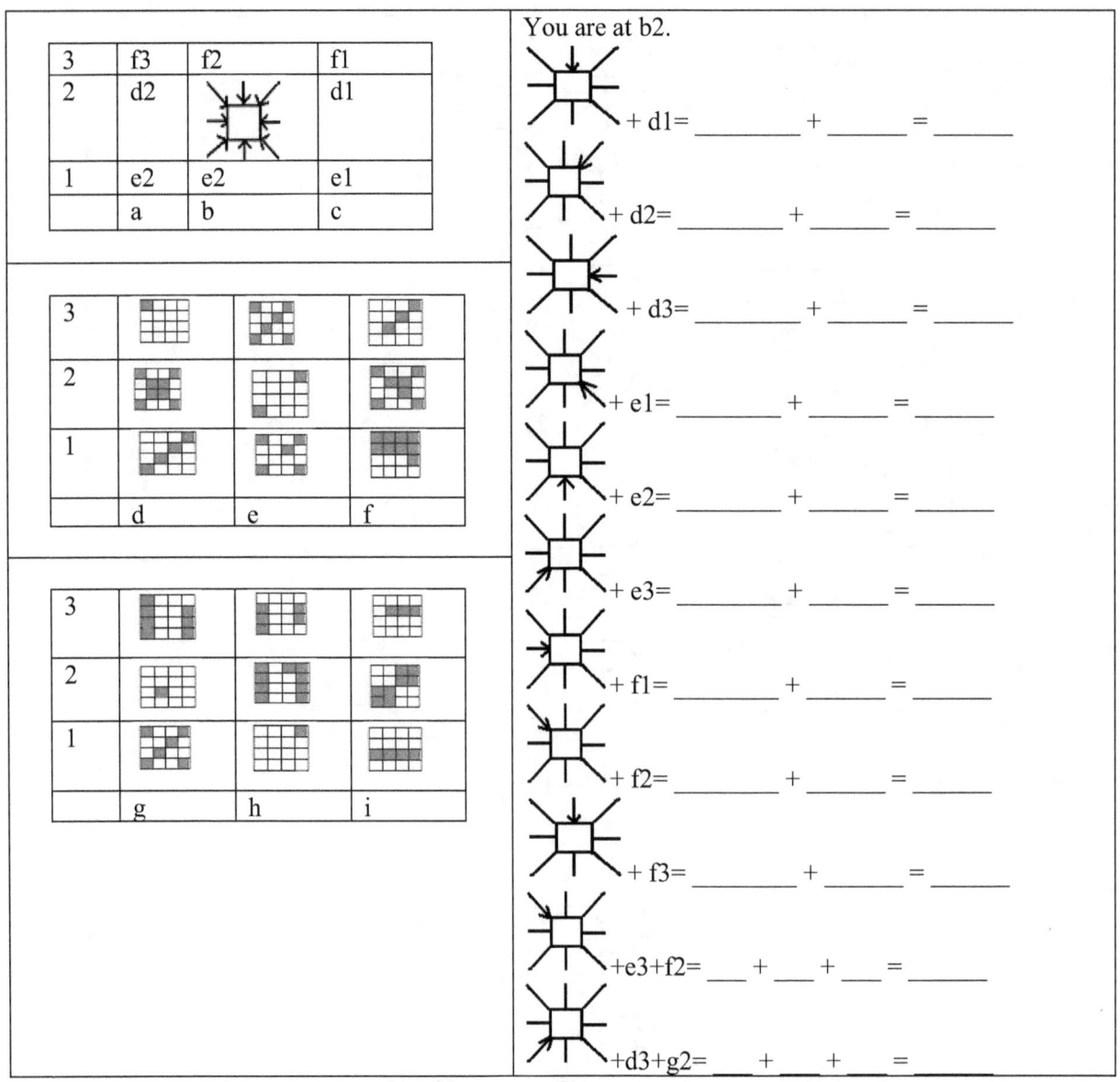

7 4 11, 9 8 17, 7 1 8, 5 5 10, 2 2 4, 2 6 8, 8 9 17, 3 7 10, 7 3 10, 7 6 7 16, 2 1 1 4

Number relationships

2 4 6, 6 8 14, 7 1 8, 3 5 8, 7 2 9, 9 6 15, 4 9 13, 5 7 12, 2 3 5, 5 6 7 18, 9 1 1 11

Number relationships

5 4 9, 6 8 14, 4 1 5, 3 5 8, 9 2 11, 7 6 13, 8 9 17, 2 7 9, 5 3 8, 2 6 7 15, 7 1 1 9

Number relationships

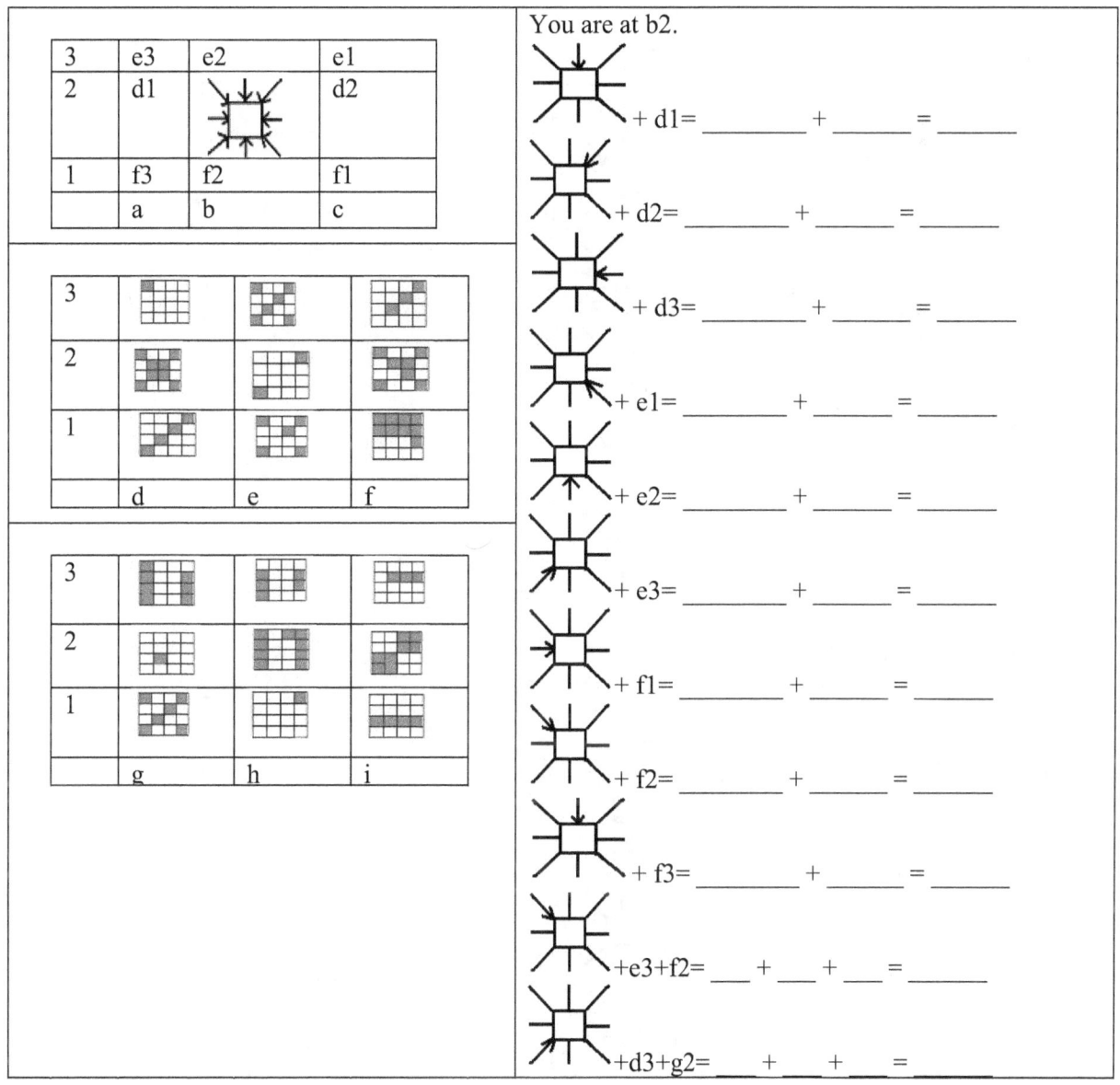

2 4 6, 5 8 13, 8 1 9, 9 5 14, 7 2 9, 3 6 9, 8 9 17, 6 7 13, 1 3 5, 6 6 8 20, 3 1 1 5

Number relationships

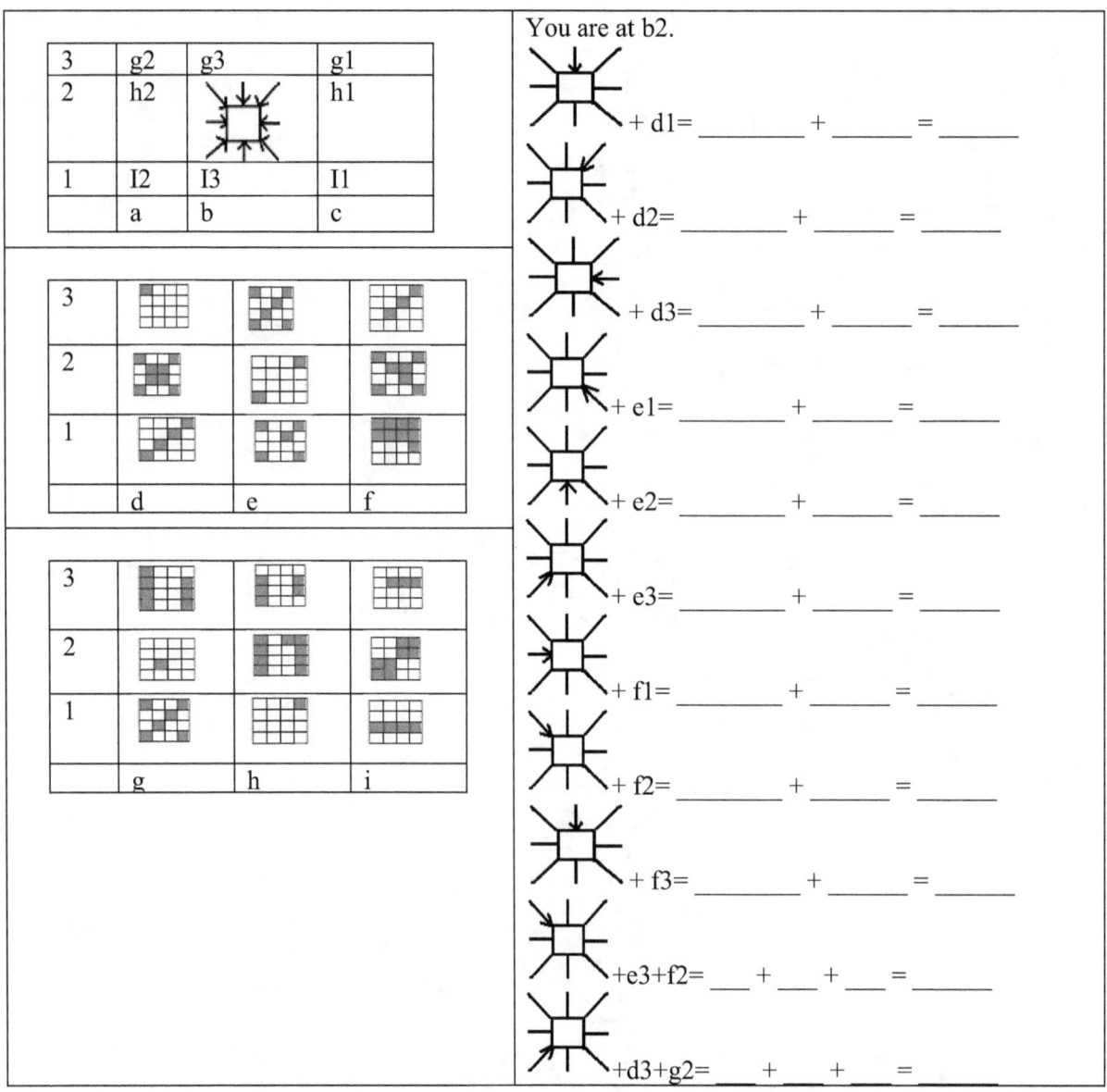

7 4 11, 6 8 14, 1 1 2, 4 5 9, 3 2 5, 8 6 14, 9 9 18, 1 7 8, 7 3 10, 1 6 7 14, 8 1 1 10

Number relationships

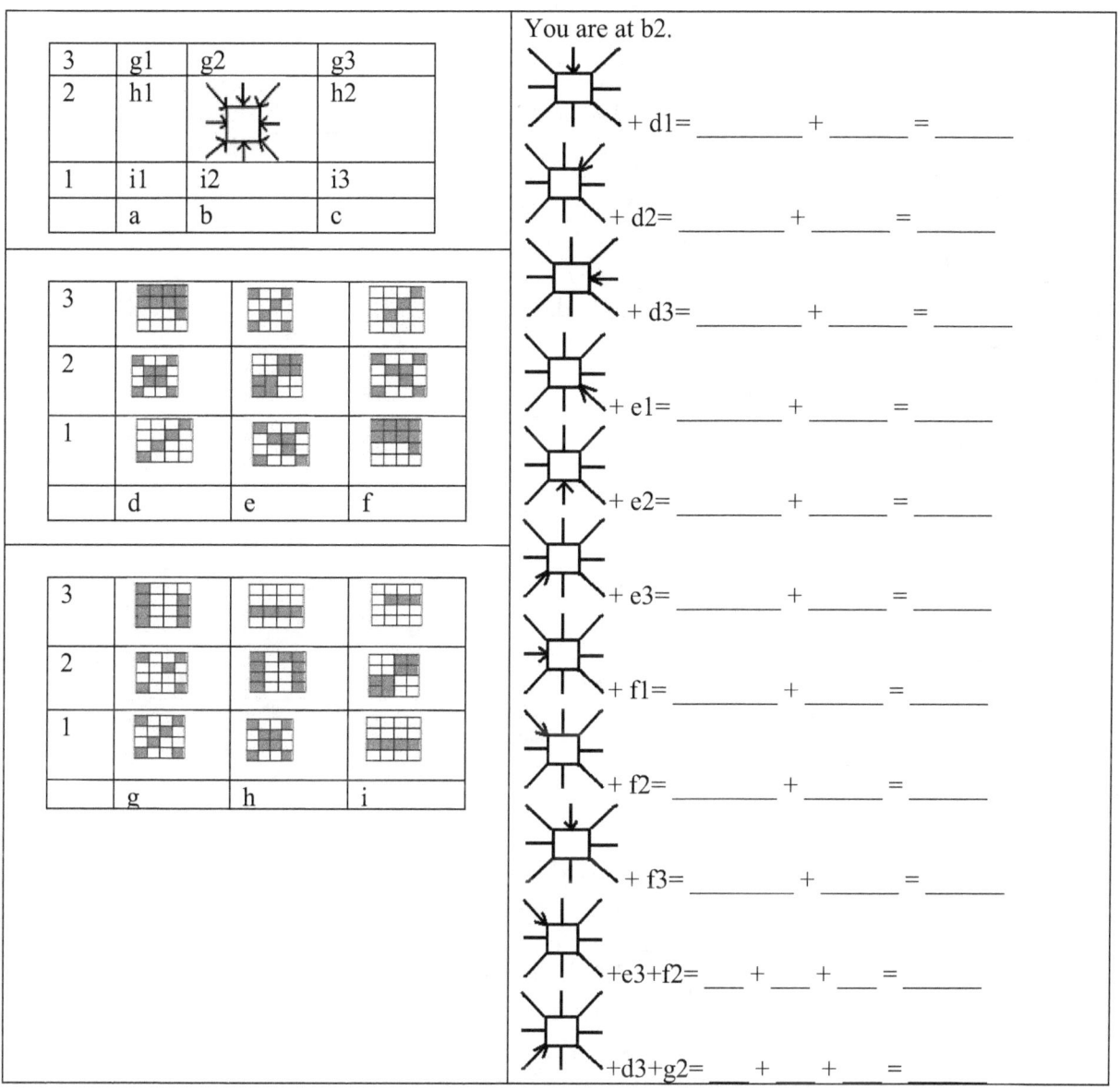

5 4 9, 7 8 15, 9 9 18, 3 7 10, 8 8 16, 4 6 10, 6 9 15, 6 7 13, 5 3 8, 6 6 8 20, 4 9 5 18

313

Number relationships

4 7 11, 4 4 8, 6 8 14, 2 6 8, 8 4 12, 7 5 12, 7 8 15, 3 8 11, 4 4 8, 3 5 8 16, 7 8 4 19

Number relationships

4 7 11, 2 2 4, 2 8 10, 4 6 10, 4 4 8, 3 5 8, 7 8 15, 7 8 15, 8 4 12, 7 5 8 20, 3 8 4 15

Number relationships

1 1 2, 7 4 11, 9 7 16, 3 6 9, 8 7 15, 4 4 8, 1 7 8, 6 3 9

Mom! I Learn Addition Using Math-Chess-Puzzles Connection

Part 4: Addition by carrying

d + dd

Wrong	Correct	Wrong	Correct
♖	₁5	7	₁7
+ 4 5	+ 4 5	+ 3 5	+ 3 5
b 4 1 0	5 0	3 1 2	4 2

6	7	♕	8
+ 6 6	+ 4 6	+ 1 6	+ 2 6
b 7 2	5 3	2 5	3 4
6	7	♕	♖
+ 5 5	+ 4 4	+ 3 3	+ 7 7
b 6 1	5 1	4 2	8 2
6	8	♕	4
+ 8 8	+ 4 8	+ 1 9	+ 2 6
b 9 4	5 6	2 8	3 0
6	♕	♖	8
+ 6 7	+ 4 6	+ 1 6	+ 2 6
b 7 3	5 5	2 1	3 4

Carrying or Regrouping for d + dd

6	4	7	♗	8
+1 4	+1 6	+1 3	+1 7	+1 2
2 0	2 0	2 0	2 0	2 0 b
2	♙	♕	♖	6
+1 8	+1 9	+1 1	+1 5	+1 4
2 0	2 0	2 0	2 0	2 0 b b
4	7	8	6	8
+1 6	+1 3	+1 2	+1 4	+1 2
2 0	2 0	2 0	2 0	2 0 b b
♕	♙	6	7	♗
+1 1	+1 9	+1 4	+1 3	+1 7
2 0	2 0	2 0	2 0	2 0 b b
♖	6	8	♕	♙
+1 5	+1 4	+1 2	+1 1	+1 9
2 0	2 0	2 0	2 0	2 0 b b

Carrying or Regrouping for d + dd

6	♖	7	4	8
+1 5	+1 6	+1 4	+1 7	+1 3
2 1	2 1	2 1	2 1	2 1 b b
♗	2	♕	6	6
+1 8	+1 9	+1 2	+1 5	+1 5
2 1	2 1	2 1	2 1	2 1 b b
6	7	8	♕	♖
+1 6	+1 7	+1 8	+1 9	+1 5
2 2	2 4	2 6	2 8	2 0 b b
♕	4	♕	♕	♕
+1 3	+1 9	+1 5	+1 6	+1 7
2 1	2 3	2 4	2 5	2 6 b b
♕	♕	8	7	6
+1 8	+1 9	+1 4	+1 4	+1 7
2 7	1 8	2 2	2 1	2 3 b b

Carrying or Regrouping for d + dd

6	4	7	♗	8
+3 4	+2 6	+2 3	+3 7	+2 2
4 0	3 0	3 0	4 0	3 0 b b
2	♙	♕	♖	6
+2 8	+3 9	+2 1	+3 5	+2 4
3 0	4 0	3 0	4 0	3 0 b b
4	7	8	6	8
+3 6	+2 3	+3 2	+2 4	+3 2
4 0	3 0	4 0	3 0	4 0 b b
♕	♙	6	7	■
+3 1	+2 9	+3 4	+2 3	+2 7
4 0	3 0	4 0	3 0	3 0 b b
♖	6	8	♕	♙
+2 5	+3 4	+2 2	+2 1	+3 9
3 0	4 0	3 0	3 0	4 0 b b

Carrying or Regrouping for d + dd

6	7	8	6	6
+1 6	+1 6	+1 6	+2 0	+1 9
2 2	2 3	2 4	2 6	2 5 b b
7	7	♕	8	8
+1 7	+1 8	+1 7	+1 8	+1 9
2 4	2 5	2 6	2 6	2 b b 7
8	♕	4	♖	♖
+1 8	+1 8	+1 4	+1 5	+1 4
2 6	2 7	2 8	2 0	1 9 b b
♖	♖	8	♕	3
+1 6	+1 7	+1 5	+1 5	+1 7
2 1	2 2	2 3	2 4	2 0 b b
7	7	8	8	4
+1 4	+1 4	+1 2	+1 3	+1 8
2 1	2 1	2 0	2 1	2 2 b b

Carrying or Regrouping for d + dd

6	4	7	3	8
+3 4	+5 6	+1 3	+4 7	+2 2
4 0	6 0	2 0	5 0	3 0 b b
2	♙	♕	♖	6
+8 8	+7 9	+1 1	+2 5	+4 4
9 0	8 0	2 0	3 0	5 0 b b
4	7	8	6	8
+6 6	+3 3	+5 2	+5 4	+8 2
7 0	4 0	6 0	6 0	9 0 b b
♕	♙	6	7	■
+1 1	+4 9	+6 4	+6 3	+7 7
2 0	5 0	7 0	7 0	8 0 b b
♖	6	8	♕	♙
+6 5	+3 4	+4 2	+5 1	+8 9
7 0	4 0	5 0	6 0	9 0 b b

dd + d

1	1	1	1	1
17	18	18	15	17
+ 6	+ ♛	+ 7	+ 8	+ ♜
23	27	25	23	22 b b
27	28	28	26	25
+ 8	+ ♛	+ 7	+ ♜	+ 7
35	37	35	31	32 b b
39	35	36	37	37
+ 8	+ 6	+ ♜	+ 8	+ ♜
47	41	41	45	42 b b
49	47	48	49	48
+ 8	+ 6	+ 7	+ 8	+ 6
57	53	55	57	54 b b
59	55	57	56	56
+ 8	+ 6	+ 6	+ 7	+ ♜
67	61	63	63	61 b b

323

dd + d

5	7	17	27	37
+ 5	+ 5	+ 5	+ ♖	+ ♖
10	12	22	32	b b 42
6	7	17	27	37
+ 6	+ 6	+ 6	+ 6	+ 6
12	13	23	33	b b 43
7	7	17	27	37
+ 7	+ 7	+ 7	+ 7	+ 7
14	14	24	34	b b 44
7	7	17	27	37
+ 7	+ 8	+ 8	+ 8	+ 8
14	15	25	35	b b 45
10	7	17	27	37
+ 7	+ ♕	+ ♕	+ ♕	+ ♕
17	16	26	36	b b 46

Ho Math Chess 何数棋谜　妈！我会棋谜式加法啦！
Mom! I Learn Addition Using Math-Chess-Puzzles Connection

Contents include both traditional and Math-Chess-Puzzles combined methods. Extra strength

©2008 － 2018 Frank Ho, Amanda Ho　　All rights reserved. www.homathchess.com

Student Name _____ Date _____

dd + d

♖	8	1 8	2 8	3 8
+ 5	+ ♖	+ ♖	+ 5	+ ♖
10	13	23	33	b b 43
6	8	1 8	2 8	3 8
+ 6	+ 6	+ 6	+ 6	+ 6
12	14	24	34	b b 44
7	8	1 8	2 8	3 8
+ 7	+ 7	+ 7	+ 7	+ 7
14	15	25	35	b b 45
8	♕	1 8	2 8	3 8
+ 8	+ 8	+ 8	+ 8	+ 8
16	17	26	36	46 b b b b
8	8	1 8	2 8	3 8
+ 8	+ ♕	+ ♕	+ ♕	+ 9
16	17	27	37	47 b b b

dd + d higher decades in pattern

5	♖	1 5	2 5	3 5
+ 5	+ 6	+ 6	+ 6	+ 6
10	11	21	31	41 b b 41
♖	1 5	2 5	3 5	4 5
+ 6	+ 6	+ 6	+ 6	+ 6
11	21	31	41	51 b b 51
1 5	2 5	3 5	4 5	5 5
+ 6	+ 6	+ 6	+ 6	+ 6
21	31	41	51	61 b b 61
2 5	3 5	4 5	5 5	6 5
+ 6	+ 6	+ 6	+ 6	+ 6
31	41	51	61	71 b b 71
3 5	4 5	5 5	6 5	7 5
+ 6	+ 6	+ 6	+ 6	+ 6
41	51	61	71	b b 81
4 5	5 5	6 5	7 5	8 5
+ 6	+ 6	+ 6	+ 6	+ 6
51	61	71	81	91 b b 91

dd + d higher decades in pattern

5	5	1 5	2 5	3 5
+ 5	+ 7	+ 7	+ 7	+ 7
10	12	22	32	b b 42
♖	1 5	2 5	3 5	4 5
+ 7	+ 7	+ 7	+ 7	+ 7
12	22	32	42	b b 52
1 5	2 5	3 5	4 5	5 5
+ 7	+ 7	+ 7	+ 7	+ 7
22	32	42	52	b b 62
2 5	3 5	4 5	5 5	6 5
+ 7	+ 7	+ 7	+ 7	+ 7
32	42	52	62	b b 72
3 5	4 5	5 5	6 5	7 5
+ 7	+ 7	+ 7	+ 7	+ 7
42	52	62	72	b b 82
4 5	5 5	6 5	7 5	8 5
+ 7	+ 7	+ 7	+ 7	+ 7
52	62	72	82	b b 92

dd + d higher decades in pattern

5	♖	1 5	2 5	3 5
+ 5	+ 8	+ 8	+ 8	+ 8
1 0	1 3	2 3	3 3	b b 4 3
5	1 5	2 5	3 5	4 5
+ 8	+ 8	+ 8	+ 8	+ 8
1 3	2 3	3 3	4 3	b b 5 3
1 5	2 5	3 5	4 5	5 5
+ 8	+ 8	+ 8	+ 8	+ 8
2 3	3 3	4 3	5 3	b b 6 3
2 5	3 5	4 5	5 5	6 5
+ 8	+ 8	+ 8	+ 8	+ 8
3 3	4 3	5 3	6 3	b b 7 3
3 5	4 5	5 5	6 5	5 5
+ 8	+ 8	+ 8	+ 8	+ 8
4 3	5 3	6 3	7 3	b b 6 3
4 5	5 5	6 5	5 5	8 5
+ 8	+ 8	+ 8	+ 8	+ 8
5 3	6 3	7 3	6 3	b b 9 3

dd + d higher decades in pattern

5 + 5 b b 10	5 + 9 14	1 5 + ♛ 2 4	2 5 + 9 3 4	3 5 + ♛ 4 4
5 + ♛ b b 14	1 5 + 9 2 4	2 5 + 9 3 4	3 5 + ♛ 4 4	4 5 + 9 5 4
1 5 + 9 b b 24	2 5 + ♛ 3 4	3 5 + 9 4 4	4 5 + ♛ 5 4	5 5 + 9 6 4
2 5 + ♛ b b 34	3 5 + 9 4 4	4 5 + ♛ 5 4	5 5 + 9 6 4	9 5 + ♛ 10 4
3 5 + 9 b b 44	4 5 + ♛ 5 4	5 5 + 9 6 4	6 5 + ♛ 7 4	7 5 + 9 8 4
4 5 + 9 b b 54	5 5 + ♛ 6 4	6 5 + 9 7 4	7 5 + ♛ 8 4	8 5 + 9 9 4

dd + d higher decades in pattern

6 + 6 b b 12	7 + 6 13	1 7 + 6 23	2 7 + 6 33	3 7 + 6 4 3
7 + 6 b b 13	1 7 + 6 23	2 7 + 6 33	3 7 + 6 43	4 7 + 6 5 3
1 7 + 6 b b 23	2 7 + 6 33	3 7 + 6 43	4 7 + 6 53	5 7 + 6 6 3
2 7 + 6 b b 33	3 7 + 6 43	4 7 + 6 53	5 7 + 6 63	6 7 + 6 7 3
3 7 + 6 b b 43	4 7 + 6 53	5 7 + 6 63	6 7 + 6 73	7 7 + 6 8 3
4 7 + 6 b b 53	5 7 + 6 63	6 7 + 6 73	7 7 + 6 83	8 7 + 6 9 3

dd + d higher decades in pattern

6	8	1 8	2 8	3 8
+ 6	+ 6	+ 6	+ 6	+ 6
b b 12	1 4	2 4	3 4	4 4
8	1 8	2 8	3 8	4 8
+ 6	+ 6	+ 6	+ 6	+ 6
b b 14	2 4	3 4	4 4	5 4
1 8	2 8	3 8	4 8	5 8
+ 6	+ 6	+ 6	+ 6	+ 6
b b 24	3 4	4 4	5 4	6 4
2 8	3 8	4 8	5 8	6 8
+ 6	+ 6	+ 6	+ 6	+ 6
b b 34	4 4	5 4	6 4	7 4
3 8	4 8	5 8	6 8	7 8
+ 6	+ 6	+ 6	+ 6	+ 6
b b 44	5 4	6 4	7 4	8 4
4 8	5 8	6 8	7 8	8 8
+ 6	+ 6	+ 6	+ 6	+ 6
b b 54	6 4	7 4	8 4	9 4

dd + d higher decades in pattern

6	♛	19	29	39
+ 6	+ 6	+ 6	+ 6	+ 6
b b 12	15	25	35	45
9	19	29	39	49
+ 6	+ 6	+ 6	+ 6	+ 6
b b 15	25	35	45	55
19	29	39	49	59
+ 6	+ 6	+ 6	+ 6	+ 6
b b 25	35	45	55	65
29	39	49	59	69
+ 6	+ 6	+ 6	+ 6	+ 6
b b 35	45	55	65	75
39	49	59	69	79
+ 6	+ 6	+ 6	+ 6	+ 6
b b 45	55	65	75	85
49	59	69	79	89
+ 6	+ 6	+ 6	+ 6	+ 6
b b 55	65	75	85	95

dd + d higher decades in pattern

7	7	17	27	37
+ 7	+ 8	+ 8	+ 8	+ 8
b b 14	15	25	35	45
7	17	27	37	47
+ 8	+ 8	+ 8	+ 8	+ 8
b b 15	25	35	45	55
17	27	37	47	57
+ 8	+ 8	+ 8	+ 8	+ 8
b b 25	35	45	55	65
27	37	47	57	87
+ 8	+ 8	+ 8	+ 8	+ 8
b b 35	45	55	65	95
37	47	57	67	77
+ 8	+ 8	+ 8	+ 8	+ 8
b b 45	55	65	75	85
47	57	67	77	87
+ 8	+ 8	+ 8	+ 8	+ 8
b b 55	65	75	85	95

dd + d higher decades in pattern

7 + 7 b b 14	7 + ♛ 16	1 7 + 9 26	2 7 + ♛ 36	3 7 + 9 4 6
7 + ♛ b b 16	1 7 + 9 26	2 7 + ♛ 36	3 7 + 9 46	4 7 + ♛ 5 6
1 7 + 9 b b 26	2 7 + ♛ 36	3 7 + 9 46	4 7 + ♛ 56	5 7 + 9 6 6
2 7 + ♛ b b 36	3 7 + 9 46	4 7 + ♛ 56	5 7 + 9 66	9 7 + ♛ 10 6
3 7 + 9 b b 46	4 7 + ♛ 56	5 7 + 9 66	6 7 + ♛ 76	7 7 + 9 8 6
4 7 + 9 b b 56	5 7 + ♛ 66	6 7 + 9 76	7 7 + ♛ 86	9 7 + 8 10 5

dd + d higher decades in pattern

8	8	1 8	2 8	3 8
+ 8	+ ♛	+ 9	+ ♛	+ 9
b b 16	17	27	37	47
8	1 8	2 8	3 8	4 8
+ ♛	+ 9	+ ♛	+ 9	+ 9
b b 17	27	37	47	57
1 8	2 8	3 8	4 8	5 8
+ 9	+ ♛	+ 9	+ ♛	+ 9
b b 27	37	47	57	67
2 8	3 8	4 8	5 8	6 8
+ ♛	+ 9	+ ♛	+ 9	+ ♛
b b 37	47	57	67	77
3 8	4 8	5 8	6 8	7 8
+ 9	+ ♛	+ 9	+ ♛	+ 9
b b 47	57	67	77	87
4 8	5 8	6 8	7 8	8 8
+ ♛	+ 9	+ ♛	+ 9	+ ♛
b b 57	67	77	87	97

Making multiples of 100

10	21	32	43	34
+ 90	+ 89	+ 78	+ 57	+ 76
b b 100	110	110	100	110
43	26	28	29	56
+ 57	+ 74	+ 82	+ 91	+ 64
b b 100	100	110	120	120
17	28	65	33	24
+ 83	+ 82	+ 45	+ 67	+ 86
b b 100	110	110	100	110
34	46	38	29	14
+ 66	+ 54	+ 72	+ 81	+ 86
b b 100	100	110	110	100
34	43	22	31	44
+ 66	+ 57	+ 78	+ 69	+ 56
b b 100	100	100	100	100
24	23	28	83	24
+ 86	+ 87	+ 82	+ 17	+ 86
b b 110	110	110	100	110

dd + d

2 8	3 8	4 8	5 8	6 8
+ 9	+ ♛	+ 9	+ ♛	+ 9
b b 37	47	57	67	77
3 8	4 8	5 8	6 8	7 8
+ ♛	+ 9	+ ♛	+ 9	+ ♛
b b 47	57	67	77	87
4 8	5 8	6 8	7 8	8 8
+ 9	+ ♛	+ 9	+ ♛	+ 9
b b 57	67	77	87	97

dd + d

67 + 6 = bb 73	69 + ♕ = 78	69 + 7 = 76	65 + 8 = 73	67 + ♖ = 72
77 + 8 = bb 85	78 + 9 = 87	78 + 7 = 85	76 + 5 = 81	75 + 7 = 82
89 + 8 = bb 97	85 + 6 = 91	86 + ♖ = 91	87 + 8 = 95	87 + 5 = 92
59 + 8 = bb 67	67 + 6 = 73	88 + 7 = 95	39 + 8 = 47	38 + 6 = 44
29 + 8 = bb 37	55 + 6 = 61	47 + 6 = 53	26 + 7 = 33	16 + ♖ = 21

338

dd + dd

11	22	33	44	10
+11	+22	+33	+44	+10
b b 22	44	66	88	20
20	30	40	10	20
+20	+30	+40	+20	+30
b b 40	60	80	30	50
30	40	12	17	14
+20	+50	+28	+23	+26
b b 50	90	40	40	40
15	26	13	18	19
+25	+34	+27	+22	+21
b b 40	60	40	40	40
35	46	37	48	39
+45	+36	+47	+38	+49
b b 80	82	84	86	88

Carrying or Regrouping for dd

Number profile	Standard Number	Write number in words
4 tens, 12 ones	☐☐ 5 2	Fifty-two
4 tens, 10 ones	☐☐50	
4 tens, 11 ones	☐☐51	
9 tens, 19 ones	☐☐☐109	
9 tens, 10 ones	☐☐☐100	
7 tens, 13 ones	☐☐83	
5 tens, 15 ones,	☐☐65	
5 tens, 29 ones,	☐☐79	
5 tens, 39 ones	☐☐89	
1 ten, 69 ones	☐☐79	
3 tens, 77 ones	☐☐☐107	

dd + ddd adding 10

19 + 91 b b 110	91 +119 210	28 + 82 110	28 +182 210	37 + 73 110
37 +173 b b 210	46 + 64 110	46 +164 210	55 + 55 110	55 +155 210
28 + 82 b b 110	28 +182 210	37 + 73 110	37 +173 210	19 + 91 110
91 +119 b b 210	55 + 55 110	55 +155 210	46 + 64 110	46 +164 210
37 + 73 b b 110	37 +173 210	28 + 82 110	28 +182 210	19 + 91 110

dd + ddd adding doubles

22 + 22 bb 44	22 +122 144	33 + 33 66	33 +133 166	44 + 44 88
44 +144 bb 188	66 +166 232	77 +177 254	55 + 55 110	55 +155 210
88 + 88 bb 176	88 +188 276	99 + 99 198	99 +199 298	99 + 99 198
44 +144 bb 188	55 + 55 110	55 +155 210	66 + 66 132	77 +177 254
77 + 77 bb 154	77 +177 254	77 + 77 154	88 +188 276	88 + 88 176

dd + ddd adding 11

29 + 92 b b 121	92 +129 221	38 + 83 121	38 +183 221	47 + 74 121
47 +174 b b 221	56 + 65 121	56 +165 221	47 + 74 121	47 +174 221
38 + 83 b b 121	38 +183 221	29 + 92 121	92 +129 221	56 + 65 121
56 +165 b b 221	47 + 74 121	47 +174 221	56 + 65 121	56 +165 221
38 +183 b b 221	38 + 83 121	29 + 92 121	92 +129 221	47 + 74 121

dd + ddd adding 12

39 + 93 b b 132	93 +139 232	48 + 84 132	48 +184 232	57 + 75 132
66 +166 b b 232	66 + 66 132	57 +175 232	48 + 84 132	39 +193 232
48 + 84 b b 132	48 +184 232	39 + 93 132	93 +139 232	66 + 66 132
66 +166 b b 232	57 + 75 132	57 +175 232	48 + 84 132	48 +184 232
48 +184 b b 232	48 + 84 132	39 + 93 132	93 +139 232	57 + 75 132

dd + ddd adding 13

49 + 94 bb 143	94 +149 243	58 + 85 143	58 +185 243	67 + 76 143
76 +167 bb 243	67 + 76 143	67 +176 243	58 + 85 143	49 +194 243
58 + 85 bb 143	58 +185 243	49 + 94 143	94 +149 243	76 + 67 143
67 +176 bb 243	67 + 76 143	67 +176 243	58 + 85 143	58 +185 243
58 +185 bb 243	58 + 85 143	49 + 94 143	94 +149 243	67 + 76 143

dd + ddd adding 14

59 + 95 bb 154	95 +159 254	68 + 86 154	68 +186 254	77 + 77 154
86 +168 bb 254	68 + 86 154	77 +177 254	68 + 86 154	59 +195 254
68 + 86 bb 154	68 +186 254	59 + 95 154	95 +159 254	68 + 86 154
86 +168 bb 254	77 + 77 154	77 +177 254	68 + 86 154	68 +186 254
68 +186 bb 254	68 + 86 154	59 + 95 154	95 +159 254	77 + 77 154

dd + ddd adding 15

69 + 96 165	96 +169 265	78 + 87 165	78 +187 265	87 + 78 165 b b
96 +169 265	69 + 96 165	87 +178 265	78 + 87 165	69 +196 265 b b
78 + 87 165	78 +187 265	69 + 96 165	96 +169 265	69 + 96 165 b b
96 +169 265	87 + 78 165	87 +178 265	78 + 87 165	78 +187 265 b b
78 +187 265	78 + 87 165	69 + 96 165	96 +169 265	87 + 78 165 b b

347

dd + ddd adding 16, 17, 18

79 + 97 176	97 +179 276	88 + 88 176	88 +188 276	97 + 79 176 b b
89 +198 287	98 + 89 187	97 +179 276	99 + 99 198	99 +199 298 b b
98 + 89 187	98 +189 287	99 + 99 198	99 +169 268	99 + 99 198 b b
88 +188 276	97 + 79 176	97 +179 276	79 + 97 176	97 +179 276 b b
88 +188 	98 + 89 187	89 + 98 187	97 +179 276	97 + 79 176 b b

dd + ddd

99 + 22 121	22 +199 221	99 + 33 132	33 +199 232	99 + 44 143 b b
44 +199 143	99 + 11 110	11 +199 210	55 + 99 154	99 +155 254 b b
66 + 99 165	77 +199 276	88 + 99 187	88 +199 287	66 + 99 165 b b
66 +199 265	77 + 77 154	99 +199 298	99 + 88 187	44 +199 243 b b
44 +199 243	99 + 88 187	77 + 99 176	77 +199 276	55 + 99 154 b b

dd + ddd

88 + 33 121	33 +188 221	88 + 33 121	33 +188 221	88 + 44 132 b b
44 +188 232	88 + 66 154	66 +188 254	55 + 88 143	88 +155 243 b b
66 + 88 154	77 +188 265	88 + 88 176	88 +188 276	66 + 88 154 b b
66 +188 254	77 + 77 154	99 +188 287	88 + 99 187	44 +188 232 b b
44 +188 232	88 + 99 187	77 + 88 165	77 +188 265	55 + 88 143 b b

dd + ddd

77 + 44 121	44 +177 221	77 + 44 121	55 +177 232	77 + 44 121 b b
55 +177 232	77 + 66 143	66 +177 243	55 + 77 132	77 +155 232 b b
66 + 77 143	77 +177 254	88 + 77 165	88 +177 265	66 + 77 143 b b
66 +177 243	77 + 77 154	77 +199 276	99 + 77 176	44 +177 221 b b
44 +177 221	77 + 88 165	77 + 77 154	77 +177 254	55 + 77 132 b b

dd + ddd

66	99	66	88	66
+ 99	+166	+ 88	+166	+ 77
165	265	154	254	143 b b
77	66	55	55	66
+166	+ 55	+166	+ 66	+155
243	121	221	121	221 b b
66	77	88	88	66
+ 66	+166	+ 66	+166	+ 66
132	243	154	254	132 b b
66	77	99	66	77
+166	+ 66	+166	+ 88	+166
232	143	265	154	243 b b
55	66	77	77	55
+166	+ 88	+ 66	+166	+ 66
221	154	143	243	121 b b

ddd + ddd

666	555	444	333	222
+766	+ 655	+544	+433	+322
1432	1210	988	766	544 b b
777	888	999	666	555
+877	+ 988	+999	+766	+ 655
1654	1876	1998	1432	1210 b b
333	222	444	555	444
+433	+322	+544	+ 655	+544
766	544	988	1210	988 b b
666	555	777	888	999
+766	+ 655	+877	+ 988	+999
1432	1210	1654	1876	1998 b b
888	999	666	555	777
+ 988	+999	+766	+ 655	+877
1876	1998	1432	1210	1654 b b

ddd + ddd

766 + 776 1542	655 + 665 1320	544 + 554 1098	433 + 443 876	322 + 332 654 b b
877 + 887 1754	988 + 998 1986	999 + 999 1998	766 + 776 1542	655 + 665 1320 b b
766 + 776 1542	877 + 887 1764	544 + 554 1098	766 + 776 1542	988 + 998 1986 b b
655 + 665 1320	544 + 554 1098	433 + 443 876	322 + 332 654	999 + 999 1998 b b
766 + 776 1542	877 + 887 1764	544 + 554 1098	766 + 776 1542	988 + 998 1986 b b

ddd + ddd

765 + 876 1641	876 + 987 1863	987 + 998 1985	598 + 599 1197	855 + 765 1620 b b
877 + 783 1660	988 + 892 1880	989 + 891 1880	756 + 664 1420	755 + 865 1620 b b
764 + 656 1420	773 + 867 1640	744 + 656 1400	668 + 572 1240	988 + 892 1880 b b
774 + 866 1640	681 + 779 1460	667 + 553 1220	578 + 662 1240	779 + 681 1460 b b
874 + 766 1640	877 + 783 1660	644 + 556 1200	766 + 674 1440	988 + 892 1880 b b

ddd + ddd

767	878	987	599	856
+ 876	+ 987	+ 999	+ 599	+ 765
1643	1865	1986	1198	1621 b b
877	988	989	756	755
+ 785	+ 895	+ 898	+ 668	+ 867
1662	1883	1887	1424	1622 b b
767	778	747	668	988
+ 656	+ 867	+ 656	+ 576	+ 897
1423	1645	1403	1244	1885 b b
779	684	667	578	779
+ 866	+ 779	+ 555	+ 665	+ 687
1645	1463	1222	1243	1466 b b
879	877	645	766	988
+ 766	+ 787	+ 556	+ 677	+ 897
1645	1664	1201	1443	1885 b b

Mom! I Learn Addition Using Math-Chess-Puzzles Connection

ddd + ddd

997	998	997	999	996
+ 873	+ 982	+ 993	+ 591	+ 764
1870	1980	1990	1590	1760 b b
993	998	999	996	995
+ 787	+ 892	+ 891	+ 664	+ 865
1780	1890	1890	1660	1860 b b
997	998	993	994	998
+ 653	+ 862	+ 657	+ 576	+ 892
1650	1860	1650	1570	1890 b b
774	682	667	575	997
+ 996	+ 998	+ 993	+ 995	+ 683
1770	1680	1660	1570	1680 b b
874	873	644	763	983
+ 996	+ 997	+ 996	+ 997	+ 997
1870	1870	1640	1760	1980 b b

Mental addition

Example

100 + 200 + 1 = 1 *hundred* + 2 *hundred* and 1 more is 301.

Complete the following problems.

100 + 200 + 2 = ___ *hundred* + ___ *hundred* and ___ more is ____.
100 + 300 + 20 = ___ *hundred* + ___ *hundred* and ___ more is ____.
100 + 500 + 30 = ___ *hundred* + ___ *hundred* and ___ more is ____.
100 + 700 + 40 = ___ *hundred* + ___ *hundred* and ___ more is ____.
100 + 800 + 50 = ___ *hundred* + ___ *hundred* and ___ more is ____.
100 + 900 + 60 = ___ *hundred* + ___ *hundred* and ___ more is ____.
200 + 400 + 80 = ___ *hundred* + ___ *hundred* and ___ more is ____.
200 + 600 + 100 = ___ *hundred* + ___ *hundred* and ___ more is ____.
200 + 800 + 2 = ___ *hundred* + ___ *hundred* and ___ more is ____.

1	2	2	302
1	3	20	420
1	5	30	630
1	7	40	840
1	8	50	950
1	9	60	1060
2	4	80	680
2	6	100	900
2	8	2	1002

Mental addition

300 + 200 + 1 = 3 *hundred* + 2 *hundred* and 1 more is 501.

400 + 200 + 2 = __4__ *hundred* + __2__ *hundred* and __2__ more is __602__.

500 + 300 + 20 = __5__ *hundred* + __3__ *hundred* and __20__ more is __820__.

600 + 500 + 30 = __6__ *hundred* + __5__ *hundred* and __30__ more is __1130__.

700 + 700 + 40 = __7__ *hundred* + __7__ *hundred* and __40__ more is __1440__.

800 + 800 + 50 = __8__ *hundred* + __8__ *hundred* and __50__ more is __1650__.

900 + 900 + 60 = __9__ *hundred* + __9__ *hundred* and __60__ more is __1860__.

1000 + 500 + 80 = __1__ *thousand* + __5__ *hundred* and __8__ more is __1580__.

1000 + 500 + 100 = __1__ *thousand* + __5__ *hundred* + __1__ *hundred* is __1600__.

__1__ __6__ __1__ __90__ __1790__

Mom! I Learn Addition Using Math-Chess-Puzzles Connection

1000 + 600 +190 = ___ thousand + ___ hundred + ___ hundred and ___ more is ___.
        1   7   1   90   1890

1000 + 700 +190 = ___ thousand + ___ hundred + ___ hundred and ___ more is ___.
        1   8   0   90   1890

1000 + 800 +90 = ___ thousand + ___ hundred + ___ hundred and ___ more is ___.

d + d + d

7 ♗ + 4	1 ♖ + 9	2 2 + 8	♖ 4 + 5	6 6 + 4
b b 14	15	12	14	16

7 3 + ♗	5 1 + 9	9 2 + 8	8 4 + 6	7 6 + 4
b b 13	15	19	18	17

7 3 + 7	1 9 + 9	8 2 + 8	6 4 + 5	6 ♗ + 7
b b 17	19	18	15	16

7 4 + 6	9 1 + 9	9 2 + 8	7 5 + ♖	6 5 + 4
b b 17	19	19	17	15

361

Mom! I Learn Addition Using Math-Chess-Puzzles Connection

d + d + d

6	8	2	♖	4
♗	♖	5	4	6
+ 4	+ 2	+ 8	+ 6	+ 4
b b 13	15	15	15	14
7	1	9	8	7
4	2	1	2	6
+ ♗	+ 9	+ 8	+ 6	+ 3
b b 14	12	18	16	16
7	2	8	5	6
3	9	4	5	♗
+ 8	+ 9	+ 8	+ 5	+ 7
b b 18	20	20	15	16
6	9	9	5	6
4	9	1	5	6
+ 6	+ 9	+ 8	+ ♖	+ 4
b b 16	27	18	15	16

d + d + d

7 ♗ + 8 —— b b 18	2 ♖ + 8 —— 15	9 2 + 8 —— 19	♖ 9 + 5 —— 19	8 6 + 4 —— 18
7 7 + ♗ —— b b 17	1 1 + 9 —— 11	9 1 + 8 —— 18	3 4 + 7 —— 14	9 6 + 4 —— 19
7 3 + 3 —— b b 13	1 1 + 9 —— 11	8 2 + 2 —— 12	6 4 + 4 —— 14	7 ♗ + 7 —— 17
7 3 + 3 —— b b 13	8 2 + 2 —— 12	9 1 + 9 —— 19	7 5 + ♖ —— 17	6 4 + 4 —— 14

363

Ho Math Chess 何数棋谜 妈！我会棋谜式加法啦！
Mom! I Learn Addition Using Math-Chess-Puzzles Connection
Contents include both traditional and Math-Chess-Puzzles combined methods. Extra strength
©2008 – 2018 Frank Ho, Amanda Ho   All rights reserved. www.homathchess.com

Student Name _____   Date _____

d + d + d

	6 ♗ + 4	♖ ♖ + 9	3 2 + 8	♖ 4 + ♖	6 ♖ + 4
	b b 13	19	13	14	15
	♖ 3 + ♗	5 1 + ♕	♕ 2 + 8	♗ 4 + 6	♕ 6 + 4
	b b 11	15	19	13	19
	7 4 + 7	2 9 + 9	8 2 + 3	6 5 + ♖	7 ♗ + 7
	b b 18	20	13	16	17
	7 5 + 6	9 2 + 9	9 2 + 9	6 5 + ♖	6 5 + 5
	b b 18	20	20	16	16

364

Number pattern

1. ♙, 4, 7, 10, 13, 16, ☐, ☐, ☐, ☐19, 22, 25, 28

2. ♗, 6, ♕, 12, 15, ☐, ☐, ☐, ☐18, 21, 24, 27

3. 2, ♜, 8, ☐, ☐, ☐, ☐11, 14, 17, 20

4. ♗, ♜, 7, ☐, ☐, ☐, ☐9, 11, 13, 15

5. 99, 96, 93, 90, 87, ☐, ☐, ☐, ☐84, 81, 78, 75

6. 105, 100, 95, 90, ☐, ☐, ☐, ☐85, 80, 75, 70

7. 100, 90, 80, 70, ☐, ☐, ☐, ☐60, 50, 40, 30

8. 104, 100, 96, 92, ☐, ☐, ☐, ☐88, 84, 80, 76

Number pattern

1. ☐, ☐, ☐, ☐, 15, 13, 11, ♛, 7, ♜

2. ♗, 6, ♛, 12, 15, ☐, ☐, ☐, ☐

3. **15, 13, 11,** ♛, ☐, ☐, ☐, ☐

4. 3, ♜, 7, ☐, ☐, ☐, ☐

5. ♔, ♜, 10, ☐, ☐, ☐, ☐

6. 98, 96, 94, 92, 90, ☐, ☐, ☐, ☐

7. 115, 110, 105, 100, ☐, ☐, ☐, ☐

8. 101, 91, 81, 71, ☐, ☐, ☐, ☐

9. 107, 109, 111, 113, ☐, ☐, ☐, ☐

23, 21, 19, 17 b b
18, 21, 24, 27 b b
7,   5,   3,   1 b b
9,   11, 13, 15 b b
15, 20, 25, 30 b b b b
 88, 86, 84, 82 b b
95, 90, 85, 80 b b
61, 51, 41, 31 b b
115, 117, 119, 221 b b

Ho Math Chess 何数棋谜   妈！我会棋谜式加法啦！
Mom! I Learn Addition Using Math-Chess-Puzzles Connection
Contents include both traditional and Math-Chess-Puzzles combined methods. Extra strength
©2008 – 2018 Frank Ho, Amanda Ho   All rights reserved. www.homathchess.com

Student Name _____ Date _____

Number pattern

1. ♙+1, 7+1, 13+1, □, □, □, □

2. ♗-1, 6-1, ♕-1, □, □, □, □

3. 2+1, ♖, 7, ♕, □, □, □

4. ♗+1, ♖+1, 7+1, □, □, □, □

5. 99-2, 96-2, 93-2, 90-2, 87-2, □, □, □, □

6. 105-3, 100-3, 95-3, 90-3, □, □, □, □

7. 100-5, 90-5, 80-5, 70-5, □, □, □, □

8. 104+1, 100+1, 96+1, 92+1, □, □, □, □

20,   26,   32,   38 b b
11,   14,   17,   20 b b
11,   13,   15 b b
10,   12,   14,   16 b b
82,   79,   76,   73 b b
82,   77,   72,   67 b b
55,   45,   35,   25 b b
89,   85,   81,   77 b b

Ho Math Chess 何数棋谜　妈！我会棋谜式加法啦！
Mom! I Learn Addition Using Math-Chess-Puzzles Connection
Contents include both traditional and Math-Chess-Puzzles combined methods. Extra strength
©2008 – 2018 Frank Ho, Amanda Ho   All rights reserved. www.homathchess.com

Student Name _____ Date _____

High performance multi-digit addition

3	889	988	898
2	999	**333**	989
1	998	888	899
	a	b	c

The original square is at b2 = ☐.

```
  090809
+ 930303
 1021112
```

b2 + ✣ = ☐☐☐ + ○○○ = 1221

☐ + ✕ = ☐☐☐ + ○○○ = 1232

b2 + ✣ = ☐☐☐ + ○○○ = 1321

☐ + ✣ = 1332, ☐☐☐ + ○○○ = 1332

b2 + ✕ = ☐☐☐ + ○○○ = 1331

☐ + ✣ = ☐☐☐ + ○○○ = 1322

b2 + ✕ = ☐☐☐ + ○○○ = 1231

☐ + ✕ = ☐☐☐ + ○○○ = 1222

High performance multi-digit addition

3	778	877	787
2	888	**444**	878
1	887	888	788
	a	b	c

The original square is at b2 = ☐.

```
  09080907
+ 94040404
---------
 103121311
```

b2 + ✥ = ☐☐☐ + ○○○ = 1332

☐ + ✕ = ☐☐☐ + ○○○ = 1232

b2 + ✥ = ☐☐☐ + ○○○ = 1321

☐ + ✥ = ☐☐☐ + ○○○ = 1332

b2 + ✕ = ☐☐☐ + ○○○ = 1331

☐ + ✥ = ☐☐☐ + ○○○ = 1322

b2 + ✕ = ☐☐☐ + ○○○ = 1231

☐ + ✕ = ☐☐☐ + ○○○ = 1222

Ho Math Chess 何数棋谜　妈！我会棋谜式加法啦！
Mom! I Learn Addition Using Math-Chess-Puzzles Connection
Contents include both traditional and Math-Chess-Puzzles combined methods. Extra strength
©2008 − 2018 Frank Ho, Amanda Ho　All rights reserved. www.homathchess.com

Student Name _____ Date _____

High performance multi-digit addition

3	666	677	676
2	777	**555**	787
1	998	888	899
	a	b	c

The original square is at b2 = ☐.

```
  0908090607
+ 9505050505
  10413141112
```

b2 + ✥ = ☐☐☐ + ○○○ = 1443

☐ + ⤫ = ☐☐☐ + ○○○ = 1454

b2 + ✥ = ☐☐☐ + ○○○ = 1232

☐ + ✥ = ☐☐☐ + ○○○ = 1332

b2 + ⤫ = ☐☐☐ + ○○○ = 1443

☐ + ✥ = ☐☐☐ + ○○○ = 1342

b2 + ⤫ = ☐☐☐ + ○○○ = 1231

☐ + ⤫ = ☐☐☐ + ○○○ = 1221

High performance multi-digit addition

3	889	988	898
2	999	666	989
1	998	888	899
	a	b	c

The original square is at b2 = ☐.

```
  0908090607
+ 9606060606
-----------
 10514151213
```

b2 + ✥ =
☐☐☐
+ ○○○
─────
1554

☐ + ⤢ =
☐☐☐
+ ○○○
─────
1565

b2 + ✥ =
☐☐☐
+ ○○○
─────
1654

☐ + ↔ =
☐☐☐
+ ○○○
─────
1665

b2 + ⤢ =
☐☐☐
+ ○○○
─────
1664

☐ + ✥ =
☐☐☐
+ ○○○
─────
1655

b2 + ⤢ =
☐☐☐
+ ○○○
─────
1564

☐ + ⤢ =
☐☐☐
+ ○○○
─────
1555

High performance multi-digit addition

3	667	766	567
2	657	**777**	756
1	998	888	899
	a	b	c

The original square is at b2 = ☐.

```
  090809050607
+ 970707070707
  1061516121314
```

b2 + ✥ =
☐☐☐
+ ○○○
1665

☐ + ⤢ =
☐☐☐
+ ○○○
1676

b2 + ✥ =
☐☐☐
+ ○○○
1543

☐ + ✥ =
☐☐☐
+ ○○○
1434

b2 + ⤢ =
☐☐☐
+ ○○○
1775

☐ + ✥ =
☐☐☐
+ ○○○
1533

b2 + ⤢ =
☐☐☐
+ ○○○
1344

☐ + ⤢ =
☐☐☐
+ ○○○
1444

Ho Math Chess 何数棋谜　妈！我会棋谜式加法啦！

Mom! I Learn Addition Using Math-Chess-Puzzles Connection

Contents include both traditional and Math-Chess-Puzzles combined methods. Extra strength

©2008 – 2018 Frank Ho, Amanda Ho　All rights reserved. www.homathchess.com

Student Name _____　Date _____

High performance multi-digit addition

3	344	454	565
2	563	**888**	678
1	789	987	976
	a	b	c

The original square is at b2 = ☐.

0809070506040302
+ 9808080808080808
10617151314121110

---

b2 + ✥ =
☐☐☐
+ ○○○
———
1875

☐ + ⤢ =
☐☐☐
+ ○○○
———
1864

---

b2 + ✥ =
☐☐☐
+ ○○○
———
1342

☐ + ✥ =
☐☐☐
+ ○○○
———
1451

---

b2 + ⤢ =
☐☐☐
+ ○○○
———
1677

☐ + ✥ =
☐☐☐
+ ○○○
———
1566

---

b2 + ⤢ =
☐☐☐
+ ○○○
———
1453

☐ + ⤢ =
☐☐☐
+ ○○○
———
1232

Mom! I Learn Addition Using Math-Chess-Puzzles Connection

Contents include both traditional and Math-Chess-Puzzles combined methods. Extra strength

©2008 – 2018 Frank Ho, Amanda Ho   All rights reserved. www.homathchess.com

Student Name _____ Date _____

Computing through function concept

3	↓	2	┼
2	4	↓	↔
1	6	7	8
	a	b	c

The original square is at b2.

a3 + b2 = ___ + ___ = ___ ↓ + ↓ = 1 + 1 = 2	b3 + b2 = ___ + ___ = ___
c3 + b2 = ___ + ___ = ___	a2 + b2 = ___ + ___ = ___
b2 + b2 = ___ + ___ = ___	c2 + b2 = ___ + ___ = ___
a1 + b2 = ___ + ___ = ___	b1 + b2 = ___ + ___ = ___
c1 + b2 = ___ + ___ = ___	a3 − b2 = ___ − ___ = ___
b3 − b2 = ___ − ___ = ___	c3 − b2 = ___ − ___ = ___
a2 − b2 = ___ − ___ = ___	b2 − b2 = ___ − ___ = ___
c2 − b2 = ___ − ___ = ___	a1 − b2 = ___ − ___ = ___
b1 − b2 = ___ − ___ = ___	c1 − b2 = ___ − ___ = ___

2, 3
4, 5
2, 6
7, 8
9, 0
1, 2
3, 0
4, 5
6, 7

Mom! I Learn Addition Using Math-Chess-Puzzles Connection

Contents include both traditional and Math-Chess-Puzzles combined methods. Extra strength
©2008 – 2018 Frank Ho, Amanda Ho   All rights reserved. www.homathchess.com

Student Name _____   Date _____

From fingering math to computing through function concept

3	↓	2	⊥
2	4	2	↔
1	6	7	8
	a	b	c

The original square is at b2.

a3 + b2 = ___ + ___ = ___	b3 + b2 = ___ + ___ = ___
c3 + b2 = ___ + ___ = ___	a2 + b2 = ___ + ___ = ___
b2 + b2 = ___ + ___ = ___	c2 + b2 = ___ + ___ = ___
a1 + b2 = ___ + ___ = ___	b1 + b2 = ___ + ___ = ___
c1 + b2 = ___ + ___ = ___	b2 − a3 = ___ − ___ = ___
b3 − b2 = ___ − ___ = ___	c3 − b2 = ___ − ___ = ___
a2 − b2 = ___ − ___ = ___	b2 − b2 = ___ − ___ = ___
c2 − b2 = ___ − ___ = ___	a1 − b2 = ___ − ___ = ___
b1 − b2 = ___ − ___ = ___	c1 − b2 = ___ − ___ = ___

3, 4
5, 6
4, 7
8, 9
10, 1
0, 1
2, 0
3, 4
5, 6

Mom! I Learn Addition Using Math-Chess-Puzzles Connection

Contents include both traditional and Math-Chess-Puzzles combined methods. Extra strength
©2008 – 2018 Frank Ho, Amanda Ho   All rights reserved. www.homathchess.com

Student Name _____ Date _____

From fingering math to computing through function concept

3	↓	2	⊥
2	5	⊥	⊕
1	6	7	8
	a	b	c

The original square is at b2.

a3 + b2 = ___ + ___ = ___	b3 + b2 = ___ + ___ = ___
c3 + b2 = ___ + ___ = ___	a2 + b2 = ___ + ___ = ___
b2 + b2 = ___ + ___ = ___	c2 + b2 = ___ + ___ = ___
a1 + b2 = ___ + ___ = ___	b1 + b2 = ___ + ___ = ___
c1 + b2 = ___ + ___ = ___	b2 − a3 = ___ − ___ = ___
b2 − b3 = ___ − ___ = ___	c3 − b2 = ___ − ___ = ___
a2 − b2 = ___ − ___ = ___	b2 − b2 = ___ − ___ = ___
c2 − b2 = ___ − ___ = ___	a1 − b2 = ___ − ___ = ___
b1 − b2 = ___ − ___ = ___	c1 − b2 = ___ − ___ = ___

4, 5
6, 8
6, 8
9, 10
11, 2
1, 0
2, 0
2, 3
4, 5

Mom! I Learn Addition Using Math-Chess-Puzzles Connection

From fingering math to computing through function concept

3	↓	2	⊥
2	4	4	↔
1	6	7	8
	a	b	c

The original square is at b2.

a3 + b2 = ___ + ___ = ___   b3 + b2 = ___ + ___ = ___
c3 + b2 = ___ + ___ = ___   a2 + b2 = ___ + ___ = ___
b2 + b2 = ___ + ___ = ___   c2 + b2 = ___ + ___ = ___
a1 + b2 = ___ + ___ = ___   b1 + b2 = ___ + ___ = ___
c1 + b2 = ___ + ___ = ___   b2 − a3 = ___ − ___ = ___
b2 − b3 = ___ − ___ = ___   b2 − c3 = ___ − ___ = ___
a2 − b2 = ___ − ___ = ___   b2 − b2 = ___ − ___ = ___
c2 − b2 = ___ − ___ = ___   a1 − b2 = ___ − ___ = ___
b1 − b2 = ___ − ___ = ___   c1 − b2 = ___ − ___ = ___

5, 6
7, 8
8, 9
10, 11
12, 3
2, 1
0, 0
1, 2
3, 4

Ho Math Chess 何数棋谜　妈！我会棋谜式加法啦！
Mom! I Learn Addition Using Math-Chess-Puzzles Connection
Contents include both traditional and Math-Chess-Puzzles combined methods. Extra strength
©2008 – 2018 Frank Ho, Amanda Ho　　All rights reserved. www.homathchess.com

Student Name _____ Date _____

From fingering math to computing through function concept

3	↔	5	↔
2	6	↔	↔
1	6	7	8
	a	b	c

The original square is at b2.

a3 + b2 = ___ + ___ = ___	b3 + b2 = ___ + ___ = ___
c3 + b2 = ___ + ___ = ___	a2 + b2 = ___ + ___ = ___
b2 + b2 = ___ + ___ = ___	c2 + b2 = ___ + ___ = ___
a1 + b2 = ___ + ___ = ___	b1 + b2 = ___ + ___ = ___
c1 + b2 = ___ + ___ = ___	a3 – b2 = ___ – ___ = ___
b3 – b2 = ___ – ___ = ___	c3 – b2 = ___ – ___ = ___
a2 – b2 = ___ – ___ = ___	b2 – b2 = ___ – ___ = ___
c2 – b2 = ___ – ___ = ___	a1 – b2 = ___ – ___ = ___
b1 – b2 = ___ – ___ = ___	c1 – b2 = ___ – ___ = ___

10, 10
10, 11
10, 10
11, 12
13, 0
0, 0
1, 0
0, 1
2, 3

Mom! I Learn Addition Using Math-Chess-Puzzles Connection

Contents include both traditional and Math-Chess-Puzzles combined methods. Extra strength

©2008 – 2018 Frank Ho, Amanda Ho   All rights reserved. www.homathchess.com

Student Name _____ Date _____

From fingering math to computing through function concept

3	7	7	8
2	8	6	✻
1	6	7	8
	a	b	c

The original square is at b2.

a3 + b2 = ___ + ___ = ___	b3 + b2 = ___ + ___ = ___
c3 + b2 = ___ + ___ = ___	a2 + b2 = ___ + ___ = ___
b2 + b2 = ___ + ___ = ___	c2 + b2 = ___ + ___ = ___
a1 + b2 = ___ + ___ = ___	b1 + b2 = ___ + ___ = ___
c1 + b2 = ___ + ___ = ___	a3 − b2 = ___ − ___ = ___
b3 − b2 = ___ − ___ = ___	c3 − b2 = ___ − ___ = ___
a2 − b2 = ___ − ___ = ___	b2 − b2 = ___ − ___ = ___
c2 − b2 = ___ − ___ = ___	a1 − b2 = ___ − ___ = ___
b1 − b2 = ___ − ___ = ___	c1 − b2 = ___ − ___ = ___

13, 13
14, 14
12, 15
12, 13
14, 1
1, 2
2, 0
3, 0
1, 2

Mom! I Learn Addition Using Math-Chess-Puzzles Connection

Contents include both traditional and Math-Chess-Puzzles combined methods. Extra strength
©2008 – 2018 Frank Ho, Amanda Ho   All rights reserved. www.homathchess.com

From fingering math to computing through function concept

3	7	✳	8
2	8	7	8
1	✳	7	8
	a	b	c

The original square is at b2.

a3 + b2 = ___ + ___ = ___	b3 + b2 = ___ + ___ = ___
c3 + b2 = ___ + ___ = ___	a2 + b2 = ___ + ___ = ___
b2 + b2 = ___ + ___ = ___	c2 + b2 = ___ + ___ = ___
a1 + b2 = ___ + ___ = ___	b1 + b2 = ___ + ___ = ___
c1 + b2 = ___ + ___ = ___	a3 − b2 = ___ − ___ = ___
b3 − b2 = ___ − ___ = ___	c3 − b2 = ___ − ___ = ___
a2 − b2 = ___ − ___ = ___	b2 − b2 = ___ − ___ = ___
c2 − b2 = ___ − ___ = ___	a1 − b2 = ___ − ___ = ___
b1 − b2 = ___ − ___ = ___	c1 − b2 = ___ − ___ = ___

14, 16
15, 15
14, 15
16, 14
15, 0
2, 1
1, 2
0, 1

Ho Math Chess 何数棋谜　妈！我会棋谜式加法啦！
Mom! I Learn Addition Using Math-Chess-Puzzles Connection

Contents include both traditional and Math-Chess-Puzzles combined methods. Extra strength

©2008 − 2018 Frank Ho, Amanda Ho　All rights reserved. www.homathchess.com

Student Name_____ Date_____

Computing additions through math and chess integrated puzzle

5			g4		
4	g3				f3
3	g1				f1
2	g2				f2
1			f4		
	a	b	c	d	e

5		5		9	
4			6	7	11
3		2			12
2	4			14	
1		7	8		15
	f	g	h	i	j

The start square is at c3 = ☐.

c3 + ⇔ = __ + __ = __	c3 + ✕ = __ + __ = __
☐ + ⇔ = __ + __ = __	☐ + ✕ = __ + __ = __
c3 + ⇔ = __ + __ = __	c3 + ✕ = __ + __ = __
☐ + ⇔ = __ + __ = __	☐ + ✕ = __ + __ = __

8, 7
12, 4
10, 5
6, 6

Ho Math Chess 何数棋谜　妈！我会棋谜式加法啦！
Mom! I Learn Addition Using Math-Chess-Puzzles Connection

Contents include both traditional and Math-Chess-Puzzles combined methods. Extra strength
©2008 – 2018 Frank Ho, Amanda Ho   All rights reserved. www.homathchess.com

Student Name _____ Date _____

Computing additions through math and chess integrated puzzle

5			g4	
4	g3			f3
3	g1			f1
2	g2			f2
1			f4	
	a	b	c	d	e

5		5		9	
4			6	7	11
3		2			12
2	4			14	
1		7	8		15
	f	g	h	i	j

The start square is at c3 = ☐.

c3 + ✥ = __ + __ = __   c3 + ⤢ = __ + __ = __

☐ + ✥ = __ + __ = __   ☐ + ⤢ = __ + __ = __

c3 + ✥ = __ + __ = __   c3 + ⤢ = __ + __ = __

☐ + ✥ = __ + __ = __   ☐ + ⤢ = __ + __ = __

14, 13
18, 10
16, 11
12, 12

Ho Math Chess 何数棋谜　妈！我会棋谜式加法啦！
Mom! I Learn Addition Using Math-Chess-Puzzles Connection

Contents include both traditional and Math-Chess-Puzzles combined methods. Extra strength

©2008 – 2018 Frank Ho, Amanda Ho　All rights reserved. www.homathchess.com

Student Name _____ Date _____

Computing additions through math and chess integrated puzzle

The start square is at c3 = ☐.

c3 + ✥ = ___ + ___ = ___

☐ + ✥ = ___ + ___ = ___

c3 + ✥ = ___ + ___ = ___

☐ + ✥ = ___ + ___ = ___

c3 + ✖ = ___ + ___ = ___

☐ + ✖ = ___ + ___ = ___

c3 + ✖ = ___ + ___ = ___

☐ + ✖ = ___ + ___ = ___

10, 9
6, 14
12, 8
7, 8

Mom! I Learn Addition Using Math-Chess-Puzzles Connection

Contents include both traditional and Math-Chess-Puzzles combined methods. Extra strength
©2008 – 2018 Frank Ho, Amanda Ho   All rights reserved. www.homathchess.com

Student Name _____ Date _____

Computing additions through math and chess integrated puzzle

5		f2	f3	g2	f1
4					
3					
2					
1	g1	f4	g3	g4	
	a	b	c	d	e

5		5		9	
4			6	7	11
3					12
2	4			14	
1		7	8		15
	f	g	h	i	j

The start square is at c3 = ☐.

c3 + ✥ = __ + __ = __   c3 + ✕ = __ + __ = __

☐ + ✥ = __ + __ = __   ☐ + ✕ = __ + __ = __

c3 + ✥ = __ + __ = __   c3 + ✕ = __ + __ = __

☐ + ✥ = __ + __ = __   ☐ + ✕ = __ + __ = __

8, 7
4, 12
10, 5
3, 6

Mom! I Learn Addition Using Math-Chess-Puzzles Connection

Contents include both traditional and Math-Chess-Puzzles combined methods. Extra strength

©2008 – 2018 Frank Ho, Amanda Ho   All rights reserved. www.homathchess.com

Student Name _____ Date _____

Computing additions through math and chess integrated puzzle

5		f2	f3	g2	f1
4		✛	✛	✛	
3		✛	✱	✛	
2		✛	✛	✛	
1	g1	f4	g3	g4	
	a	b	c	d	e

5	✕	5	✱	9	✱
4	✱	✕	6	7	11
3	↓	2	✛	✕	12
2	4	✛		14	✕
1	✢	7	8	✕	15
	f	g	h	i	j

The start square is at c3 = ☐.

c3 + ✢ = __ + __ = __   c3 + ✕ = __ + __ = __

☐ + ✢ = __ + __ = __   ☐ + ✕ = __ + __ = __

c3 + ✢ = __ + __ = __   c3 + ✕ = __ + __ = __

☐ + ✢ = __ + __ = __   ☐ + ✕ = __ + __ = __

c3 + ✢ = __ + __ = __   c3 + ✕ = __ + __ = __

☐ + ✢ = __ + __ = __   ☐ + ✕ = __ + __ = __

14, 10, 16, 11, 16, 11
13, 18, 12, 12, 12, 12

Ho Math Chess 何数棋谜 妈！我会棋谜式加法啦！
Mom! I Learn Addition Using Math-Chess-Puzzles Connection
Contents include both traditional and Math-Chess-Puzzles combined methods. Extra strength
©2008 — 2018 Frank Ho, Amanda Ho   All rights reserved. www.homathchess.com

Student Name _____   Date _____

Computing additions through math and chess integrated puzzle

5			g4	
4	g3	┬	┬	┬	f3
3	g1	┬	✥	┬	f1
2	g2	┬	┬	┬	f2
1			f4	
	a	b	c	d	e

5	✕	5	✸	9	✸
4	✸	✕	6	7	11
3	↓	2	┬	✕	12
2	4	┬		14	✕
1	✥	7	8	✕	15
	f	g	h	i	j

The start square is at c3 = ☐.

c3 + ✥ = __ + __ = ___   c3 + ✕ = __ + __ = ___

☐ + ✥ = __ + __ = ___   ☐ + ✕ = __ + __ = ___

c3 + ✥ = __ + __ = ___   c3 + ✕ = __ + __ = ___

☐ + ✥ = __ + __ = ___   ☐ + ✕ = __ + __ = ___

10, 9
14, 6
12, 7
8, 8

Ho Math Chess 何数棋谜 妈！我会棋谜式加法啦！

Mom! I Learn Addition Using Math-Chess-Puzzles Connection

Contents include both traditional and Math-Chess-Puzzles combined methods. Extra strength

©2008 – 2018 Frank Ho, Amanda Ho   All rights reserved. www.homathchess.com

Student Name _____ Date _____

Computing additions through math and chess integrated puzzle

5				g4	
4	g3	┼	┼	┼	f3
3	g1	┼	✕	┼	f1
2	g2	┼	┼	┼	f2
1			f4		
	a	b	c	d	e

5	✕	5	✻	9	✻
4	✻	✕	6	7	11
3	↓	2	┼	✕	12
2	4	┼		14	✕
1	✚	7	8	✕	15
	f	g	h	i	j

The start square is at c3 = ☐.

c3 + ✚ = ___ + ___ = ___

c3 + ✕ = ___ + ___ = ___

☐ + ✚ = ___ + ___ = ___

☐ + ✕ = ___ + ___ = ___

c3 + ✚ = ___ + ___ = ___

c3 + ✕ = ___ + ___ = ___

☐ + ✚ = ___ + ___ = ___

☐ + ✕ = ___ + ___ = ___

8, 7
12, 4
10, 5
6, 6

Mom! I Learn Addition Using Math-Chess-Puzzles Connection

Contents include both traditional and Math-Chess-Puzzles combined methods. Extra strength
©2008 − 2018 Frank Ho, Amanda Ho   All rights reserved. www.homathchess.com

Student Name _____ Date _____

Computing additions through math and chess integrated puzzle

The start square is at c3 = ☐.

c3 + ✥ = ___ + ___ = ___

☐ + ✥ = ___ + ___ = ___

c3 + ✥ = ___ + ___ = ___

☐ + ✥ = ___ + ___ = ___

☐ + ✥ = ___ + ___ = ___

c3 + ✕ = ___ + ___ = ___

☐ + ✕ = ___ + ___ = ___

c3 + ✕ = ___ + ___ = ___

☐ + ✕ = ___ + ___ = ___

☐ + ✕ = ___ + ___ = ___

14, 14
18, 10
16, 11
12, 12
12, 12

100 table and pattern

10	10	20	30	40	50	60	70	80	90	100
9	9	18	27	36	45	54	63	72	81	90
8	8	16	24	32	40	48	56	64	72	80
7	7	14	21	28	35	42	49	56	63	70
6	6	12	18	24	30	36	42	48	54	60
5	5	10	15	20	25	30	35	40	45	50
4	4	8	12	16	20	24	28	32	36	40
3	3	6	9	12	15	18	21	24	27	30
2	2	4	6	8	10	12	14	16	18	20
1	1	2	3	4	5	6	7	8	9	10
	a	b	c	d	e	f	g	h	i	j

The pattern rule of b1 to b10 is _____.

Observe one's place; the core of the pattern is Subtract at 2 and add 2 each time.

The pattern rule of c1 to c10 is _____.

Start at 3 and then count by 2.

The pattern rule of d1 to d10 is _____.

Observe one's place; the core of the pattern is Start at 4 and then count (add) by 4.

The pattern rule of e1 to e10 is _____.

Start at 5 and then count by 5.

The pattern rule of f1 to f10 is_____.

Start at 6 and then count by 6.

The pattern rule of g1 to g10 is _____.

Start at 7 and then count by 7.

The pattern rule of h1 to h10 is _____.

Start at 8 and then count by 8.

Ho Math Chess 何数棋谜   妈！我会棋谜式加法啦！
Mom! I Learn Addition Using Math-Chess-Puzzles Connection
Contents include both traditional and Math-Chess-Puzzles combined methods. Extra strength
©2008 — 2018 Frank Ho, Amanda Ho   All rights reserved. www.homathchess.com

Student Name _____ Date _____

The pattern rule of i1 to i10 is _____.

Start at 9 then count by 9.

The pattern rule of j1 to j10 is _____.

Start at 10 and then add by 10.

Ho Math Chess 何数棋谜　妈！我会棋谜式加法啦！
Mom! I Learn Addition Using Math-Chess-Puzzles Connection
Contents include both traditional and Math-Chess-Puzzles combined methods. Extra strength
©2008 — 2018 Frank Ho, Amanda Ho   All rights reserved. www.homathchess.com

Student Name _____ Date _____

100 table and pattern

10	91	92	93	94	95	96	97	98	99	100
9	81	82	83	84	85	86	87	88	89	90
8	71	72	73	74	75	76	77	78	79	80
7	61	62	63	64	65	66	67	68	69	70
6	51	52	53	54	55	56	57	58	59	60
5	41	42	43	44	45	46	47	48	49	50
4	31	32	33	34	35	36	37	38	39	40
3	21	22	23	24	25	26	27	28	29	30
2	11	12	13	14	15	16	17	18	19	20
1	1	2	3	4	5	6	7	8	9	10
	a	b	c	d	e	f	g	h	i	j

Start at f7 ⤫, the pattern rule is to subtract ____ to get the next number.

Start at f7 ⤫, the pattern rule is to subtract ____ to get the next number.

Start at e1 ⤫, the pattern rule is to add ____ to get the next number.

Start at e1 ⤫, the pattern rule is to add ____ to get the next number.

Start at a5 ✛, the pattern rule is to add ____ to get the next number.

Start at i1 ✛, the pattern rule is to add ____ to get the next number.

11
9
9

11
1
10

Ho Math Chess 何数棋谜   妈！我会棋谜式加法啦！
Mom! I Learn Addition Using Math-Chess-Puzzles Connection
Contents include both traditional and Math-Chess-Puzzles combined methods. Extra strength
©2008 – 2018 Frank Ho, Amanda Ho   All rights reserved. www.homathchess.com

Student Name _____ Date _____

100 table and pattern

10	91	92	93	94	95	96	97	98	99	100
9	81	82	83	84	85	86	87	88	89	90
8	71	72	73	74	75	76	77	78	79	80
7	61	62	63	64	65	66	67	68	69	70
6	51	52	53	54	55	56	57	58	59	60
5	41	42	43	44	45	46	47	48	49	50
4	31	32	33	34	35	36	37	38	39	40
3	21	22	23	24	25	26	27	28	29	30
2	11	12	13	14	15	16	17	18	19	20
1	1	2	3	4	5	6	7	8	9	10
	a	b	c	d	e	f	g	h	i	j

Start at a1, count every 2, you will get _____.

Start at b1, count every 2, you will get _____.

Start at g1, move along, you will get _____.

Start at e1, move along, you will get _____.

Start at d10, move along, you will get _____.

3, 5, 7, 9
4, 6, 8, 10
28
15
75

Mom! I Learn Addition Using Math-Chess-Puzzles Connection

Contents include both traditional and Math-Chess-Puzzles combined methods. Extra strength

©2008 – 2018 Frank Ho, Amanda Ho   All rights reserved. www.homathchess.com

Student Name _____ Date _____

Using dd + d or d + dd adding up to 10 concept using chess moves

3	5	15	35
2	25	5	45
1	65	75	55
	a	b	c

You are a chess piece and located at b2.

The original square is at b2.	Horizontal from	Vertical form	The original square is at b2.	Horizontal from	Vertical form
b2+ ↕↔	5 + 15 = 20	5 + 15 20	↕↔ +b2	15 + 5 = 20	15 + 5 20
b2+ ↕↔	5 + 75 = 80	5 + 75 80	↕↔ +b2	75 + 5 = 80	75 + 5 80

Mom! I Learn Addition Using Math-Chess-Puzzles Connection

Using adding up to 10 concept

3	5	15	35
2	25	5	45
1	65	75	55
	a	b	c

You are a chess piece and located at b2.

The original square is at b2.	Horizontal from	Vertical form	The original square is at b2.	Horizontal from	Vertical form
b2+ (↔↕)			(↔↕) +b2		
b2+ (↔)			(↔) +b2		

30
80

Mom! I Learn Addition Using Math-Chess-Puzzles Connection

Contents include both traditional and Math-Chess-Puzzles combined methods. Extra strength
©2008 – 2018 Frank Ho, Amanda Ho   All rights reserved. www.homathchess.com

Student Name _____ Date _____

Using adding up to 10 concept

3	5	15	35
2	25	5	45
1	65	75	55
	a	b	c

You are a chess piece and located at b2.

The original square is at b2.	Horizontal from	Vertical form	The original square is at b2.	Horizontal from	Vertical form
b2+✕			✕+b2		
b2+✕			✕+b2		

10
60

Mom! I Learn Addition Using Math-Chess-Puzzles Connection

Contents include both traditional and Math-Chess-Puzzles combined methods. Extra strength
©2008 − 2018 Frank Ho, Amanda Ho   All rights reserved. www.homathchess.com

Student Name _____ Date _____

Using adding up to 10 concept

3	5	15	35
2	25	5	45
1	65	75	55
	a	b	c

You are a chess piece and located at b2.

The original square is at b2.	Horizontal from	Vertical form	The original square is at b2.	Horizontal from	Vertical form
b2+✕			✕+b2		
b2+✕			✕+b2		

40
70

Ho Math Chess 何数棋谜 妈！我会棋谜式加法啦！
Mom! I Learn Addition Using Math-Chess-Puzzles Connection

Contents include both traditional and Math-Chess-Puzzles combined methods. Extra strength
©2008 – 2018 Frank Ho, Amanda Ho   All rights reserved. www.homathchess.com

Student Name _____ Date _____

Using adding up to 10 concept

3	7	17	37
2	27	3	47
1	67	77	57
	a	b	c

You are a chess piece and located at b2.

The original square is at b2.	Horizontal from	Vertical form	The original square is at b2.	Horizontal from	Vertical form
b2+			+b2		
b2+			+b2		

20
80

Using adding up to 10 concept

3	7	17	37
2	27	3	47
1	67	77	57
	a	b	c

You are a chess piece and located at b2.

The original square is at b2.	Horizontal from	Vertical form	The original square is at b2.	Horizontal from	Vertical form
b2+			+b2		
b2+			+b2		

30
50

398

Ho Math Chess 何数棋谜   妈！我会棋谜式加法啦！
Mom! I Learn Addition Using Math-Chess-Puzzles Connection
Contents include both traditional and Math-Chess-Puzzles combined methods. Extra strength
©2008 – 2018 Frank Ho, Amanda Ho   All rights reserved. www.homathchess.com

Student Name _____ Date _____

Using adding up to 10 concept

3	7	17	37	You are a chess piece and located at b2.
2	27	3	47
1	67	77	57
	a	b	c

The original square is at b2.	Horizontal from	Vertical form	The original square is at b2.	Horizontal from	Vertical form
b2+⤫			⤫+b2		
b2+⤫			⤫+b2		

10
60

399

Using adding up to 10 concept

3	7	17	37
2	27	3	47
1	67	77	57
	a	b	c

You are a chess piece and located at b2.

The original square is at b2.	Horizontal from	Vertical form	The original square is at b2.	Horizontal from	Vertical form
b2+✕			✕ +b2		
b2+✕			✕ +b2		

40
70

Mom! I Learn Addition Using Math-Chess-Puzzles Connection

Contents include both traditional and Math-Chess-Puzzles combined methods. Extra strength

©2008 – 2018 Frank Ho, Amanda Ho   All rights reserved. www.homathchess.com

Student Name _____ Date _____

Using adding up to 10 concept

3	6	16	36
2	26	4	46
1	66	76	56
	a	b	c

You are a chess piece and located at b2.

The original square is at b2.	Horizontal from	Vertical form	The original square is at b2.	Horizontal from	Vertical form
b2+ ⇕			⇕ +b2		
b2+ ⇕			⇕ +b2		

20
80

Ho Math Chess 何数棋谜   妈！我会棋谜式加法啦！

Mom! I Learn Addition Using Math-Chess-Puzzles Connection

Contents include both traditional and Math-Chess-Puzzles combined methods. Extra strength

©2008 – 2018 Frank Ho, Amanda Ho   All rights reserved. www.homathchess.com

Student Name _____   Date _____

Using adding up to 10 concept

3	6	16	36	You are a chess piece and located at b2.
2	26	4	46	
1	66	76	56	
	a	b	c	

The original square is at b2.	Horizontal from	Vertical form	The original square is at b2.	Horizontal from	Vertical form
b2+			+b2		
b2+			+b2		

30
50

Ho Math Chess 何数棋谜　妈！我会棋谜式加法啦！
Mom! I Learn Addition Using Math-Chess-Puzzles Connection

Contents include both traditional and Math-Chess-Puzzles combined methods. Extra strength
©2008 − 2018 Frank Ho, Amanda Ho　All rights reserved. www.homathchess.com

Student Name _____　Date_____

Using adding up to 10 concept

3	6	16	36	You are a chess piece and located at b2.
2	26	4	46	
1	66	76	56	
	a	b	c	

The original square is at b2.	Horizontal from	Vertical form	The original square is at b2.	Horizontal from	Vertical form
b2+✕			✕+b2		
b2+✕			✕+b2		

10
60

403

Mom! I Learn Addition Using Math-Chess-Puzzles Connection

Using adding up to 10 concept

3	6	16	36	You are a chess piece and located at b2.
2	26	4	46	
1	66	76	56	
	a	b	c	

The original square is at b2.	Horizontal from	Vertical form	The original square is at b2.	Horizontal from	Vertical form
b2+ ✕			✕ +b2		
b2+ ✕			✕ +b2		

40
70

Ho Math Chess 何数棋谜 妈！我会棋谜式加法啦！
Mom! I Learn Addition Using Math-Chess-Puzzles Connection
Contents include both traditional and Math-Chess-Puzzles combined methods. Extra strength
©2008 – 2018 Frank Ho, Amanda Ho   All rights reserved. www.homathchess.com

Student Name _____ Date _____

Adding up to 10

3	6	16	36
2	26	4	46
1	66	76	56
	a	b	c

You are a chess piece and located at b2.

The original square is at b2.	Horizontal from	Vertical form	The original square is at b2.	Horizontal from	Vertical form
b2+			+b2		
b2+			+b2		

20
80

Mom! I Learn Addition Using Math-Chess-Puzzles Connection

Contents include both traditional and Math-Chess-Puzzles combined methods. Extra strength
©2008 − 2018 Frank Ho, Amanda Ho   All rights reserved. www.homathchess.com

Student Name _____   Date _____

Using adding up to 10 concept

3	8	18	38
2	28	2	48
1	68	78	58
	a	b	c

You are a chess piece and located at b2.

The original square is at b2.	Horizontal from	Vertical form	The original square is at b2.	Horizontal from	Vertical form
b2+ ✥			✥ +b2		
b2+ ✥			✥ +b2		

30
50

Ho Math Chess 何数棋谜　妈！我会棋谜式加法啦！
Mom! I Learn Addition Using Math-Chess-Puzzles Connection

Contents include both traditional and Math-Chess-Puzzles combined methods. Extra strength

©2008 − 2018 Frank Ho, Amanda Ho   All rights reserved. www.homathchess.com

Student Name _____ Date _____

Using adding up to 10 concept

3	8	18	38
2	28	2	48
1	68	78	58
	a	b	c

You are a chess piece and located at b2.

The original square is at b2.	Horizontal from	Vertical form	The original square is at b2.	Horizontal from	Vertical form
b2+⤢	2+8=10	2 + 8 10	⤢+b2	8+2=10	8 + 2 10
b2+⤢	2+58=60	2 + 58 60	⤢+b2	58+2=60	58 + 2 60

Ho Math Chess 何数棋谜   妈！我会棋谜式加法啦！
Mom! I Learn Addition Using Math-Chess-Puzzles Connection
Contents include both traditional and Math-Chess-Puzzles combined methods. Extra strength
©2008 – 2018 Frank Ho, Amanda Ho   All rights reserved. www.homathchess.com

Student Name _____ Date _____

Using adding up to 10 concept

3	8	18	38
2	28	2	48
1	68	78	58
	a	b	c

You are a chess piece and located at b2.

The original square is at b2.	Horizontal from	Vertical form	The original square is at b2.	Horizontal from	Vertical form
b2+✕	40		✕ +b2		
b2+✕	70		✕ +b2		

408

Ho Math Chess 何数棋谜　妈！我会棋谜式加法啦！
Mom! I Learn Addition Using Math-Chess-Puzzles Connection
Contents include both traditional and Math-Chess-Puzzles combined methods. Extra strength
©2008 — 2018 Frank Ho, Amanda Ho   All rights reserved. www.homathchess.com

Student Name _____ Date _____

Using adding up to 10 concept

3	9	19	39	You are a chess piece and located at b2.
2	29	1	49	
1	66	79	59	
	a	b	c	

The original square is at b2.	Horizontal from	Vertical form	The original square is at b2.	Horizontal from	Vertical form
b2+ ✥	20		✥ +b2		
b2+ ✥	80		✥ +b2		

Using adding up to 10 concept

3	9	19	39
2	29	1	49
1	66	79	59
	a	b	c

You are a chess piece and located at b2.

The original square is at b2.	Horizontal from	Vertical form	The original square is at b2.	Horizontal from	Vertical form
b2+	30		+b2		
b2+	50		+b2		

410

Mom! I Learn Addition Using Math-Chess-Puzzles Connection

Contents include both traditional and Math-Chess-Puzzles combined methods. Extra strength
©2008 – 2018 Frank Ho, Amanda Ho   All rights reserved. www.homathchess.com

Student Name _____ Date _____

Using adding up to 10 concept

3	9	19	39	You are a chess piece and located at b2.
2	29	1	49	
1	66	79	59	
	a	b	c	

The original square is at b2.	Horizontal from	Vertical form	The original square is at b2.	Horizontal from	Vertical form
b2+✕	10		✕+b2		
b2+✕	60		✕+b2		

411

Ho Math Chess 何数棋谜　妈！我会棋谜式加法啦！
Mom! I Learn Addition Using Math-Chess-Puzzles Connection

Contents include both traditional and Math-Chess-Puzzles combined methods. Extra strength
©2008 – 2018 Frank Ho, Amanda Ho   All rights reserved. www.homathchess.com

Student Name _____ Date _____

Using adding up to 10 concept

3	9	19	39
2	29	1	49
1	66	79	59
	a	b	c

You are a chess piece and located at b2.

The original square is at b2.	Horizontal from	Vertical form	The original square is at b2.	Horizontal from	Vertical form
b2+✕	40		✕ +b2		
b2+✕	67		✕ +b2		

412

Ho Math Chess 何数棋谜   妈！我会棋谜式加法啦！
Mom! I Learn Addition Using Math-Chess-Puzzles Connection
Contents include both traditional and Math-Chess-Puzzles combined methods. Extra strength
©2008 – 2018 Frank Ho, Amanda Ho   All rights reserved. www.homathchess.com

Student Name _____ Date _____

Addition using chess moves

e	2	2	3	4	7
d	6	4	8	9	0
c	1	2	2	3	4
b	5	6	7	6	9
a	0	1	2	3	8
	1	2	3	4	5

Move by one square at a time from the original square.

The original square is at c3.

c3 + ✥ = ____ + ____ = ____   2 + 3 = 5

The original square is at c3.

c3 + ✥ = ____ + ____ = ____ 10

The original square is at c3.

c3 + ✥ = ____ + ____ = ____ 4

The original square is at c3.

c3 + ✥ = ____ + ____ = ____ 9

Ho Math Chess 何数棋谜 妈！我会棋谜式加法啦！
Mom! I Learn Addition Using Math-Chess-Puzzles Connection

Contents include both traditional and Math-Chess-Puzzles combined methods. Extra strength
©2008 — 2018 Frank Ho, Amanda Ho   All rights reserved. www.homathchess.com

Student Name _____ Date _____

Addition

e	1	2	3	4	5
d	6	7	8	9	0
c	1	2	5	3	4
b	5	6	7	8	9
a	0	1	2	3	4
	1	2	3	4	5

Move by one square at a time from the original square.

The original square is at c3.

C3 + ✥ + ✥ = ____ + ____ + ____ = _____ 12

The original square is at c3.

C3 + ✥ + ✥ = ____ + ____ + ____ = _____ 16

The original square is at c3.

C3 + ✥ + ✥ = ____ + ____ + ____ = _____ 8

The original square is at c3.

C3 + ✥ + ✥ = ____ + ____ + ____ = _____ 14

Mom! I Learn Addition Using Math-Chess-Puzzles Connection

Addition

e	1	2	3	4	5
d	6	7	8	19	2
🍦	1	2	7	3	4
b	5	6	7	8	9
a	3	3	3	3	4
	1	2	3	4	5

Move by one square at a time from the original square.

The original square is at c3.

🍦₃ + ⬌ + ⬍ +

= __7__ + __3__ + __4__ = 14_____

The original square is at c3.

C3 + ⬆ + ⬇ = ____ + ____ + ____ = _7+8+3=18_____

The original square is at c3.

C3 + ⬆ + ⬇ = ____ + ____ + ____ = _____
7+2+1=10

The original square is at c3.

C3 + ⬌ + ⬍ = ____ + ____ + ____ = __7+7+3=17_____

Ho Math Chess 何数棋谜   妈！我会棋谜式加法啦！
Mom! I Learn Addition Using Math-Chess-Puzzles Connection

Contents include both traditional and Math-Chess-Puzzles combined methods. Extra strength
©2008 – 2018 Frank Ho, Amanda Ho   All rights reserved. www.homathchess.com

Student Name _____   Date _____

Addition

e	11	12	13	14	15
d	16	17	18	19	20
c	21	22	2	23	24
b	25	26	27	28	29
a	30	31	32	33	34
	1	2	3	4	5

Move by one square at a time from the original square.

The original square is at c3.

c3 + ✕ = ____ + ____ = ____

The original square is at c3.

c3 + ✕ = ____ + ____ = ____

The original square is at c3.

c3 + ✕ = ____ + ____ = ____

The original square is at c3.

c3 + ✕ = ____ + ____ = ____

21, 30, 28, 19

Mom! I Learn Addition Using Math-Chess-Puzzles Connection

Contents include both traditional and Math-Chess-Puzzles combined methods. Extra strength
©2008 – 2018 Frank Ho, Amanda Ho   All rights reserved. www.homathchess.com

Student Name _____ Date _____

Addition

e	11	12	13	14	15
d	16	17	18	19	20
c	21	22	2	23	24
b	25	26	27	28	29
a	30	31	32	33	34
	1	2	3	4	5

Move by one square at a time from the original square.

The original square is at c3.

c3 + ✥ = ____ + ____ = _____

The original square is at c3.

c3 + ⬌ = ____ + ____ = _____

The original square is at c3.

c3 + ⬍ = ____ + ____ = _____

The original square is at c3.

c3 + ⬍ = ____ + ____ = _____

25, 20, 24, 29

Addition

e	11	12	13	14	15
d	16	17	18	19	20
c	21	22	2	23	24
b	25	26	27	28	29
a	30	31	32	33	34
	1	2	3	4	5

Move by one square at a time from where you are.

The original square is at c3.

C3 + ⬌ + ⬌ = ____ + ____ + ____ = ____

The original square is at c3.

C3 + ⬌ + ⬌ = ____ + ____ + ____ = ____

The original square is at c3.

C3 + ⬌ + ⬌ = ____ + ____ + ____ = ____

The original square is at c3.

C3 + ⬌ + ⬌ = ____ + ____ + ____ = ____

49,33,45,61

Addition

e	11	12	13	14	15
d	16	17	18	19	20
🍦	21	22	2	23	24
b	25	26	27	28	29
a	30	31	32	33	34
	1	2	3	4	5

Move by one square at a time from where you are.

The original square is at c3.

🍦₃ + ⬌ + ⬌

= ____ + ____ + ____ = ____

The original square is at c3.

c3 + ⬍ + ⬍ = ____ + ____ + ____ = ____

The original square is at c3.

c3 + ⬌ + ⬍ = ____ + ____ + ____ = ____

The original square is at c3.

c3 + ⬍ + ⬍ = ____ + ____ + ____ = ____

49, 33, 45, 61

Addition

e	11	12	13	14	15
d	16	17	18	19	20
c	21	22	2	23	24
b	25	26	27	28	29
a	30	31	32	33	34
	1	2	3	4	5

Move by one square at a time from the original square.

The original square is at c3.

c3 + ↗↙ = ____ + _____ = _____

The original square is at c3.

c3 + ↖↘ = ____ + _____ = _____

The original square is at c3.

c3 + ↗↘ = ____ + _____ = _____

The original square is at c3.

c3 + ↖↙ = ____ + _____ = _____

21, 30, 28, 19

Addition

e	11	12	13	14	15
d	16	17	18	19	20
c	21	22	2	23	24
b	25	26	27	28	29
a	30	31	32	33	34
	1	2	3	4	5

Move by one square at a time from the original square.

The original square is at c3.

c3 + ↖↘ + ↗↙ = ___ + ___ + ___ = ___

The original square is at c3.

c3 + ↖↘ + ↗↙ = ___ + ___ + ___ = ___

The original square is at c3.

c3 + ↖↘ + ↗↙ = ___ + ___ + ___ = ___

The original square is at c3.

c3 + ↖↘ + ↗↙ = ___ + ___ + ___ = ___

36, 64, 58, 30

Addition

e	11	12	13	14	15
d	16	17	18	19	20
c	21	22	2	23	24
b	25	26	27	28	29
a	30	31	32	33	34
	1	2	3	4	5

Move by one square at a time from the original square.

The original square is at c3.

c3 + ↗↙ + ↖↘ = ____ + ____ + ____ = ____

The original square is at c3.

c3 + ↗↙ + ↖↘ = ____ + ____ + ____ = ____

The original square is at c3.

c3 + ↗↙ + ↖↘ = ____ + ____ + ____ = ____

The original square is at c3.

c3 + ↗↙ + ↖↘ = ____ + ____ + ____ = ____

36, 64, 58, 30

Ho Math Chess 何数棋谜　妈！我会棋谜式加法啦！
Mom! I Learn Addition Using Math-Chess-Puzzles Connection

Contents include both traditional and Math-Chess-Puzzles combined methods. Extra strength
©2008 – 2018 Frank Ho, Amanda Ho   All rights reserved. www.homathchess.com

Student Name _____ Date _____

Addition

e	11	12	13	14	15
d	16	17	18	19	20
c	21	22	2	23	24
b	25	26	27	28	29
a	30	31	32	33	34
	1	2	3	4	5

Move by one square at a time from the original square.

The original square is at c3.

C3 + ┼ = _____ + _____ = _____

The original square is at c3.

C3 + ┼ = _____ + _____ = _____

The original square is at c3.

C3 + ┼ = _____ + _____ = _____

The original square is at c3.

C3 + ┼ = _____ + _____ = _____

14, 22, 31, 35

Addition

e	11	12	13	14	15
d	16	17	18	19	20
c	21	22	2	23	24
b	25	26	27	28	29
a	30	31	32	33	34
	1	2	3	4	5

Move by one square at a time from the original square.

The original square is at c3.

C3 + ⊤ + ⊤ = ____ + ____ + ____ = ____

The original square is at c3.

C3 + ⊤ + ⊤ = ____ + ____ + ____ = ____

The original square is at c3.

C3 + ⊤ + ⊤ = ____ + ____ + ____ = ____

The original square is at c3.

C3 + ⊤ + ⊤ = ____ + ____ + ____ = ____

45, 53, 62, 51

Addition

e	11	12	13	14	15
d	16	17	18	19	20
c	21	22	2	23	24
b	25	26	27	28	29
a	30	31	32	33	34
	1	2	3	4	5

Move by one square at a time from the original square.

The original square is at c3.

C3 + ┼ = ____ + ____ = ____

The original square is at c3.

C3 + ┼ = ____ + ____ = ____

The original square is at c3.

C3 + ┼ = ____ + ____ = ____

The original square is at c3.

C3 + ┼ = ____ + ____ = ____

33, 27, 18, 14

Addition

e	11	12	13	14	15
d	16	17	18	19	20
c	21	22	2	23	24
b	25	26	27	28	29
a	30	31	32	33	34
	1	2	3	4	5

Move by one square at a time from the original square.

The original square is at c3.

c3 + [plus-shape] + [plus-shape] = ____ + ____ + ____ = ____

The original square is at c3.

c3 + [plus-shape] + [move-arrows] = ____ + ____ + ____ = ____

The original square is at c3.

c3 + [plus-shape] + [plus-shape] = ____ + ____ + ____ = ____

The original square is at c3.

c3 + [plus-shape] + [plus-shape] = ____ + ____ + ____ = ____

49, 50, 32, 39

Addition

e	11	12	13	14	15
d	16	17	18	19	20
c	21	22	3	23	24
b	25	26	27	28	29
a	30	31	32	33	34
	1	2	3	4	5

Move by one square at a time from the original square.

The original square is at c3.

C3 + ✥ = _____ + _____ = _____

The original square is at c3.

C3 + ⇔ = _____ + _____ = _____

The original square is at c3.

C3 + ⇔ = _____ + _____ = _____

The original square is at c3.

C3 + ⇔ = _____ + _____ = _____

26, 21, 25, 30

Addition

e	11	12	13	14	15
d	16	17	18	19	20
c	21	22	3	23	24
b	25	26	27	28	29
a	30	31	32	33	34
	1	2	3	4	5

Move by one square at a time from the original square.

The original square is at c3.

c3 + ⬌(↕) + ⬌(↓) = ____ + ____ + ____ = ____

The original square is at c3.

c3 + ⚓(↔) + ⚓(↔) = ____ + ____ + ____ = ____

The original square is at c3.

c3 + ⬌(⚓) + ⬌(⚓) = ____ + ____ + ____ = ____

The original square is at c3.

c3 + ⬌(⚓↓) + ⬌(⚓↓) = ____ + ____ + ____ = ____

50, 34, 46, 62

Mom! I Learn Addition Using Math-Chess-Puzzles Connection

Contents include both traditional and Math-Chess-Puzzles combined methods. Extra strength

©2008 – 2018 Frank Ho, Amanda Ho   All rights reserved. www.homathchess.com

Student Name _____ Date _____

Addition

e	11	12	13	14	15
d	16	17	18	19	20
c	21	22	3	23	24
b	25	26	27	28	29
a	30	31	32	33	34
	1	2	3	4	5

Move by one square at a time from the original square.

The original square is at c3.

c3 + ✥ + ✥ = ____ + ____ + ____ = ____ 50

The original square is at c3.

c3 + ✥ + ✥ = ____ + ____ + ____ = ____ 34

The original square is at c3.

c3 + ✥ + ✥ = ____ + ____ + ____ = ____ 46

The original square is at c3.

c3 + ✥ + ✥ = ____ + ____ + ____ = ____ 62

Addition

e	11	12	13	14	15
d	16	17	18	19	20
c	21	22	3	23	24
b	25	26	27	28	29
a	30	31	32	33	34
	1	2	3	4	5

Move by one square at a time from the original square.

The original square is at c3.

c3 + ⤢⤡ = _____ + _____ = _____ 22

The original square is at c3.

c3 + ⤢⤡ = _____ + _____ = _____ 31

The original square is at c3.

c3 + ⤢⤡ = _____ + _____ = _____ 29

The original square is at c3.

c3 + ⤢⤡ = _____ + _____ = _____ 20

Addition

e	11	12	13	14	15
d	16	17	18	19	20
c	21	22	3	23	24
b	25	26	27	28	29
a	30	31	32	33	34
	1	2	3	4	5

Move by one square at a time from the original square.

The original square is at c3.

c3 + ⤢ + ⤢ = ____ + ____ + ____ = ____ 37

The original square is at c3.

c3 + ⤢ + ⤢ = ____ + ____ + ____ = ____ 65

The original square is at c3.

c3 + ⤢ + ⤢ = ____ + ____ + ____ = ____ 59

The original square is at c3.

c3 + ⤢ + ⤢ = ____ + ____ + ____ = ____ 31

Addition

e	11	12	13	14	15
d	16	17	18	19	20
c	21	22	3	23	24
b	25	26	27	28	29
a	30	31	32	33	34
	1	2	3	4	5

Move by one square at a time from the original square.

The original square is at c3.

c3 + ✕ + ✕ = ____ + ____ + ____ = ____ 37

The original square is at c3.

c3 + ✕ + ✕ = ____ + ____ + ____ = ____ 65

The original square is at c3.

c3 + ✕ + ✕ = ____ + ____ + ____ = ____ 59

The original square is at c3.

c3 + ✕ + ✕ = ____ + ____ + ____ = ____ 31

Addition

e	11	12	13	14	15
d	16	17	18	19	20
c	21	22	3	23	24
b	25	26	27	28	29
a	30	31	32	33	34
	1	2	3	4	5

Move by one square at a time from the original square.

The original square is at c3.

C3 + ✛ = _____ + _____ = _____ 17

The original square is at c3.

C3 + ✛ = _____ + _____ = _____ 23

The original square is at c3.

C3 + ✛ = _____ + _____ = _____ 32

The original square is at c3.

C3 + ✛ = _____ + _____ = _____ 34

433

Addition

e	11	12	13	14	15
d	16	17	18	19	20
c	21	22	3	23	24
b	25	26	27	28	29
a	30	31	32	33	34
	1	2	3	4	5

Move by one square at a time from the original square.

The original square is at c3.

c3 + ⊕ = _____ + _____ = _____ 34

The original square is at c3.

c3 + ⊕ = _____ + _____ = _____ 28

The original square is at c3.

c3 + ⊕ = _____ + _____ = _____ 19

The original square is at c3.

c3 + ⊕ = _____ + _____ = _____ 15

Addition

e	11	12	13	14	15
d	16	17	18	19	20
c	21	22	4	23	24
b	25	26	27	28	29
a	30	31	32	33	34
	1	2	3	4	5

Move by one square at a time from the original square.

The original square is at c3.

C3 + ✥ = ____ + ____ = _____ 27

The original square is at c3.

C3 + ⇩ = ____ + ____ = _____ 22

The original square is at c3.

C3 + ⇔ = ____ + ____ = _____ 26

The original square is at c3.

C3 + ⇔ = ____ + ____ = _____ 31

Addition

e	11	12	13	14	15
d	16	17	18	19	20
c	21	22	2	23	24
b	25	26	27	28	29
a	30	31	32	33	34
	1	2	3	4	5

Move by one square at a time from the original square.

The original square is at c3.

c3 + ✥ + ✥ = ____ + ____ + ____ = ____ 49

The original square is at c3.

c3 + ⚓ + ⚓ = ____ + ____ + ____ = ____ 33

The original square is at c3.

c3 + ✥ + ✥ = ____ + ____ + ____ = ____ 45

The original square is at c3.

c3 + ⚓ + ⚓ = ____ + ____ + ____ = ____ 56

Mom! I Learn Addition Using Math-Chess-Puzzles Connection

Contents include both traditional and Math-Chess-Puzzles combined methods. Extra strength
©2008 – 2018 Frank Ho, Amanda Ho   All rights reserved. www.homathchess.com

Student Name _____   Date _____

Addition

e	11	12	13	14	15
d	16	17	18	19	20
c	21	22	4	23	24
b	25	26	27	28	29
a	30	31	32	33	34
	1	2	3	4	5

Move by one square at a time from the original square.

The original square is at c3.

c3 + ✥ + ✥ = ____ + ____ + ____ = _____ 51

The original square is at c3.

c3 + ✥ + ✥ = ____ + ____ + ____ = _____ 35

The original square is at c3.

c3 + ✥ + ✥ = ____ + ____ + ____ = _____ 47

The original square is at c3.

c3 + ✥ + ✥ = ____ + ____ + ____ = _____ 63

Mom! I Learn Addition Using Math-Chess-Puzzles Connection

Contents include both traditional and Math-Chess-Puzzles combined methods. Extra strength

©2008 – 2018 Frank Ho, Amanda Ho   All rights reserved. www.homathchess.com

Student Name _____   Date _____

Addition

e	11	12	13	14	15
d	16	17	18	19	20
c	21	22	4	23	24
b	25	26	27	28	29
a	30	31	32	33	34
	1	2	3	4	5

Move by one square at a time from the original square.

The original square is at c3.

c3 + ✕ = ____ + ____ = _____ 23

The original square is at c3.

c3 + ✕ = ____ + ____ = _____ 32

The original square is at c3.

c3 + ✕ = ____ + ____ = _____ 30

The original square is at c3.

c3 + ✕ = ____ + ____ = _____ 21

Ho Math Chess 何数棋谜 妈！我会棋谜式加法啦！
Mom! I Learn Addition Using Math-Chess-Puzzles Connection

Contents include both traditional and Math-Chess-Puzzles combined methods. Extra strength

©2008 – 2018 Frank Ho, Amanda Ho   All rights reserved. www.homathchess.com

Student Name _____   Date _____

Addition

e	11	12	13	14	15
d	16	17	18	19	20
c	21	22	4	23	24
b	25	26	27	28	29
a	30	31	32	33	34
	1	2	3	4	5

Move by one square at a time from the original square.

The original square is at c3.

c3 + ⤢⤡ + ⤢⤡ = ____ + ____ + ____ = _____ 38

The original square is at c3.

c3 + ⤢⤡ + ⤢⤡ = ____ + ____ + ____ = _____ 66

The original square is at c3.

c3 + ⤢⤡ + ⤢⤡ = ____ + ____ + ____ = _____ 60

The original square is at c3.

c3 + ⤢⤡ + ⤢⤡ = ____ + ____ + ____ = _____ 32

Ho Math Chess 何数棋谜   妈！我会棋谜式加法啦！
Mom! I Learn Addition Using Math-Chess-Puzzles Connection

Contents include both traditional and Math-Chess-Puzzles combined methods. Extra strength
©2008 – 2018 Frank Ho, Amanda Ho   All rights reserved. www.homathchess.com

Student Name _____   Date _____

Addition

e	11	12	13	14	15
d	16	17	18	19	20
c	21	22	4	23	24
b	25	26	27	28	29
a	30	31	32	33	34
	1	2	3	4	5

Move by one square at a time from the original square.

The original square is at c3.

c3 + ⤢ + ⤢ = ____ + ____ + ____ = _____ 38

The original square is at c3.

c3 + ⤢ + ⤢ = ____ + ____ + ____ = _____ 66

The original square is at c3.

c3 + ⤢ + ⤢ = ____ + ____ + ____ = _____ 60

The original square is at c3.

c3 + ⤢ + ⤢ = ____ + ____ + ____ = _____ 33

Ho Math Chess 何数棋谜　妈！我会棋谜式加法啦！
Mom! I Learn Addition Using Math-Chess-Puzzles Connection

Contents include both traditional and Math-Chess-Puzzles combined methods. Extra strength
©2008 — 2018 Frank Ho, Amanda Ho　　All rights reserved. www.homathchess.com

Student Name _____ Date _____

Addition

e	11	12	13	14	15
d	16	17	18	19	20
c	21	22	4	23	24
b	25	26	27	28	29
a	30	31	32	33	34
	1	2	3	4	5

Move by one square at a time from the original square.

The original square is at c3.

c3 + ╋ = _____ + _____ = _____ 18

The original square is at c3.

c3 + ╋ = _____ + _____ = _____ 24

The original square is at c3.

c3 + ╋ = _____ + _____ = _____ 33

The original square is at c3.

c3 + ╋ = _____ + _____ = _____ 35

Ho Math Chess 何数棋谜 妈！我会棋谜式加法啦！
Mom! I Learn Addition Using Math-Chess-Puzzles Connection

Contents include both traditional and Math-Chess-Puzzles combined methods. Extra strength
©2008 – 2018 Frank Ho, Amanda Ho   All rights reserved. www.homathchess.com

Student Name _____ Date _____

Addition

e	11	12	13	14	15
d	16	17	18	19	20
c	21	22	4	23	24
b	25	26	27	28	29
a	30	31	32	33	34
	1	2	3	4	5

Move by one square at a time from the original square.

The original square is at c3.

C3 ✚ ┼ = _____ ✚ _____ = _____ 35

The original square is at c3.

C3 ✚ ┼ = _____ ✚ _____ = _____ 29

The original square is at c3.

C3 ✚ ┼ = _____ ✚ _____ = _____ 20

The original square is at c3.

C3 ✚ ┼ = _____ ✚ _____ = _____ 16

Ho Math Chess 何数棋谜   妈！我会棋谜式加法啦！
Mom! I Learn Addition Using Math-Chess-Puzzles Connection
Contents include both traditional and Math-Chess-Puzzles combined methods. Extra strength
©2008 — 2018 Frank Ho, Amanda Ho   All rights reserved. www.homathchess.com

Student Name_____ Date_____

Addition

e	11	12	13	14	15
d	16	17	18	19	20
c	21	22	5	23	24
b	25	26	27	28	29
a	30	31	32	33	34
	1	2	3	4	5

Move by one square at a time from the original square.

The original square is at c3.

c3 + ✥ = _____ + _____ = _____ 28

The original square is at c3.

c3 + ⇔ = _____ + _____ = _____ 23

The original square is at c3.

c3 + ⇔ = _____ + _____ = _____ 27

The original square is at c3.

c3 + ✥ = _____ + _____ = _____ 32

Addition

e	11	12	13	14	15
d	16	17	18	19	20
c	21	22	5	23	24
b	25	26	27	28	29
a	30	31	32	33	34
	1	2	3	4	5

Move by one square at a time from the original square.

The original square is at c3.

c3 + ✕ = _____ + _____ = _____ 24

The original square is at c3.

c3 + ✕ = _____ + _____ = _____ 33

The original square is at c3.

c3 + ✕ = _____ + _____ = _____ 31

The original square is at c3.

c3 + ✕ = _____ + _____ = _____ 22

Ho Math Chess 何数棋谜   妈！我会棋谜式加法啦！
Mom! I Learn Addition Using Math-Chess-Puzzles Connection
Contents include both traditional and Math-Chess-Puzzles combined methods. Extra strength
©2008 – 2018 Frank Ho, Amanda Ho   All rights reserved. www.homathchess.com

Student Name _____   Date _____

Addition

e	11	12	13	14	15
d	16	17	18	19	20
c	21	22	5	23	24
b	25	26	27	28	29
a	30	31	32	33	34
	1	2	3	4	5

Move by one square at a time from the original square.

The original square is at c3.

c3 + ⤢ + ⤢ = ____ + ____ + ____ = ____ 39

The original square is at c3.

c3 + ⤢ + ⤢ = ____ + ____ + ____ = ____ 67

The original square is at c3.

c3 + ⤢ + ⤢ = ____ + ____ + ____ = ____ 61

The original square is at c3.

c3 + ⤢ + ⤢ = ____ + ____ + ____ = ____ 33

Ho Math Chess 何数棋谜　妈！我会棋谜式加法啦！
Mom! I Learn Addition Using Math-Chess-Puzzles Connection
Contents include both traditional and Math-Chess-Puzzles combined methods. Extra strength
©2008 — 2018 Frank Ho, Amanda Ho　　All rights reserved. www.homathchess.com

Student Name _____ Date _____

Addition

e	11	12	13	14	15
d	16	17	18	19	20
c	21	22	5	23	24
b	25	26	27	28	29
a	30	31	32	33	34
	1	2	3	4	5

Move by one square at a time from the original square.

The original square is at c3.

c3 + ↗↘ + ↖↘ = ____ + ____ + ____ = _____ 39

The original square is at c3.

c3 + ↗↙ + ↖↘ = ____ + ____ + ____ = _____ 67

The original square is at c3.

c3 + ↖↘ + ↗↙ = ____ + ____ + ____ = _____ 61

The original square is at c3.

c3 + ↖↗ + ↖↗ = ____ + ____ + ____ = _____ 33

446

Ho Math Chess 何数棋谜 妈！我会棋谜式加法啦！
Mom! I Learn Addition Using Math-Chess-Puzzles Connection

Contents include both traditional and Math-Chess-Puzzles combined methods. Extra strength

©2008 – 2018 Frank Ho, Amanda Ho   All rights reserved. www.homathchess.com

Student Name _____ Date _____

Addition

e	11	12	13	14	15
d	16	17	18	19	20
c	21	22	5	23	24
b	25	26	27	28	29
a	30	31	32	33	34
	1	2	3	4	5

Move by one square at a time from the original square.

The original square is at c3.

C3 + ☐ = _____ + _____ = _____ 19

The original square is at c3.

C3 + ☐ = _____ + _____ = _____ 25

The original square is at c3.

C3 + ☐ = _____ + _____ = _____ 34

The original square is at c3.

C3 + ☐ = _____ + _____ = _____ 38

Addition

e	11	12	13	14	15
d	16	17	18	19	20
c	21	22	5	23	24
b	25	26	27	28	29
a	30	31	32	33	34
	1	2	3	4	5

Move by one square at a time from the original square.

The original square is at c3.

C3 + ☐ = _____ + _____ = _____ 36

The original square is at c3.

C3 + ☐ = _____ + _____ = _____ 30

The original square is at c3.

C3 + ☐ = _____ + _____ = _____ 21

The original square is at c3.

C3 + ☐ = _____ + _____ = _____ 17

Addition

e	11	12	13	14	15
d	16	17	18	19	20
c	21	22	6	23	24
b	25	26	27	28	29
a	30	31	32	33	34
	1	2	3	4	5

Move by one square at a time from the original square.

The original square is at c3.

c3 + ✦ = _____ + _____ = _____ 29

The original square is at c3.

c3 + ↔ = _____ + _____ = _____ 24

The original square is at c3.

c3 + ↕ = _____ + _____ = _____ 28

The original square is at c3.

c3 + ✦ = _____ + _____ = _____ 33

Ho Math Chess 何数棋谜　妈！我会棋谜式加法啦！
Mom! I Learn Addition Using Math-Chess-Puzzles Connection

Contents include both traditional and Math-Chess-Puzzles combined methods. Extra strength

©2008 – 2018 Frank Ho, Amanda Ho　　All rights reserved. www.homathchess.com

Student Name _____ Date _____

Addition

e	11	12	13	14	15
d	16	17	18	19	20
c	21	22	6	23	24
b	25	26	27	28	29
a	30	31	32	33	34
	1	2	3	4	5

Move by one square at a time from the original square.

The original square is at c3.

c3 + ✥ + ✥ = ____ + ____ + ____ = ____ 53

The original square is at c3.

c3 + ✥ + ✥ = ____ + ____ + ____ = ____ 37

The original square is at c3.

c3 + ✥ + ✥ = ____ + ____ + ____ = ____ 49

The original square is at c3.

c3 + ✥ + ✥ = ____ + ____ + ____ = ____ 66

Ho Math Chess 何数棋谜 妈！我会棋谜式加法啦！
Mom! I Learn Addition Using Math-Chess-Puzzles Connection
Contents include both traditional and Math-Chess-Puzzles combined methods. Extra strength
©2008 – 2018 Frank Ho, Amanda Ho   All rights reserved. www.homathchess.com

Student Name_____ Date_____

Addition

e	11	12	13	14	15
d	16	17	18	19	20
c	21	22	6	23	24
b	25	26	27	28	29
a	30	31	32	33	34
	1	2	3	4	5

Move by one square at a time from the original square.

The original square is at c3.

C3 + ✥ + ✥ = ____ + ____ + ____ = _____ 53

The original square is at c3.

C3 + ⚓ + ⚓ = ____ + ____ + ____ = _____ 37

The original square is at c3.

C3 + ⚓ + ⚓ = ____ + ____ + ____ = _____ 49

The original square is at c3.

C3 + ⚓ + ⚓ = ____ + ____ + ____ = _____ 65

451

Addition

e	11	12	13	14	15
d	16	17	18	19	20
c	21	22	6	23	24
b	25	26	27	28	29
a	30	31	32	33	34
	1	2	3	4	5

Move by one square at a time from the original square.

The original square is at c3.

c3 + ✕ = __6__ + __19__ = __25__

The original square is at c3.

c3 + ✕ = __6__ + __28__ = __34__

The original square is at c3.

c3 + ✕ = __6__ + __26__ = __32__

The original square is at c3.

c3 + ✕ = __6__ + __17__ = __23__

Addition

e	11	12	13	14	15
d	16	17	18	19	20
c	21	22	6	23	24
b	25	26	27	28	29
a	30	31	32	33	34
	1	2	3	4	5

Move by one square at a time from the original square.

The original square is at c3.

c3 + ↘↙ + ↘↗ = ___ + ___ + ___ = ___ 40

The original square is at c3.

c3 + ↙↗ + ↖↗ = ___ + ___ + ___ = ___ 68

The original square is at c3.

c3 + ↗↙ + ↙↗ = ___ + ___ + ___ = ___ 62

The original square is at c3.

c3 + ↖↘ + ↙↗ = ___ + ___ + ___ = ___ 34

Mom! I Learn Addition Using Math-Chess-Puzzles Connection

Contents include both traditional and Math-Chess-Puzzles combined methods. Extra strength
©2008 – 2018 Frank Ho, Amanda Ho   All rights reserved. www.homathchess.com

Student Name _____   Date _____

Addition

e	11	12	13	14	15
d	16	17	18	19	20
c	21	22	6	23	24
b	25	26	27	28	29
a	30	31	32	33	34
	1	2	3	4	5

Move by one square at a time from the original square.

The original square is at c3.

c3 + ✕ + ✕ = __6__ + __19__ + __15__ = __40__

The original square is at c3.

c3 + ✕ + ✕ = __6__ + __28__ + __34__ = __68__

The original square is at c3.

c3 + ✕ + ✕ = __6__ + __26__ + __30__ = __62__

The original square is at c3.

c3 + ✕ + ✕ = __6__ + __17__ + __11__ = __34__

Addition

e	11	12	13	14	15
d	16	17	18	19	20
c	21	22	6	23	24
b	25	26	27	28	29
a	30	31	32	33	34
	1	2	3	4	5

Move by one square at a time from the original square.

The original square is at c3.

c3 + ⊞ = _____ + _____ = _____ 20

The original square is at c3.

c3 + ⊞ = _____ + _____ = _____ 26

The original square is at c3.

c3 + ⊞ = _____ + _____ = _____ 35

The original square is at c3.

c3 + ⊞ = _____ + _____ = _____ 39

Addition

e	11	12	13	14	15
d	16	17	18	19	20
c	21	22	6	23	24
b	25	26	27	28	29
a	30	31	32	33	34
	1	2	3	4	5

Move by one square at a time from the original square.

The original square is at c3.

C3 + ✚ = _____ + _____ = _____ 37

The original square is at c3.

C3 + ✚ = _____ + _____ = _____ 31

The original square is at c3.

C3 + ✚ = _____ + _____ = _____ 22

The original square is at c3.

C3 + ✚ = _____ + _____ = _____ 18

Mom! I Learn Addition Using Math-Chess-Puzzles Connection

Addition

e	11	12	13	14	15
d	16	17	18	19	20
c	21	22	7	23	24
b	25	26	27	28	29
a	30	31	32	33	34
	1	2	3	4	5

Move by one square at a time from the original square.

The original square is at c3.

c3 + ⬌ = _____ + _____ = _____ 30

The original square is at c3.

c3 + ⬌ = _____ + _____ = _____ 25

The original square is at c3.

c3 + ⬌ = _____ + _____ = _____ 29

The original square is at c3.

c3 + ⬌ = _____ + _____ = _____ 34

Addition

e	11	12	13	14	15
d	16	17	18	19	20
c	21	22	7	23	24
b	25	26	27	28	29
a	30	31	32	33	34
	1	2	3	4	5

Move by one square at a time from the original square.

The original square is at c3.

c3 + ✕ = _____ + _____ = _____ 26

The original square is at c3.

c3 + ✕ = _____ + _____ = _____ 35

The original square is at c3.

c3 + ✕ = _____ + _____ = _____ 33

The original square is at c3.

c3 + ✕ = _____ + _____ = _____ 24

Ho Math Chess 何数棋谜　妈！我会棋谜式加法啦！
Mom! I Learn Addition Using Math-Chess-Puzzles Connection
Contents include both traditional and Math-Chess-Puzzles combined methods. Extra strength
©2008 – 2018 Frank Ho, Amanda Ho   All rights reserved. www.homathchess.com

Student Name_____ Date_____

Addition

e	11	12	13	14	15
d	16	17	18	19	20
c	21	22	7	23	24
b	25	26	27	28	29
a	30	31	32	33	34
	1	2	3	4	5

Move by one square at a time from the original square.

The original square is at c3.

c3 + ⤢ + ⤢ = ____ + ____ + ____ = _____ 41

The original square is at c3.

c3 + ✕ + ✕ = ____ + ____ + ____ = _____ 69

The original square is at c3.

c3 + ✕ + ✕ = ____ + ____ + ____ = _____ 63

The original square is at c3.

c3 + ✕ + ✕ = ____ + ____ + ____ = _____ 35

Addition

e	11	12	13	14	15
d	16	17	18	19	20
c	21	22	7	23	24
b	25	26	27	28	29
a	30	31	32	33	34
	1	2	3	4	5

Move by one square at a time from the original square.

The original square is at c3.

c3 + ↗↙ + ↖↘ = ____ + ____ + ____ = ____ 41

The original square is at c3.

c3 + ↗↙ + ↖↘ = ____ + ____ + ____ = ____ 69

The original square is at c3.

c3 + ↗↙ + ↖↘ = ____ + ____ + ____ = ____ 63

The original square is at c3.

c3 + ↗↙ + ↖↘ = ____ + ____ + ____ = ____ 35

Ho Math Chess 何数棋谜   妈！我会棋谜式加法啦！
Mom! I Learn Addition Using Math-Chess-Puzzles Connection
Contents include both traditional and Math-Chess-Puzzles combined methods. Extra strength
©2008 – 2018 Frank Ho, Amanda Ho   All rights reserved. www.homathchess.com

Student Name _____   Date _____

Addition

e	11	12	13	14	15
d	16	17	18	19	20
c	21	22	7	23	24
b	25	26	27	28	29
a	30	31	32	33	34
	1	2	3	4	5

Move by one square at a time from the original square.

The original square is at c3.

C3 + ✚ = _____ + _____ = _____ 21

The original square is at c3.

C3 + ✚ = _____ + _____ = _____ 27

The original square is at c3.

C3 + ✚ = _____ + _____ = _____ 36

The original square is at c3.

C3 + ✚ = _____ + _____ = _____ 40

461

Ho Math Chess 何数棋谜 妈！我会棋谜式加法啦！
Mom! I Learn Addition Using Math-Chess-Puzzles Connection

Contents include both traditional and Math-Chess-Puzzles combined methods. Extra strength

©2008 — 2018 Frank Ho, Amanda Ho   All rights reserved. www.homathchess.com

Student Name_____ Date_____

Addition

e	11	12	13	14	15
d	16	17	18	19	20
c	21	22	7	23	24
b	25	26	27	28	29
a	30	31	32	33	34
	1	2	3	4	5

Move by one square at a time from the original square.

The original square is at c3.

c3 + ┼ + ┼ = _____ + _____ = _____ 38

The original square is at c3.

c3 + ┼ + ┼ = _____ + _____ = _____ 32

The original square is at c3.

c3 + ┼ + ┼ = _____ + _____ = _____ 23

The original square is at c3.

c3 + ┼ + ┼ = _____ + _____ = _____ 19

Ho Math Chess 何数棋谜 妈！我会棋谜式加法啦！
Mom! I Learn Addition Using Math-Chess-Puzzles Connection
Contents include both traditional and Math-Chess-Puzzles combined methods. Extra strength
©2008 — 2018 Frank Ho, Amanda Ho   All rights reserved. www.homathchess.com

Student Name _____ Date _____

Addition

e	11	12	13	14	15
d	16	17	18	19	20
c	21	22	8	23	24
b	25	26	27	28	29
a	30	31	32	33	34
	1	2	3	4	5

Move by one square at a time from the original square.

The original square is at c3.

c3 + ✥ = _____ + _____ = _____ 31

The original square is at c3.

c3 + ⚓ = _____ + _____ = _____ 26

The original square is at c3.

c3 + ⚓ = _____ + _____ = _____ 30

The original square is at c3.

c3 + ⚓ = _____ + _____ = _____ 35

Ho Math Chess 何数棋谜   妈！我会棋谜式加法啦！

Mom! I Learn Addition Using Math-Chess-Puzzles Connection

Contents include both traditional and Math-Chess-Puzzles combined methods. Extra strength

©2008 – 2018 Frank Ho, Amanda Ho   All rights reserved. www.homathchess.com

Student Name _____   Date _____

Addition

e	11	12	13	14	15
d	16	17	18	19	20
c	21	22	8	23	24
b	25	26	27	28	29
a	30	31	32	33	34
	1	2	3	4	5

Move by one square at a time from the original square.

The original square is at c3.

c3 + ✛ ✕ = _____ + _____ = _____ 27

The original square is at c3.

c3 + ✛ ✕ = _____ + _____ = _____ 36

The original square is at c3.

c3 + ✛ ✕ = _____ + _____ = _____ 34

The original square is at c3.

c3 + ✛ ✕ = _____ + _____ = _____ 25

Ho Math Chess 何数棋谜 妈！我会棋谜式加法啦！
Mom! I Learn Addition Using Math-Chess-Puzzles Connection
Contents include both traditional and Math-Chess-Puzzles combined methods. Extra strength
©2008 – 2018 Frank Ho, Amanda Ho   All rights reserved. www.homathchess.com

Student Name _____ Date _____

Addition

e	11	12	13	14	15
d	16	17	18	19	20
c	21	22	8	23	24
b	25	26	27	28	29
a	30	31	32	33	34
	1	2	3	4	5

Move by one square at a time from the original square.

The original square is at c3.

c3 + ⤢ + ⤢ = ____ + ____ + ____ = _____ 42

The original square is at c3.

c3 + ⤢ + ⤢ = ____ + ____ + ____ = _____ 70

The original square is at c3.

c3 + ⤢ + ⤢ = ____ + ____ + ____ = _____ 64

The original square is at c3.

c3 + ⤢ + ⤢ = ____ + ____ + ____ = _____ 36

Addition

e	11	12	13	14	15
d	16	17	18	19	20
c	21	22	8	23	24
b	25	26	27	28	29
a	30	31	32	33	34
	1	2	3	4	5

Move by one square at a time from the original square.

The original square is at c3.

c3 + ⤢⤡ + ⤢⤡ = ____ + ____ + ____ = ____ 42

The original square is at c3.

c3 + ⤢⤡ + ⤢⤡ = ____ + ____ + ____ = ____ 70

The original square is at c3.

c3 + ⤢⤡ + ⤢⤡ = ____ + ____ + ____ = ____ 64

The original square is at c3.

c3 + ⤢⤡ + ⤢⤡ = ____ + ____ + ____ = ____ 36

Addition

e	11	12	13	14	15
d	16	17	18	19	20
c	21	22	8	23	24
b	25	26	27	28	29
a	30	31	32	33	34
	1	2	3	4	5

Move by one square at a time from the original square.

The original square is at c3.

C3 + ⊕ = _____ + _____ = _____ 22

The original square is at c3.

C3 + ⊕ = _____ + _____ = _____ 28

The original square is at c3.

C3 + ⊕ = _____ + _____ = _____ 37

The original square is at c3.

C3 + ⊕ = _____ + _____ = _____ 41

Addition

e	11	12	13	14	15
d	16	17	18	19	20
c	21	22	8	23	24
b	25	26	27	28	29
a	30	31	32	33	34
	1	2	3	4	5

Move by one square at a time from the original square.

The original square is at c3.

c3 + ⊤⊢ = _____ + _____ = _____ 39

The original square is at c3.

c3 + ⊤⊢ = _____ + _____ = _____ 33

The original square is at c3.

c3 + ⊤⊢ = _____ + _____ = _____ 24

The original square is at c3.

c3 + ⊤⊢ = _____ + _____ = _____ 20

Addition

e	11	12	13	14	15
d	16	17	18	19	20
c	21	22	9	23	24
b	25	26	27	28	29
a	30	31	32	33	34
	1	2	3	4	5

Move by one square at a time from the original square.

The original square is at c3.

c3 + ✥ = _____ + _____ = _____ 32

The original square is at c3.

c3 + ↔ = _____ + _____ = _____ 27

The original square is at c3.

c3 + ↔ = _____ + _____ = _____ 31

The original square is at c3.

c3 + ↔ = _____ + _____ = _____ 36

Mom! I Learn Addition Using Math-Chess-Puzzles Connection

Contents include both traditional and Math-Chess-Puzzles combined methods. Extra strength
©2008 − 2018 Frank Ho, Amanda Ho   All rights reserved. www.homathchess.com

Student Name_____ Date_____

Addition

e	11	12	13	14	15
d	16	17	18	19	20
c	21	22	9	23	24
b	25	26	27	28	29
a	30	31	32	33	34
	1	2	3	4	5

Move by one square at a time from the original square.

The original square is at c3.

c3 + ✥ + ✥ = ____ + ____ + ____ = _____ 56

The original square is at c3.

c3 + ✥ + ✥ = ____ + ____ + ____ = _____ 40

The original square is at c3.

c3 + ✥ + ✥ = ____ + ____ + ____ = _____ 52

The original square is at c3.

c3 + ✥ + ✥ = ____ + ____ + ____ = _____ 68

Mom! I Learn Addition Using Math-Chess-Puzzles Connection

Contents include both traditional and Math-Chess-Puzzles combined methods. Extra strength

©2008 – 2018 Frank Ho, Amanda Ho   All rights reserved. www.homathchess.com

Student Name_____ Date_____

Addition

e	11	12	13	14	15
d	16	17	18	19	20
c	21	22	9	23	24
b	25	26	27	28	29
a	30	31	32	33	34
	1	2	3	4	5

Move by one square at a time from the original square.

The original square is at c3.

C3 + ⬌ + ⬌ = ____ + ____ + ____ = _____ 56

The original square is at c3.

C3 + ⬌ + ⬌ = ____ + ____ + ____ = _____ 40

The original square is at c3.

C3 + ⬌ + ⬌ = ____ + ____ + ____ = _____ 52

The original square is at c3.

C3 + ⬌ + ⬌ = ____ + ____ + ____ = _____ 68

Ho Math Chess 何数棋谜　妈！我会棋谜式加法啦！
Mom! I Learn Addition Using Math-Chess-Puzzles Connection
Contents include both traditional and Math-Chess-Puzzles combined methods. Extra strength
©2008 – 2018 Frank Ho, Amanda Ho   All rights reserved. www.homathchess.com

Student Name _____ Date _____

Addition

e	11	12	13	14	15
d	16	17	18	19	20
c	21	22	9	23	24
b	25	26	27	28	29
a	30	31	32	33	34
	1	2	3	4	5

Move by one square at a time from the original square.

The original square is at c3.

c3 + ✕ = ____ + ____ = ____ 28

The original square is at c3.

c3 + ✕ = ____ + ____ = ____ 37

The original square is at c3.

c3 + ✕ = ____ + ____ = ____ 35

The original square is at c3.

c3 + ✕ = ____ + ____ = ____ 26

Mom! I Learn Addition Using Math-Chess-Puzzles Connection

Contents include both traditional and Math-Chess-Puzzles combined methods. Extra strength
©2008 – 2018 Frank Ho, Amanda Ho   All rights reserved. www.homathchess.com

Student Name _____   Date _____

Addition

e	11	12	13	14	15
d	16	17	18	19	20
c	21	22	9	23	24
b	25	26	27	28	29
a	30	31	32	33	34
	1	2	3	4	5

Move by one square at a time from the original square.

The original square is at c3.

c3 + ⤢ + ⤫ = ____ + ____ + ____ = ____ 43

The original square is at c3.

c3 + ⤢ + ⤫ = ____ + ____ + ____ = ____ 71

The original square is at c3.

c3 + ⤢ + ⤫ = ____ + ____ + ____ = ____ 65

The original square is at c3.

c3 + ⤢ + ⤫ = ____ + ____ + ____ = ____ 37

Ho Math Chess 何数棋谜 妈！我会棋谜式加法啦！
Mom! I Learn Addition Using Math-Chess-Puzzles Connection
Contents include both traditional and Math-Chess-Puzzles combined methods. Extra strength
©2008 – 2018 Frank Ho, Amanda Ho   All rights reserved. www.homathchess.com

Student Name _____   Date _____

Addition

e	11	12	13	14	15
d	16	17	18	19	20
c	21	22	9	23	24
b	25	26	27	28	29
a	30	31	32	33	34
	1	2	3	4	5

Move by one square at a time from the original square.

The original square is at c3.

c3 + ✕ + ✕ = __9__ + __19__ + __15__ = __43__ 43

The original square is at c3.

c3 + ✕ + ✕ = __9__ + __28__ + __34__ = __71__ 71

The original square is at c3.

c3 + ✕ + ✕ = __9__ + __26__ + __30__ = __65__ 65

The original square is at c3.

c3 + ✕ + ✕ = __9__ + __17__ + __11__ = __37__ 37

Ho Math Chess 何数棋谜　妈！我会棋谜式加法啦！
Mom! I Learn Addition Using Math-Chess-Puzzles Connection
Contents include both traditional and Math-Chess-Puzzles combined methods. Extra strength
©2008 – 2018 Frank Ho, Amanda Ho   All rights reserved. www.homathchess.com

Student Name _____   Date _____

Addition

e	11	12	13	14	15
d	16	17	18	19	20
c	21	22	9	23	24
b	25	26	27	28	29
a	30	31	32	33	34
	1	2	3	4	5

Move by one square at a time from the original square.

The original square is at c3.

c3 ✚ ┼ = _____ ✚ _____ = _____ 23

The original square is at c3.

c3 ✚ ┼ = _____ ✚ _____ = _____ 29

The original square is at c3.

c3 ✚ ┼ = _____ ✚ _____ = _____ 38

The original square is at c3.

c3 ✚ ┼ = _____ ✚ _____ = _____ 42

475

Addition

e	11	12	13	14	15
d	16	17	18	19	20
c	21	22	9	23	24
b	25	26	27	28	29
a	30	31	32	33	34
	1	2	3	4	5

Move by one square at a time from the original square.

The original square is at c3.

c3 + ⊥⊤ = ____ + _____ = _____ 40

The original square is at c3.

c3 + ⊥⊤ = ____ + _____ = _____ 34

The original square is at c3.

c3 + ⊥⊤ = ____ + _____ = _____ 25

The original square is at c3.

c3 + ⊥⊤ = ____ + _____ = _____ 21

Addition and subtraction concept by link

3	6	3	7
2	2	11	9
1	4	8	5
	a	b	c

You are a chess piece and located at b2.

The original square is at the first term

b2= ↕ + ↕ ____ = __+__
11=9+2

↔ = b2 − ↕ ____ = __−__ 9=11-2

b2= ↕ + ↕ ____ = __+__ 11=3+8

↔ = b2 − ↕ ____ = __−__ 3=11-8

The original square is at the first term

b2= ↔ + ↔ ____ = __+__
11=2+9

↕ = b2 − ↕ ____ = __−__ 2 = 11-9

b2= ↕ + ↕ ____ = __+__ 11=8+3

↕ = b2 − ↕ ____ = __−__ 8=11-3

Ho Math Chess 何数棋谜　妈！我会棋谜式加法啦！
Mom! I Learn Addition Using Math-Chess-Puzzles Connection
Contents include both traditional and Math-Chess-Puzzles combined methods. Extra strength
©2008 — 2018 Frank Ho, Amanda Ho   All rights reserved. www.homathchess.com

Student Name _____   Date _____

Addition and subtraction concept

3	6	5	7
2	8	14	6
1	7	9	8
	a	b	c

You are a chess piece and located at b2.

The original square is at the first term	The original square is at the first term
b2= ⬇ + ⬇ ____ = __+__	b2= ⬇ + ⬇ ____ = __+__
⬆ = b2 – ⬆ ____ = __-__	⬆ = b2 – ⬆ ____ = __-__
b2= ⬆ + ⬆ ____ = __+__	b2= ⬇ + ⬇ ____ = __+__
⬆ = b2 – ⬆ ____ = __-__	⬅ = b2 – ⬅ ____ = __-__

14=6+8　14 = 8 + 6
6=14-8　8 = 14 – 6
14 = 5 + 9　14 = 9 + 5
5 = 14 – 9　9 = 14 – 5

Ho Math Chess 何数棋谜 妈！我会棋谜式加法啦！
Mom! I Learn Addition Using Math-Chess-Puzzles Connection

Contents include both traditional and Math-Chess-Puzzles combined methods. Extra strength

©2008 — 2018 Frank Ho, Amanda Ho   All rights reserved. www.homathchess.com

Student Name _____ Date _____

Addition and subtraction concept

3	9	6	5
2	8	13	5
1	8	7	4
	a	b	c

You are a chess piece and located at b2.

The original square is at the first term	The original square is at the first term
b2= ⇕ + ⇕ _____ = __+__	b2= ⇕ + ⇕ _____ = __+__
⇕ = b2 − ⇕ _____ = __−__	⇕ = b2 − ⇕ _____ = __−__
b2= ⇔ + ⇔ _____ = __+__	b2= ⇔ + ⇔ _____ = __+__
⇔ = b2 − ⇔ _____ = __−__	⇔ = b2 − ⇔ _____ = __−__

13 = 5 + 8   13 = 8 + 5
5 = 13 − 8   8 = 13 − 5
13 = 6 + 7   13 = 7 + 6
6 = 13 − 7   7 = 13 − 6

Ho Math Chess 何数棋谜 妈！我会棋谜式加法啦！
Mom! I Learn Addition Using Math-Chess-Puzzles Connection

Contents include both traditional and Math-Chess-Puzzles combined methods. Extra strength
©2008 – 2018 Frank Ho, Amanda Ho   All rights reserved. www.homathchess.com

Student Name _____ Date _____

Addition and subtraction concept

3	9	7	6
2	8	12	4
1	6	5	3
	a	b	c

You are a chess piece and located at b2.

The original square is at the first term	The original square is at the first term
b2= ⊕ + ⊕ _____ = __+__	b2= ⊕ + ⊕ _____ = __+__
⊕ = b2 – ⊕ _____ = __–__	⊕ = b2 – ⊕ _____ = __–__
b2= ⊕ + ⊕ _____ = __+__	b2= ⊕ + ⊕ _____ = __+__
⊕ = b2 – ⊕ _____ = __–__	⊕ = b2 – ⊕ _____ = __–__

12 = 4 + 8 12 = 8+4
4 = 12 – 8 8 = 12 – 4
12 = 7 + 5 12 = 5 + 7
7 = 12 – 5 5 = 12 – 7

Ho Math Chess 何数棋谜　妈！我会棋谜式加法啦！

Mom! I Learn Addition Using Math-Chess-Puzzles Connection

Contents include both traditional and Math-Chess-Puzzles combined methods. Extra strength

©2008 − 2018 Frank Ho, Amanda Ho　All rights reserved. www.homathchess.com

Student Name _____ Date _____

Addition and subtraction concept

3	6	6	9
2	7	15	8
1	6	9	9
	a	b	c

You are a chess piece and located at b2.

The original square is at the first term	The original square is at the first term
b2= ⇔ + ⇔ ____ = __+__	b2= ⇔ + ⇔ ____ = __+__
⇔ = b2 − ⇔ ____ = __−__	⇔ = b2 − ⇔ ____ = __−__
b2= ⇔ + ⇔ ____ = __+__	b2= ⇔ + ⇔ ____ = __+__
⇔ = b2 − ⇔ ____ = __−__	⇔ = b2 − ⇔ ____ = __−__

15 = 8 + 7　15 = 7 + 8
8 = 15 − 7　7 = 15 − 8
15 = 6 + 9　15 = 9 + 6
6 = 15 − 9　9 = 15 − 6

Ho Math Chess 何数棋谜　妈！我会棋谜式加法啦！
Mom! I Learn Addition Using Math-Chess-Puzzles Connection
Contents include both traditional and Math-Chess-Puzzles combined methods. Extra strength
©2008 – 2018 Frank Ho, Amanda Ho   All rights reserved. www.homathchess.com

Student Name _____   Date _____

Addition and subtraction concept

3	9	9	8
2	8	17	9
1	9	8	8
	a	b	c

You are a chess piece and located at b2.

The original square is at the first term	The original square is at the first term
b2= ⬍ + ⬍ ____ = __+__	b2= ⬍ + ⬍ ____ = __+__
⬍ = b2 – ⬍ ____ = __-__	⬍ = b2 – ⬍ ____ = __-__
b2= ⬌ + ⬌ ____ = __+__	b2= ⬌ + ⬌ ____ = __+__
⬌ = b2 – ⬌ ____ = __-__	⬌ = b2 – ⬌ ____ = __-__

17=9+8   17 = 8 + 9
9 = 17 – 8   8 = 17-9
17 = 9 + 8   17 = 8 + 9
9 = 17 -8   8 = 17 - 9

Ho Math Chess 何数棋谜 妈！我会棋谜式加法啦！
Mom! I Learn Addition Using Math-Chess-Puzzles Connection
Contents include both traditional and Math-Chess-Puzzles combined methods. Extra strength
©2008 – 2018 Frank Ho, Amanda Ho   All rights reserved. www.homathchess.com

Student Name _____ Date _____

Addition and subtraction concept

3	9	7	8
2	9	16	7
1	8	9	7
	a	b	c

You are a chess piece and located at b2.

The original square is at the first term	The original square is at the first term
b2= ⬍ + ⬍ ____ = __+__	b2= ⬍ + ⬍ ____ = __+__
⬍ = b2 − ⬍ ____ = __−__	⬍ = b2 − ⬍ ____ = __−__
b2= ⬍ + ⬍ ____ = __+__	b2= ⬍ + ⬍ ____ = __+__
⬍ = b2 − ⬍ ____ = __−__	⬍ = b2 − ⬍ ____ = __−__

16 = 7 + 9   16 = 9 + 7
7 = 16 − 9   9 = 16 − 7
16 = 7 + 9   16 = 9 + 7
7 = 16 − 9   9 = 16 − 7

Ho Math Chess 何数棋谜　妈！我会棋谜式加法啦！
Mom! I Learn Addition Using Math-Chess-Puzzles Connection

Contents include both traditional and Math-Chess-Puzzles combined methods. Extra strength
©2008 – 2018 Frank Ho, Amanda Ho   All rights reserved. www.homathchess.com

Student Name _____ Date _____

Addition and subtraction concept

3	9	9	9
2	9	18	9
1	9	9	9
	a	b	c

You are a chess piece and located at b2.

The original square is at the first term	The original square is at the first term
b2= ⬍ + ⬍ ____ = __+__	b2= ⬍ + ⬍ ____ = __+__
⬍ = b2 − ⬍ ____ = __−__	⬍ = b2 − ⬍ ____ = __−__
b2= ⬍ + ⬍ ____ = __+__	b2= ⬍ + ⬍ ____ = __+__
⬍ = b2 − ⬍ ____ = __−__	⬍ = b2 − ⬍ ____ = __−__

18 = 9 + 9   18 = 9 + 9
9 = 18 − 9   9 = 18 − 9
18 = 9 + 9   18 = 9 + 9
9 = 18 − 9   9 = 18 − 9

Mom! I Learn Addition Using Math-Chess-Puzzles Connection

Addition and subtraction concept

3	6	3	7
2	2	11	9
1	4	8	5
	a	b	c

You are a chess piece and located at b2.

The original square is at the first term	The original square is at the first term
b2 = ↖↘ + ↗↙ ____ = __ + __	b2 = ↖↘ + ↗↙ ____ = __ + __
↖↘ = b2 − ↗↙ ____ = __ − __	↖↘ = b2 − ↗↙ ____ = __ − __
b2 = ↖↙ + ↖↗ ____ = __ + __	b2 = ↖↙ + ↖↗ ____ = __ + __
↖↙ = b2 − ↖↗ ____ = __ − __	↖↙ = b2 − ↖↗ ____ = __ − __

11=7+4   11 = 4+7
7=11-4   4=11-7
11=6+5   11=5+6
6=11-5   5=11-6

Ho Math Chess 何数棋谜　妈！我会棋谜式加法啦！
Mom! I Learn Addition Using Math-Chess-Puzzles Connection
Contents include both traditional and Math-Chess-Puzzles combined methods. Extra strength
©2008 — 2018 Frank Ho, Amanda Ho   All rights reserved. www.homathchess.com

Student Name _____ Date _____

Addition and subtraction concept

3	6	5	7
2	8	14	6
1	7	9	8
	a	b	c

You are a chess piece and located at b2.

The original square is at the first term	The original square is at the first term
b2 = ⤩ + ⤨　____ = __ + __	b2 = ⤩ + ⤨　____ = __ + __
⤨ = b2 – ⤩　____ = __ – __	⤨ = b2 – ⤩　____ = __ – __
b2 = ⤩ + ⤨　____ = __ + __	b2 = ⤩ + ⤨　____ = __ + __
⤨ = b2 – ⤩　____ = __ – __	⤨ = b2 – ⤩　____ = __ – __

14==7+7　　　14==7+7
7 = 14 – 7　　7 = 14 – 7
14==7+7　　　14==7+7
7 = 14 – 7　　7 = 14 – 7

Mom! I Learn Addition Using Math-Chess-Puzzles Connection

Contents include both traditional and Math-Chess-Puzzles combined methods. Extra strength
©2008 – 2018 Frank Ho, Amanda Ho   All rights reserved. www.homathchess.com

Student Name_____ Date_____

Addition and subtraction concept

3	7	6	5
2	4	13	9
1	8	7	6
	a	b	c

You are a chess piece and located at b2.

The original square is at the first term	The original square is at the first term
b2 = ⭨ + ⭩ ____ = __ + __	b2 = ⭨ + ⭩ ____ = __ + __
⭨ = b2 − ⭩ ____ = __ − __	⭨ = b2 − ⭩ ____ = __ − __
b2 = ⭨ + ⭩ ____ = __ + __	b2 = ⭨ + ⭩ ____ = __ + __
⭨ = b2 − ⭩ ____ = __ − __	⭨ = b2 − ⭩ ____ = __ − __

13 = 5+8   13 = 8 + 5
5 = 13 – 8   8 = 13 – 5
13 = 7 + 6   13 = 6 + 7
7 = 13 – 6   6 = 13 - 7

Mom! I Learn Addition Using Math-Chess-Puzzles Connection

Contents include both traditional and Math-Chess-Puzzles combined methods. Extra strength
©2008 – 2018 Frank Ho, Amanda Ho   All rights reserved. www.homathchess.com

Student Name _____ Date _____

Addition and subtraction concept

3	9	7	6
2	8	12	4
1	6	5	3
	a	b	c

You are a chess piece and located at b2.

The original square is at the first term	The original square is at the first term
b2 = ⤨ + ⤨ ____ = __ + __	b2 = ⤨ + ⤨ ____ = __ + __
⤨ = b2 – ⤨ ____ = __ – __	⤨ = b2 – ⤨ ____ = __ – __
b2 = ⤨ + ⤨ ____ = __ + __	b2 = ⤨ + ⤨ ____ = __ + __
⤨ = b2 – ⤨ ____ = __ – __	⤨ = b2 – ⤨ ____ = __ – __

12 = 6 + 6   12 = 6 + 6
6 = 12 - 6   6 = 12 - 6
12 = 9 + 3   12 = 3 + 9
9 = 12 - 3   3 = 12 - 9

Addition and subtraction concept

3	6	6	9
2	7	15	8
1	6	9	9
	a	b	c

You are a chess piece and located at b2.

The original square is at the first term	The original square is at the first term
b2 = ⤢ + ⤢ ____ = __ + __	b2 = ⤢ + ⤢ ____ = __ + __
⤢ = b2 − ⤢ ____ = __ − __	⤢ = b2 − ⤢ ____ = __ − __
b2 = ⤢ + ⤢ ____ = __ + __	b2 = ⤢ + ⤢ ____ = __ + __
⤢ = b2 − ⤢ ____ = __ − __	⤢ = b2 − ⤢ ____ = __ − __

15 = 9 + 6   15 = 6 + 9
9 = 15 − 6   6 = 15 − 9
15 = 6 + 9   15 = 9 + 6
6 = 15 − 9   9 = 15 − 6

Ho Math Chess 何数棋谜   妈！我会棋谜式加法啦！
Mom! I Learn Addition Using Math-Chess-Puzzles Connection
Contents include both traditional and Math-Chess-Puzzles combined methods. Extra strength
©2008 — 2018 Frank Ho, Amanda Ho   All rights reserved. www.homathchess.com

Student Name _____ Date _____

Addition and subtraction concept

3	7	8	9
2	8	16	8
1	7	9	9
	a	b	c

You are a chess piece and located at b2.

The original square is at the first term	The original square is at the first term
b2 = ⤢ + ⤢ ____ = __ + __	b2 = ⤢ + ⤢ ____ = __ + __
⤢ = b2 − ⤢ ____ = __ − __	⤢ = b2 − ⤢ ____ = __ − __
b2 = ⤢ + ⤢ ____ = __ + __	b2 = ⤢ + ⤢ ____ = __ + __
⤢ = b2 − ⤢ ____ = __ − __	⤢ = b2 − ⤢ ____ = __ − __

16 = 9 + 7   16 = 7 + 9
9 = 16 − 7   7 = 16 − 9
16 = 7 + 9   16 = 9 + 7
7 = 16 − 9   9 = 16 - 7

Mom! I Learn Addition Using Math-Chess-Puzzles Connection

Addition and subtraction concept

3	8	8	9
2	9	17	8
1	8	9	9
	a	b	c

You are a chess piece and located at b2.

The original square is at the first term	The original square is at the first term
b2 = ↙ + ↙ ___ = __ + __	b2 = ↙ + ↙ ___ = __ + __
↙ = b2 − ↙ ___ = __ − __	↙ = b2 − ↙ ___ = __ − __
b2 = ↙ + ↙ ___ = __ + __	b2 = ↙ + ↙ ___ = __ + __
↙ = b2 − ↙ ___ = __ − __	↙ = b2 − ↙ ___ = __ − __

17 = 9 + 8   17 = 8 + 9
9 = 17 − 8   8 = 17 − 9
17 = 8 + 9   17 = 9 + 8
8 = 17 − 9   9 = 17 − 8

Mom! I Learn Addition Using Math-Chess-Puzzles Connection

Addition and subtraction concept

3	9	9	9
2	9	18	9
9	6	9	9
	a	b	c

You are a chess piece and located at b2.

The original square is at the first term	The original square is at the first term
b2 = ⤡ + ⤡ ____ = __ + __	b2 = ⤡ + ⤡ ____ = __ + __
⤡ = b2 – ⤡ ____ = __ – __	⤡ = b2 – ⤡ ____ = __ – __
b2 = ⤡ + ⤡ ____ = __ + __	b2 = ⤡ + ⤡ ____ = __ + __
⤡ = b2 – ⤡ ____ = __ – __	⤡ = b2 – ⤡ ____ = __ – __

18 = 9 + 9   18 = 9 + 9
9 = 18 – 9   9 = 18 – 9
18 = 9 + 9   18 = 9 + 9
9 = 18 – 9   9 = 18 – 9

Ho Math Chess 何数棋谜   妈！我会棋谜式加法啦！
Mom! I Learn Addition Using Math-Chess-Puzzles Connection
Contents include both traditional and Math-Chess-Puzzles combined methods. Extra strength
©2008 – 2018 Frank Ho, Amanda Ho   All rights reserved. www.homathchess.com

Student Name _____ Date _____

Addition and subtraction by link using dots

5		•• •• -1		•• ••	
4	••• • +1	••• ••	••	••	•• ••
3		••• +2	2+1 5+1 6+2 7+3 3+2 9+2 8+1 5+4 4+3	• +3	
2	• +2	••• 	•• •• -1	•• ••	•••
1		•		• •• +3	
	a	b	c	d	e

You are a chess piece located at c3

• = 1

5, 13
13, 12
13, 13
8, 10
11, 12

Mom! I Learn Addition Using Math-Chess-Puzzles Connection

Contents include both traditional and Math-Chess-Puzzles combined methods. Extra strength
©2008 – 2018 Frank Ho, Amanda Ho   All rights reserved. www.homathchess.com

Student Name_____ Date_____

Addition and subtraction by link

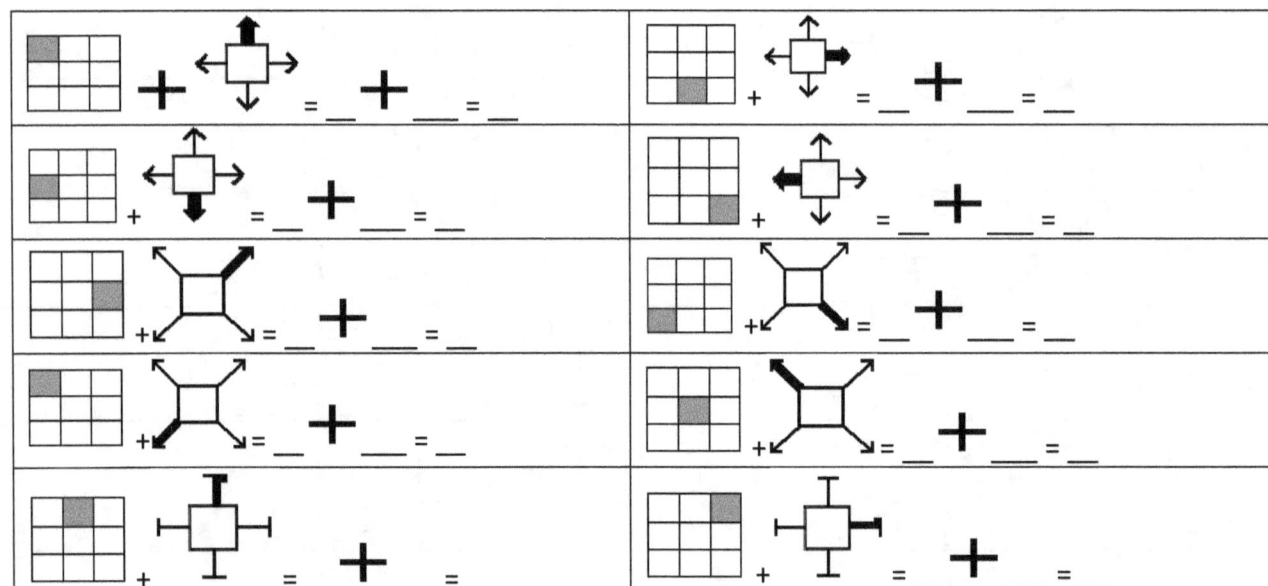

7, 10
20, 8
5, 11
7, 12
13, 10

Ho Math Chess 何数棋谜 妈！我会棋谜式加法啦！
Mom! I Learn Addition Using Math-Chess-Puzzles Connection
Contents include both traditional and Math-Chess-Puzzles combined methods. Extra strength
©2008 – 2018 Frank Ho, Amanda Ho   All rights reserved. www.homathchess.com

Student Name_____ Date_____

Addition and subtraction by link

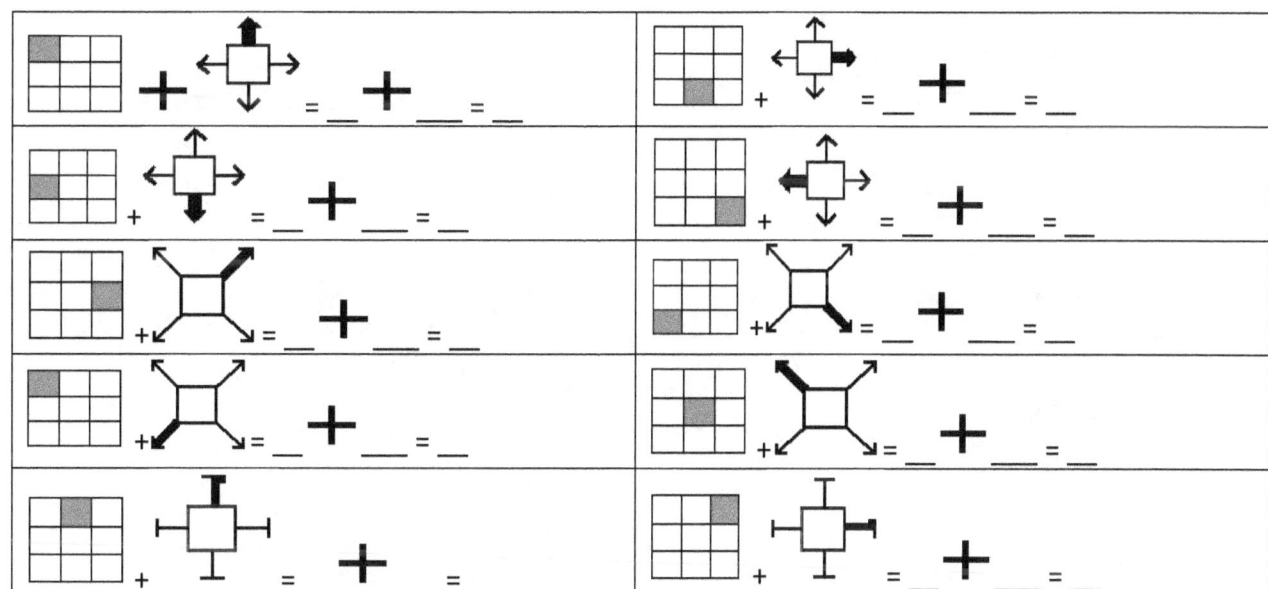

15, 3
18, 10
11. 11
19, 6
11, 9

Ho Math Chess 何数棋谜　妈！我会棋谜式加法啦！
Mom! I Learn Addition Using Math-Chess-Puzzles Connection

Contents include both traditional and Math-Chess-Puzzles combined methods. Extra strength
©2008 – 2018 Frank Ho, Amanda Ho   All rights reserved. www.homathchess.com

Student Name _____ Date _____

Addition and subtraction by link

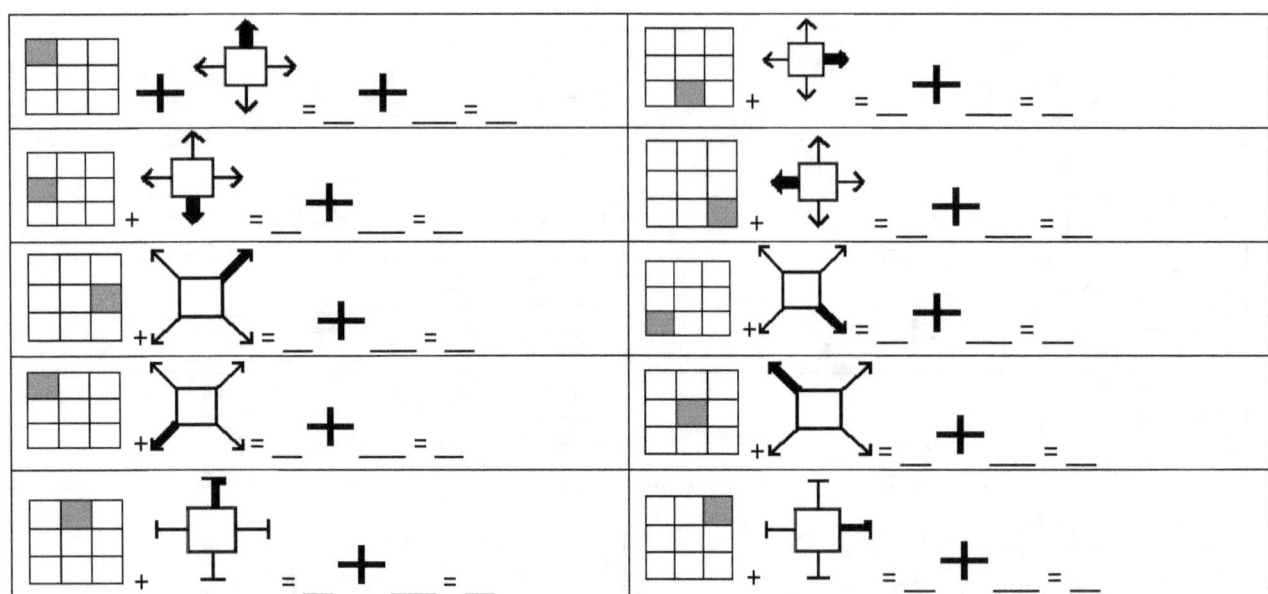

4, 8
13, 9
9, 12
8, 10
16, 16

Ho Math Chess 何数棋谜 妈！我会棋谜式加法啦！
Mom! I Learn Addition Using Math-Chess-Puzzles Connection

Contents include both traditional and Math-Chess-Puzzles combined methods. Extra strength

©2008 — 2018 Frank Ho, Amanda Ho   All rights reserved. www.homathchess.com

Student Name _____ Date _____

Addition and subtraction by link

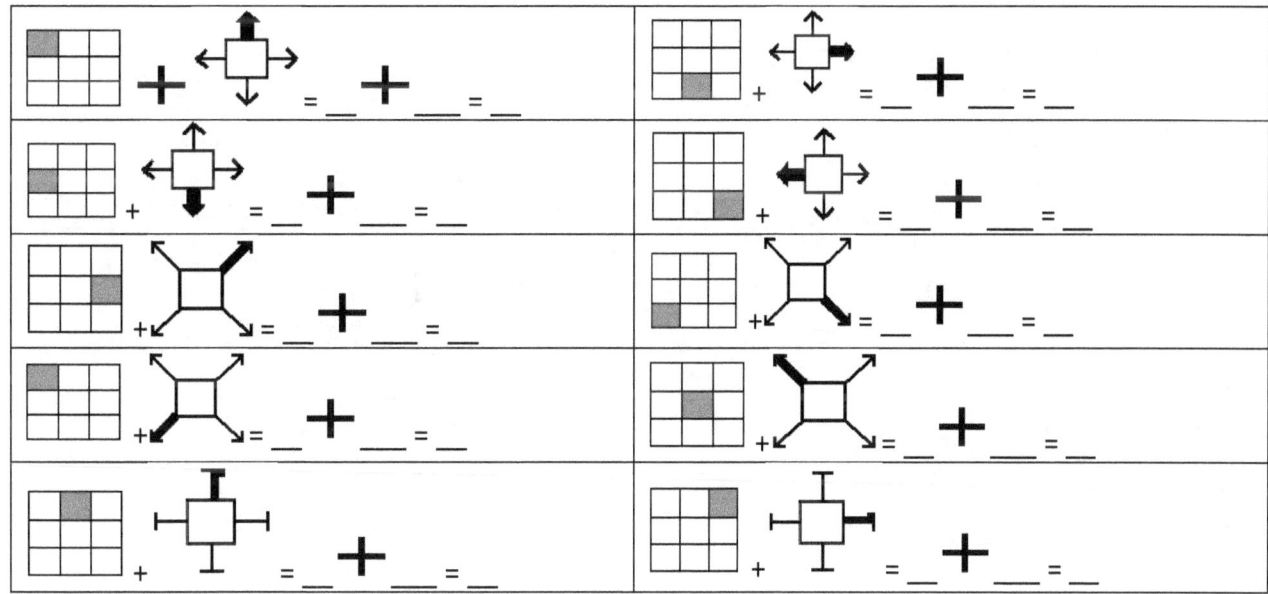

9, 7
17, 3
11, 16
10, 16
12, 10

Ho Math Chess 何数棋谜 妈！我会棋谜式加法啦！
Mom! I Learn Addition Using Math-Chess-Puzzles Connection
Contents include both traditional and Math-Chess-Puzzles combined methods. Extra strength
©2008 – 2018 Frank Ho, Amanda Ho   All rights reserved. www.homathchess.com

Student Name_____ Date_____

Addition and subtraction by link

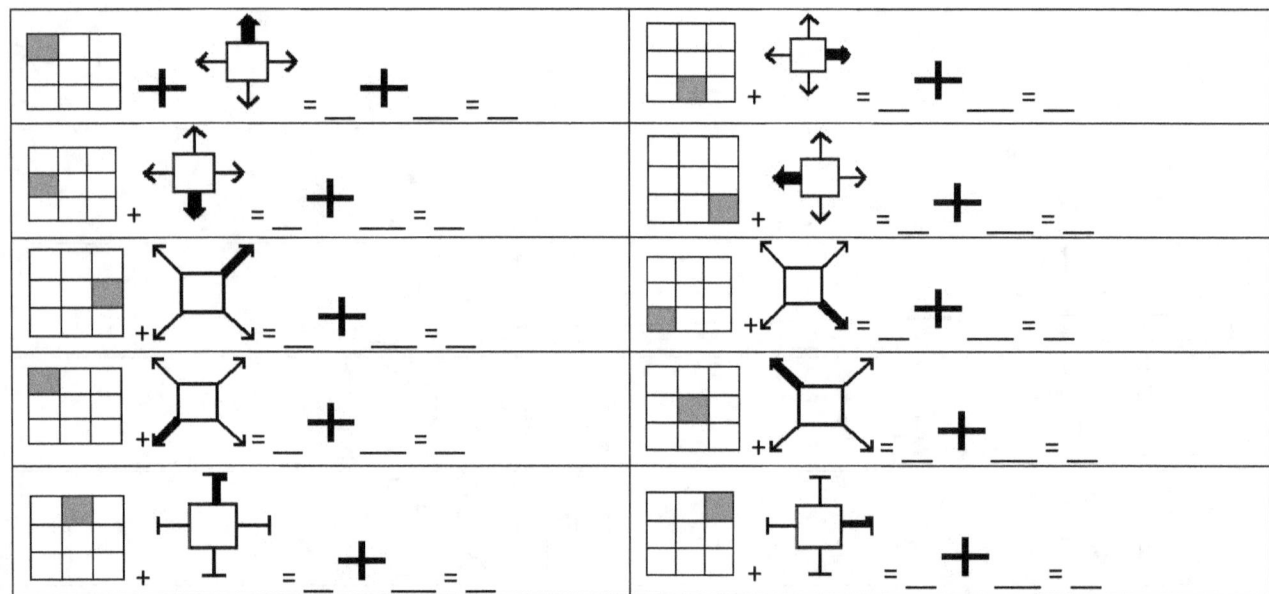

17, 2
10, 12
10, 9
7, 5
10, 6

Mom! I Learn Addition Using Math-Chess-Puzzles Connection

Contents include both traditional and Math-Chess-Puzzles combined methods. Extra strength

©2008 – 2018 Frank Ho, Amanda Ho   All rights reserved. www.homathchess.com

Student Name _____   Date _____

Addition and subtraction by link

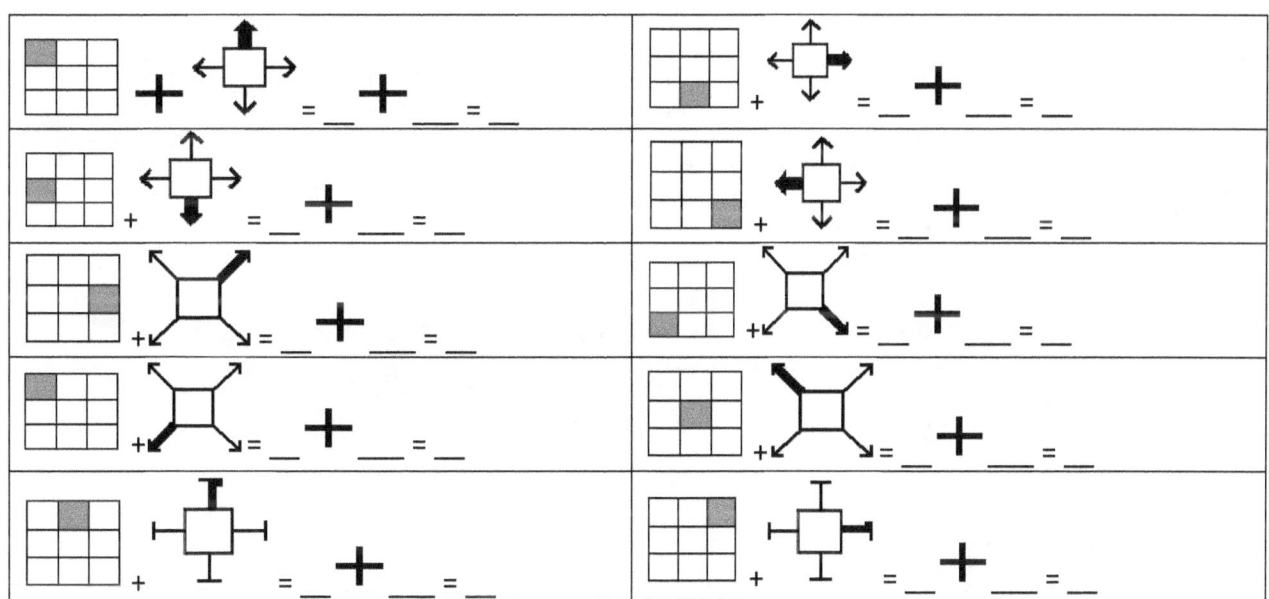

12, 8
15, 11
4, 6
15, 6
6, 5

Ho Math Chess 何数棋谜   妈！我会棋谜式加法啦！

Mom! I Learn Addition Using Math-Chess-Puzzles Connection

Contents include both traditional and Math-Chess-Puzzles combined methods. Extra strength

©2008 – 2018 Frank Ho, Amanda Ho   All rights reserved. www.homathchess.com

Student Name _____ Date _____

Addition and subtraction by link

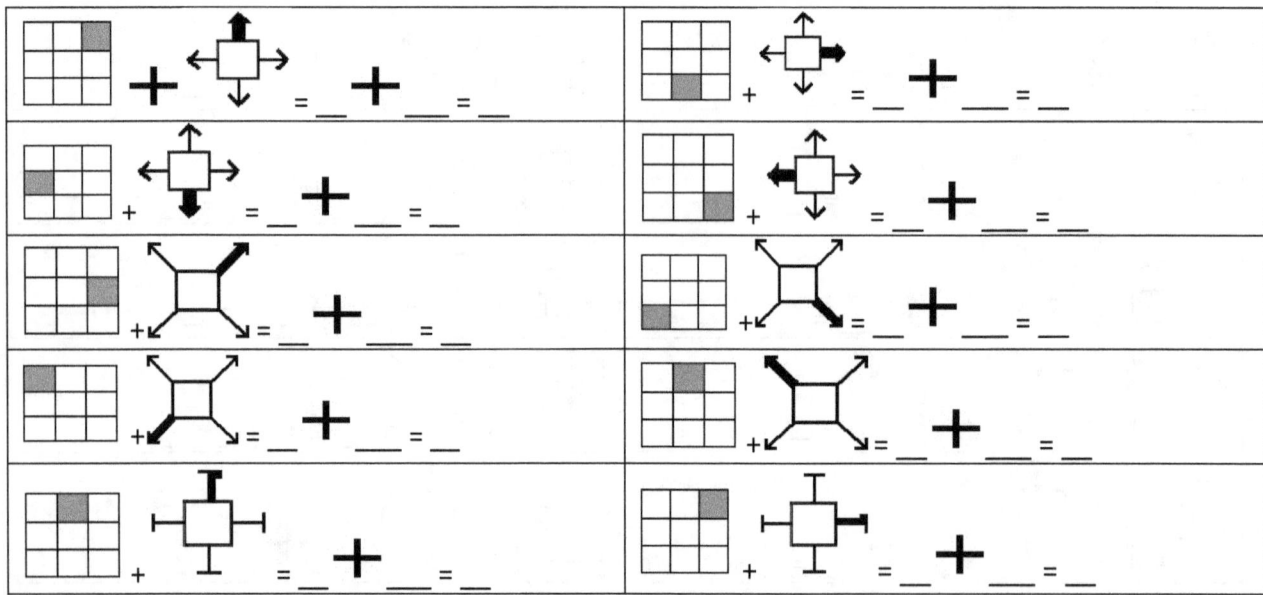

13 10
15 12
12 6
19 18
14 14

500

d + d of math and chess integrated problems

One-digit 5 addition

		3		4	
4	5	3	2	4	5
3		2	5	3	
2	3	4	3	5	4
1		5		2	
	a	b	c	d	e

You are at square c3 = ☐.

6 addition

	a	b	c	d	e
5		4		3	
4	5	5	2	4	5
3		4	6	5	
2	5	6	3	6	2
1		4		3	

You are at square c3 = ☐.

= ____8

= ____10

= ____11

= ____12

= ____9

= ____12

= ____10

= ____11

= ____9

= ____11

= ____8

= ____9

= ____10

= ____11

= ____11

= ____10

502

Ho Math Chess 何数棋谜　妈！我会棋谜式加法啦！
Mom! I Learn Addition Using Math-Chess-Puzzles Connection

Contents include both traditional and Math-Chess-Puzzles combined methods. Extra strength

©2008 – 2018 Frank Ho, Amanda Ho　All rights reserved. www.homathchess.com

Student Name_____ Date_____

7 addition

	5		4		3	
	4	5	5	2	4	5
	3		4	7	5	
	2	5	6	3	6	2
	1		4		3	
		a	b	c	d	e

You are at square c3 = ☐.

503

8 addition

5		4		3	
4	5	5	2	4	5
3		4	8	5	
2	5	6	3	6	2
1		4		3	
	a	b	c	d	e

You are at square c3 = ☐.

+ = ____ 10

+ = ____ 12

+ = ____ 13

+ = ____ 14

+ = ____ 11

+ = ____ 14

+ = ____ 12

+ = ____ 13

+ = ____ 11

+ = ____ 13

+ = ____ 14

+ = ____ 11

+ = ____ 12

+ = ____ 13

+ = ____ 13

+ = ____ 12

504

9 addition

	a	b	c	d	e
5	5		3		4
4	4	5	6	3	2	5
3	3		2	9	3
2	2	3	5	4	4	4
1	1		5		2

You are at square c3 = ☐.

11 addition

5		13		14	
4	15	16	13	12	15
3		19	11	13	
2	13	18	14	4	14
1		15		19	
	a	b	c	d	e

You are at square c3 = ☐.

Ho Math Chess 何数棋谜 妈！我会棋谜式加法啦！
Mom! I Learn Addition Using Math-Chess-Puzzles Connection
Contents include both traditional and Math-Chess-Puzzles combined methods. Extra strength
©2008 – 2018 Frank Ho, Amanda Ho   All rights reserved. www.homathchess.com

Student Name_____ Date_____

12 addition

5		13		14	
4	15	14	12	13	15
3		19	12	15	
2	13	18	16	4	14
1		17		19	
	a	b	c	d	e

You are at square c3 = ☐.

13 addition

5		13		14	
4	15	16	13	12	15
3		19	13	13	
2	13	18	14	4	14
1		15		19	
	a	b	c	d	e

You are at square c3 = ☐.

= ____ 26   = ____ 25

= ____ 26   = ____ 17

= ____ 27   = ____ 31

= ____ 32   = ____ 29

= ____ 27   = ____ 28

= ____ 27   = ____ 32

= ____ 28   = ____ 26

= ____ 28   = ____ 26

Mom! I Learn Addition Using Math-Chess-Puzzles Connection

Contents include both traditional and Math-Chess-Puzzles combined methods. Extra strength
©2008 – 2018 Frank Ho, Amanda Ho   All rights reserved. www.homathchess.com

Student Name _____ Date _____

Adding (be patient with your writing.)

3	3090807	4060706	4070409
2	3040506	3060504	3040708
1	4090709	2070608	3090709
	a	b	c

You are at square b2 = ☐.

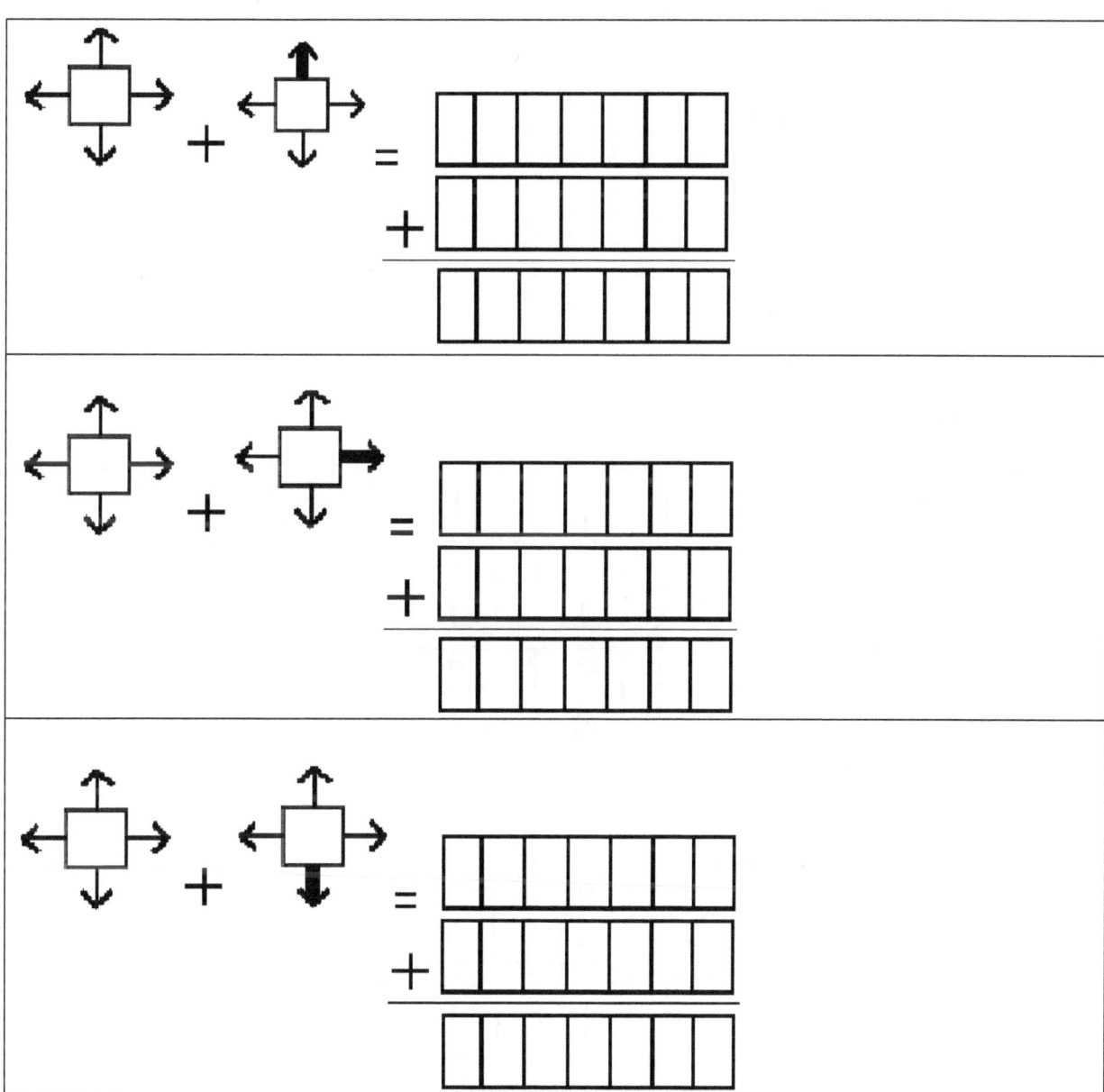

7121210, 6101212, 5131112

Ho Math Chess 何数棋谜 妈！我会棋谜式加法啦！
Mom! I Learn Addition Using Math-Chess-Puzzles Connection
Contents include both traditional and Math-Chess-Puzzles combined methods. Extra strength
©2008 — 2018 Frank Ho, Amanda Ho All rights reserved. www.homathchess.com

Student Name _____ Date _____

Adding (be patient with your writing.)

3	3090807	4060706	4070409
2	3040506	3060504	3040708
1	4090709	2070608	3090709
	a	b	c

You are at square b2 = ☐.

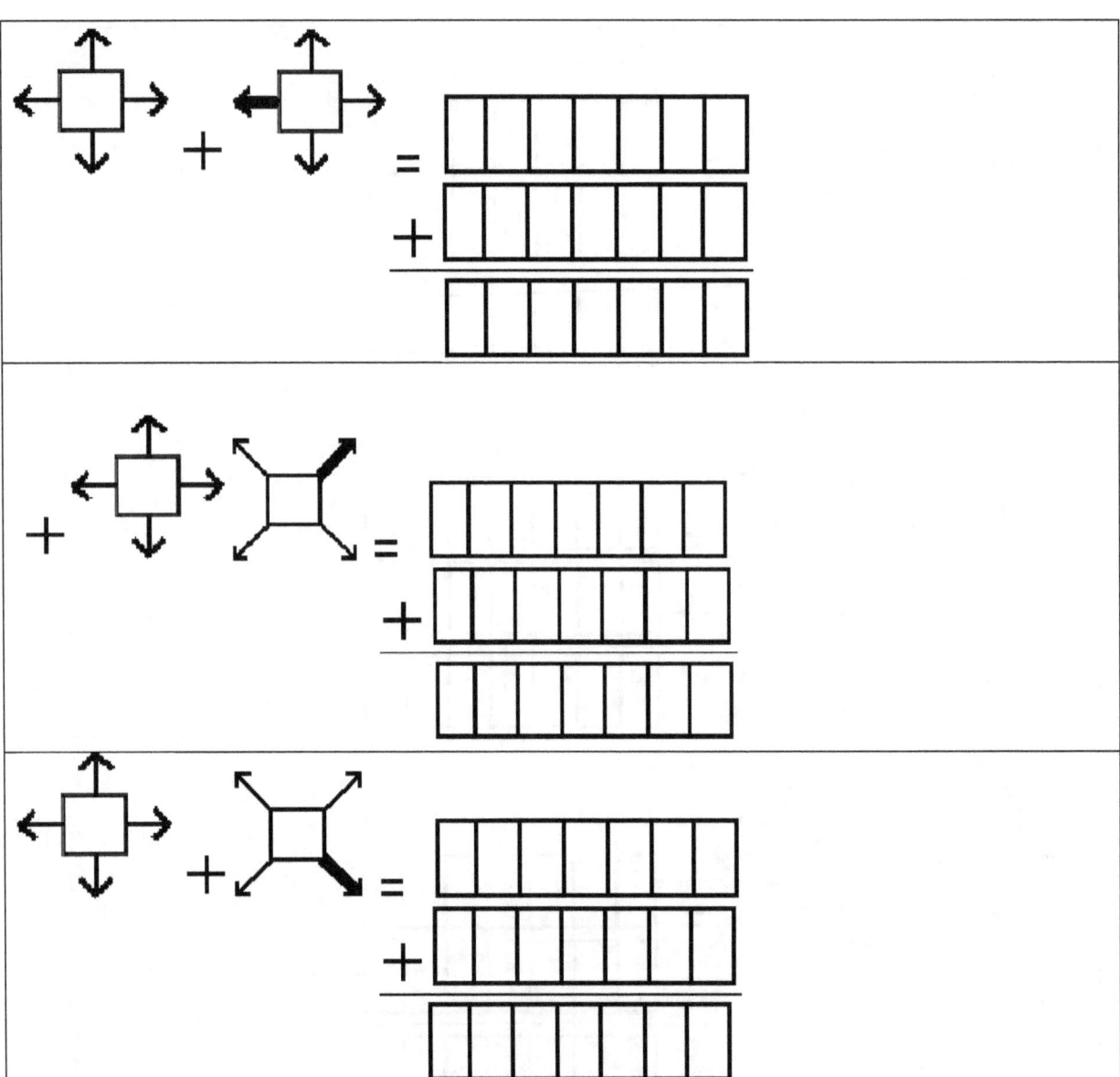

6101010, 7130913, 6151213

510

Ho Math Chess 何数棋谜　妈！我会棋谜式加法啦！
Mom! I Learn Addition Using Math-Chess-Puzzles Connection
Contents include both traditional and Math-Chess-Puzzles combined methods. Extra strength
©2008 － 2018 Frank Ho, Amanda Ho　All rights reserved. www.homathchess.com

Student Name _____ Date _____

Adding (be patient with your writing.)

3	3090807	4060706	4070409
2	3040506	3060504	3040708
1	4090709	2070608	3090709
	a	b	c

You are at square b2 = ☐.

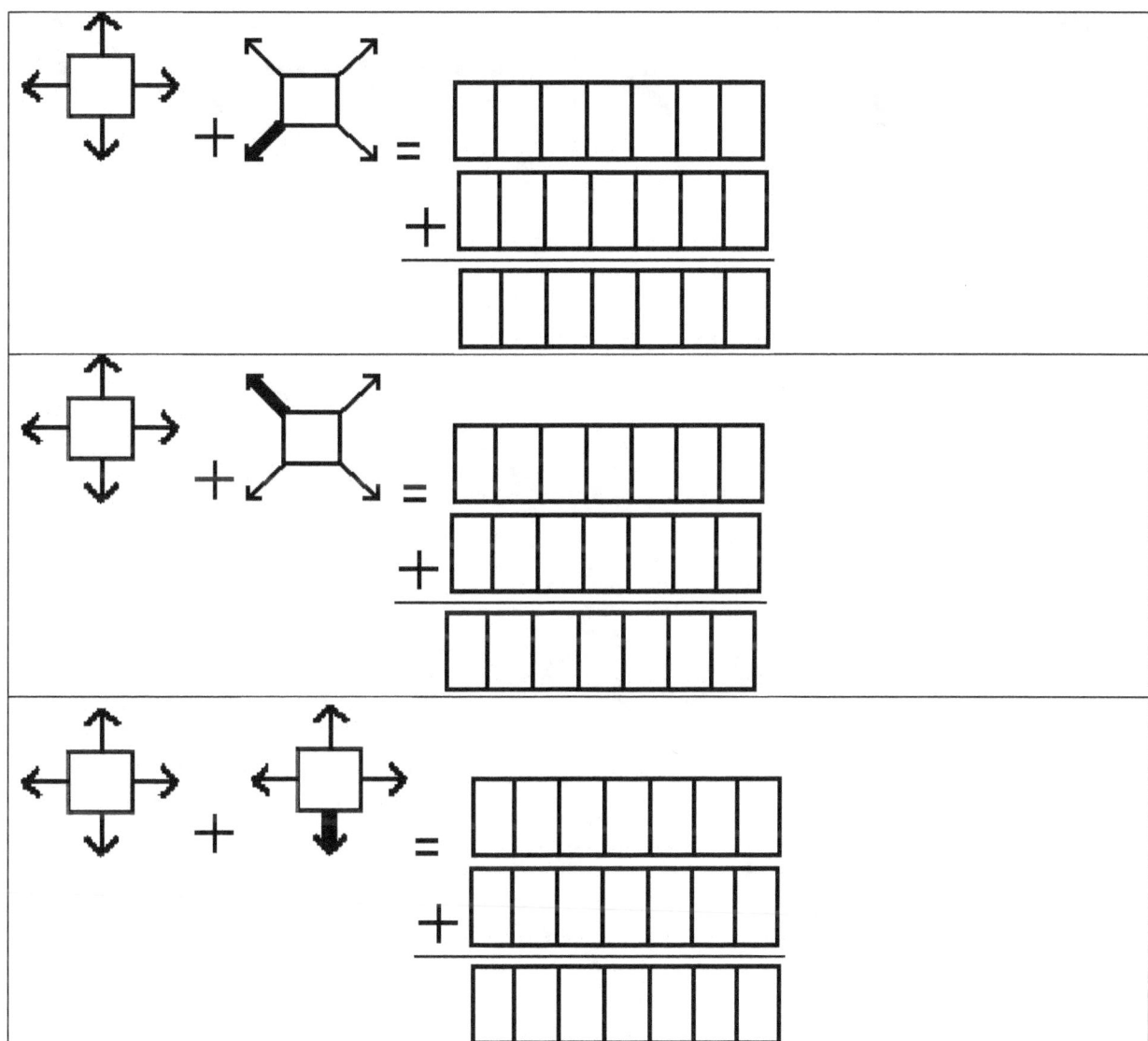

7151213, 6151311, 5131112

Part 5: Cognitive ability 思唯意識

Puzzle # 1

Complete the sequence.

Puzzle # 2

Complete the sequence.

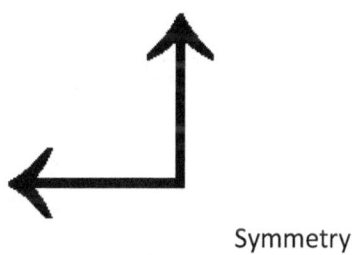

Symmetry

Puzzle # 3

□ = 5

△ = 3

◯ − □ = 5

◯ is a whole number from 0 to 9.

What is ◯ ?

7- (5-3) =5 because the circle is limited from 0 to 9

Puzzle # 4

Replace each ? with a number.

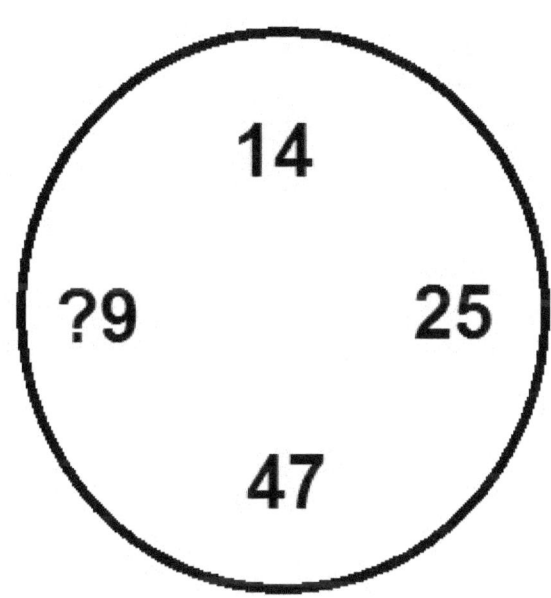

69 because the ones digit - the tens digit is 3.

Puzzle # 5

Replace ? by a number.

1, 1, 2, 3, ?, 8, 13

5

Ho Math Chess 何数棋谜　妈！我会棋谜式加法啦！
Mom! I Learn Addition Using Math-Chess-Puzzles Connection

Contents include both traditional and Math-Chess-Puzzles combined methods. Extra strength

©2008 – 2018 Frank Ho, Amanda Ho　　All rights reserved. www.homathchess.com

Student Name _____ Date _____

Puzzle # 6

Replace ? by a number.

1.

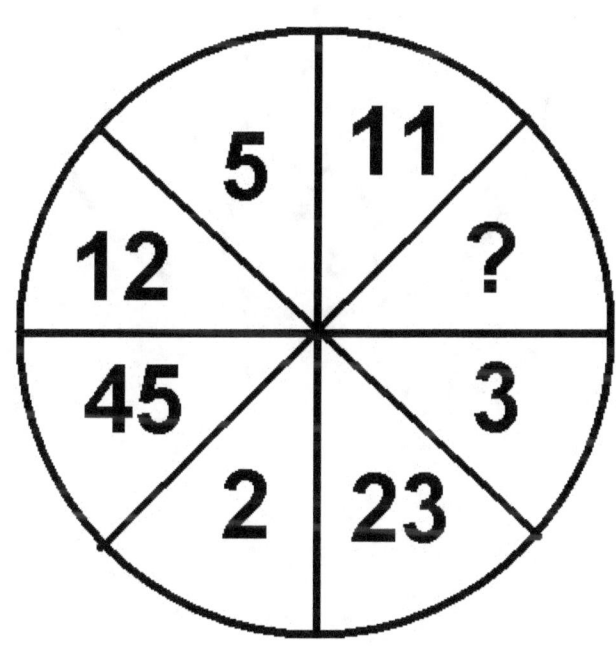

4 + 5　　　　　　　　　　　　　　　　　　　　　　= 9

517

Puzzle # 7

Replace ? by a number.

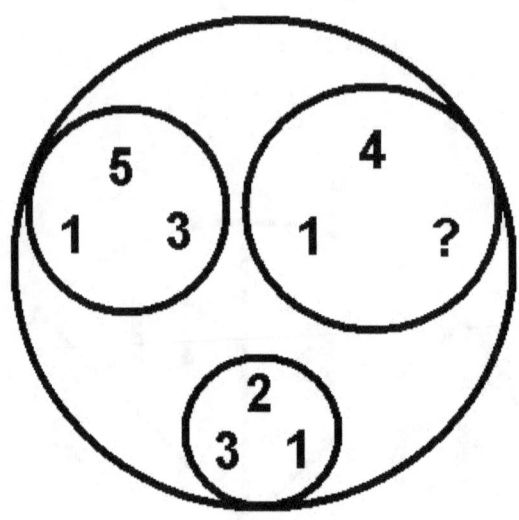

7

Puzzle # 8

Replace each ? with a number.

1, 22, 333, 444?, 55?55, ?

4, 5, 666666

Puzzle # 9

Replace each ? with a number.

11

Replace each ? with a letter.

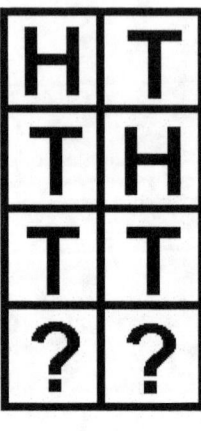
HH

Puzzle # 10

Replace ? by a number.

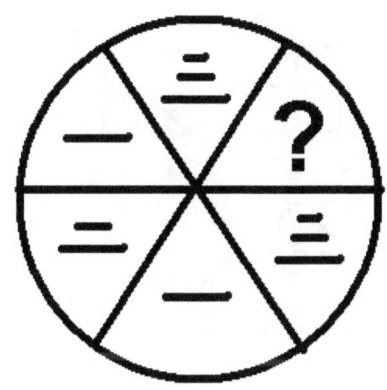

The lines add to 4 so the answer is 2 horizontal lines.

Puzzle # 11

What is the next term in the following sequence?

3 + 1, 6 + 3 + 1, 10 + 6 + 3 + 1, _____

15 + 10 + 6 + 3 + 1= 35

3.

Ho Math Chess 何数棋谜   妈！我会棋谜式加法啦！
Mom! I Learn Addition Using Math-Chess-Puzzles Connection
Contents include both traditional and Math-Chess-Puzzles combined methods. Extra strength
©2008 — 2018 Frank Ho, Amanda Ho   All rights reserved. www.homathchess.com

Student Name_____ Date_____

One-drawable figure

Circle the following figure which can be traced with one continuous line without lifting your pencil or retracing it.

answer

Find the answer for the ?

answer

Ho Math Chess 何数棋谜　妈！我会棋谜式加法啦！
Mom! I Learn Addition Using Math-Chess-Puzzles Connection

Contents include both traditional and Math-Chess-Puzzles combined methods. Extra strength

©2008 － 2018 Frank Ho, Amanda Ho　　All rights reserved. www.homathchess.com

Student Name_____ Date_____

Find the answer for the ?

9.89	6.86
1.06	? 1.09

Find the answer for them ?

Find the answer for the ?

Matrix Reasoning

Number or figure pattern Completion

Draw the missing piece.

Answer _____

answer

Reasoning by Analogy (row or column change)

Find the answer for ? by examining how the objects change across the row, column, or diagonal.

Answer _____

answer

Matrix Reasoning

Serial Reasoning (row, column, and diagonal changes)

Find the answer for ? by examining how the objects change across the row and/or column and/or diagonal.

Answer _____

Observe diagonally.

Spatial Reasoning (partial figures or numbers to make a whole or vice versa. Shape transformation may be involved.)

Answer _____

Observe diagonally.

525

Ho Math Chess 何数棋谜　妈！我会棋谜式加法啦！
Mom! I Learn Addition Using Math-Chess-Puzzles Connection
Contents include both traditional and Math-Chess-Puzzles combined methods. Extra strength
©2008 — 2018 Frank Ho, Amanda Ho　All rights reserved. www.homathchess.com

Student Name _____ Date _____

Matrix Reasoning

Number or figure pattern Completion

Draw the missing piece.

Answer _____

Reasoning by Analogy (row or column change)

Find the answer for ? by examining how the objects change across the row, column, or diagonal.

Answer _____

answer

Ho Math Chess 何数棋谜   妈！我会棋谜式加法啦！
Mom! I Learn Addition Using Math-Chess-Puzzles Connection

Contents include both traditional and Math-Chess-Puzzles combined methods. Extra strength

©2008 – 2018 Frank Ho, Amanda Ho   All rights reserved. www.homathchess.com

Student Name _____ Date _____

Matrix Reasoning

Serial Reasoning (row, column, and diagonal changes)

Find the answer for ? by examining how the objects change across the row and/or column and/or diagonal.

Answer _____

clockwise 90-degree circulation
Observe row to row and see which arrow is missing?

Spatial Reasoning (partial figures or numbers to make a whole or vice versa. Shape transformation may be involved.)

Answer _____

answer

Ho Math Chess 何数棋谜 妈！我会棋谜式加法啦！
Mom! I Learn Addition Using Math-Chess-Puzzles Connection
Contents include both traditional and Math-Chess-Puzzles combined methods. Extra strength
©2008 — 2018 Frank Ho, Amanda Ho   All rights reserved. www.homathchess.com
Student Name _____ Date _____

Matrix Reasoning

Number or figure pattern Completion

Draw the missing piece.

Answer _____

Reasoning by Analogy (row or column change)

Find the answer for ? by examining how the objects change across the row, column, or diagonal.

Answer _____

Turn 45 degree clockwise.

Ho Math Chess 何数棋谜　妈！我会棋谜式加法啦！
Mom! I Learn Addition Using Math-Chess-Puzzles Connection

Contents include both traditional and Math-Chess-Puzzles combined methods. Extra strength
©2008 – 2018 Frank Ho, Amanda Ho   All rights reserved. www.homathchess.com

Student Name _____ Date _____

Matrix Reasoning

Serial Reasoning (row, column, and diagonal changes)

Find the answer for ? by examining how the objects change across the row and/or column and/or diagonal.

Answer _____

Flip each figure 90 degree clockwise.

Spatial Reasoning (partial figures or numbers to make a whole or vice versa. Shape transformation may be involved.)

Answer _____

answer

529

Matrix reasoning

Find an answer to each question mark.

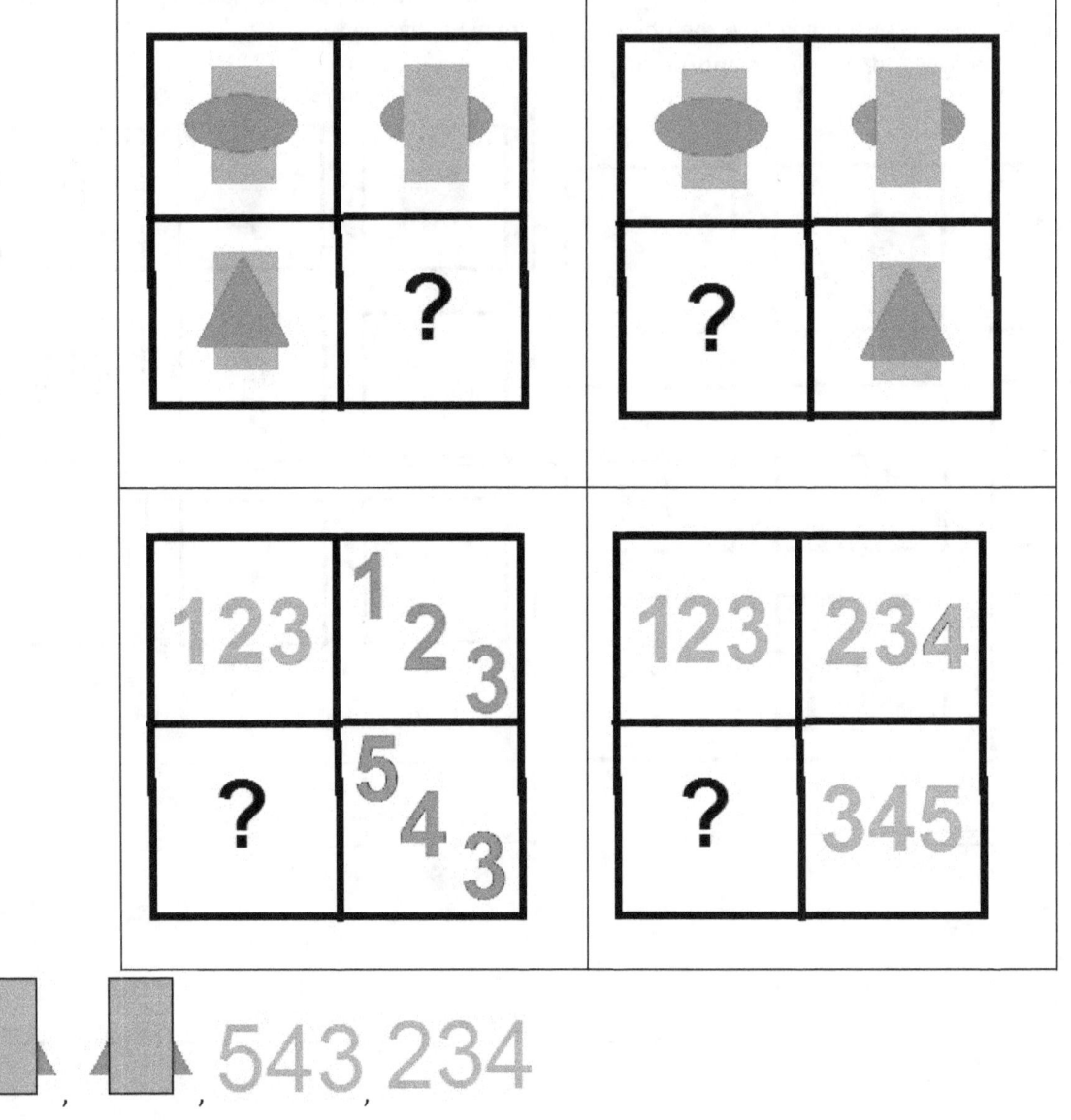

Matrix reasoning

Find an answer to each question mark.

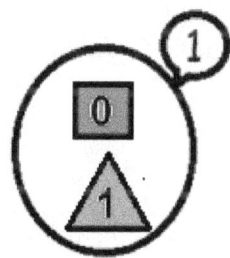
answer

Matrix reasoning

Find an answer to each question mark.

 answer

Matrix reasoning

Find an answer to each question mark.

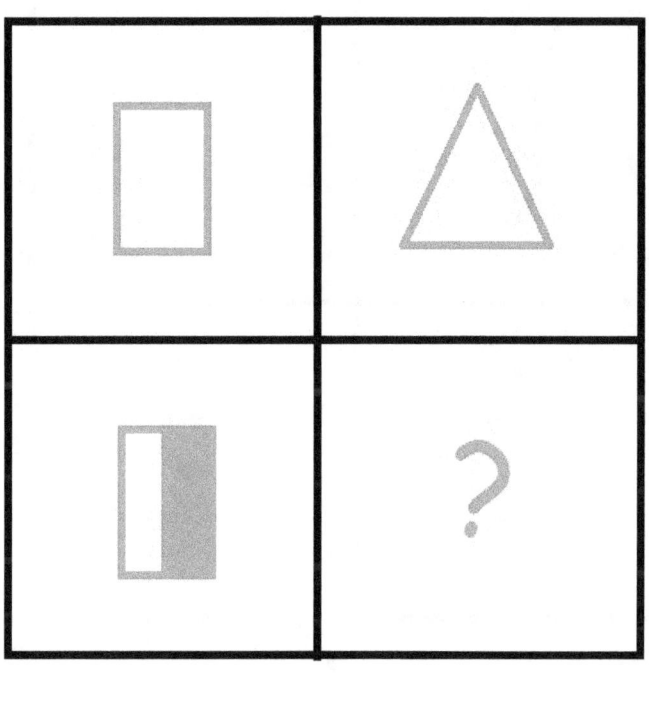

answer

Matrix reasoning

Find an answer to each question mark.

   answer

Matrix reasoning

Find an answer to each question mark.

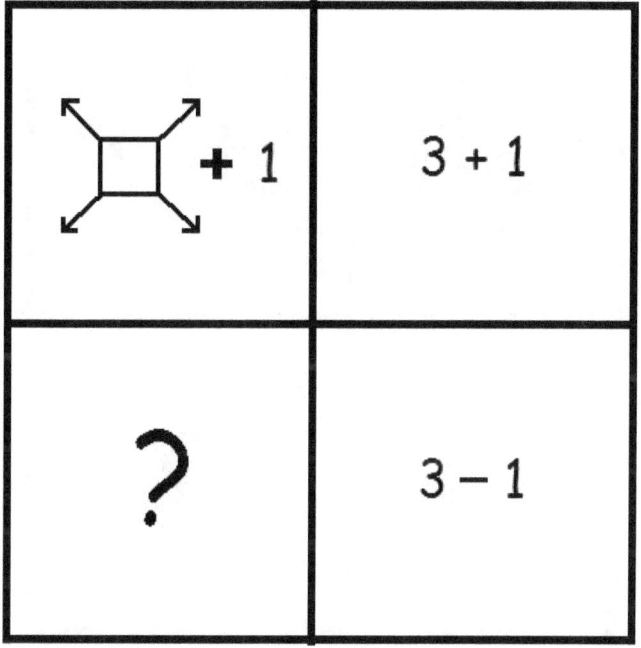

answer

Matrix reasoning

Find an answer to each question mark.

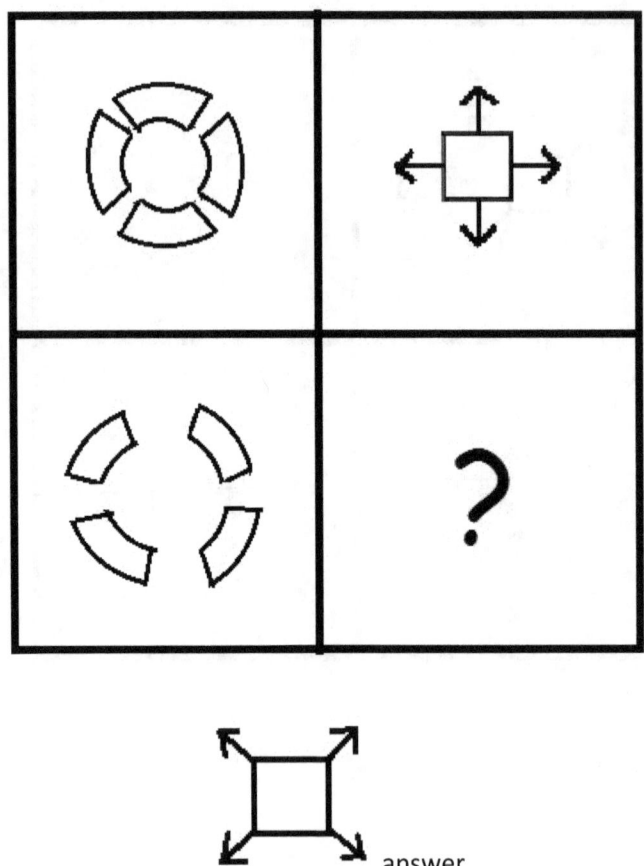

answer

Part 6: Pattern, geometry, time, calendar, a fraction

Find the pattern and fill in each blank.

1	2	3	___	___	___	4,5,6
2	4	6	___	___	___	8,10,12
5	10	15	___	___	___	20,25,30
10	100	1000	___	___	___	10000,100000,1000000
325	335	345	___	___	___	355,365,375
4236	5236	6236	___	___	___	7236,8236,9236
270	280	290	___	___	___	300,310,320,
1150	1200	1250	___	___	___	1300,1350,1400
7385	7375	7365	___	___	___	7355,7345,7335
4350	4250	4150	___	___	___	4050,3950,3850
1007	1008	1009	___	___	___	1010,1011,1012
4256	4506	4756	___	___	___	5006,5256
21	32	43	___	___	___	54,65,76
613	524	435	___	___	___	346,257,168
987	876	765	___	___	___	654,543,432

Mom! I Learn Addition Using Math-Chess-Puzzles Connection

Looking for patterns

3	_____	15	31	63	127	255	7
10	20	40	80	160	_____		320
2	4	16	_____	65536			256
16	8	_____	_____	1			4, 2
_____	12	36	108	324	972		4
_____	_____	20	24	96	100	400	1, 5
600	600	300	100	25	_____		5
784529	78452	7452	452	_____	_____		42, 2
11	18	25	32	39	_____	53	46
810	270	_____	30	10			90
1	4	16	64	256	_____		1024
$400	$200	$100	$50	_____			$25
100	99	97	94	90	_____	_____	85, 79
15	12	14	11	13	_____	_____	10, 12
1	3	7	15	31	_____	_____	63, 127
67	35	19	11	7	_____	_____	5, 4

Looking for patterns.

```
                    1
                   2  2
                  3  4  3
                 4  7  __  4                    7
                5  11  __  11  5                14
               6  __  25  __  16  6             16, 25
              7  22  41  __  41  __  7          50
```

```
                    1
                   2  4
                  3  9  27
                 4  16  __  256                 64
                5  __  __  625  3125            25, 125
```

```
                    1
                   2  4
                  3  6  9
                 4  8  __  16                   12
                5  __  15  __  __               10, 20, 25
               6  __  __  __  __  __            12, 18, 24, 30, 36
```

Complete the following Pattern.

Part 7: Geometry

Lines

straight line   curve   parallel lines

Matching.

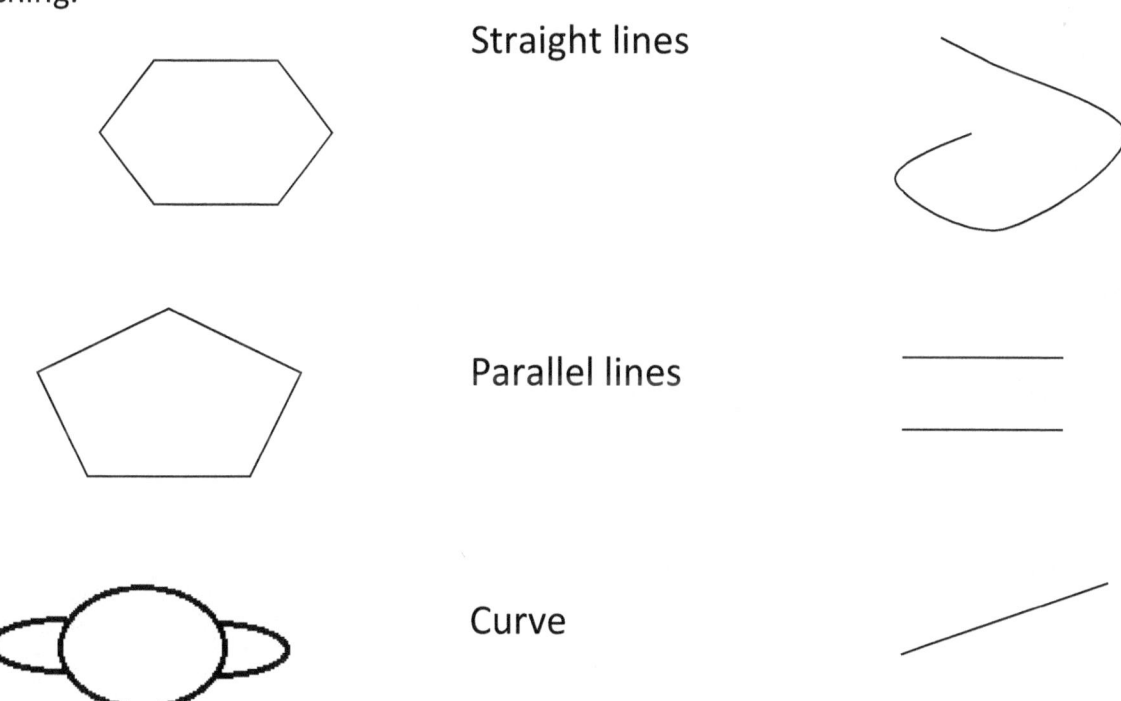

Straight lines

Parallel lines

Curve

Ho Math Chess 何数棋谜　妈！我会棋谜式加法啦！
Mom! I Learn Addition Using Math-Chess-Puzzles Connection
Contents include both traditional and Math-Chess-Puzzles combined methods. Extra strength
©2008 – 2018 Frank Ho, Amanda Ho　All rights reserved. www.homathchess.com

Student Name _____ Date _____

What kind of lines does each picture have?

	Straight line	Curve	Parallel lines
(bell)		√	
(ruler)	√		√
(star)	√		
(mushroom)		√	

Draw line(s) through the following dots.

Straight line	Parallel lines	Curve
· B · A	· C · D	· E · F

Name of lines

horizontal lines	vertical lines	parallel lines
intersecting lines	perpendicular lines	

What types of lines can you find in each figure?

	horizontal lines	vertical lines	parallel lines	intersecting lines	perpendicular lines
#	√	√	√	√	√
(diagonal parallel)	√		√	√	
(diagonal cross)	√	√		√	√

Angles

An angle is formed by two rays with a common endpoint. An angle with a measure of 90 degrees is a right angle.

1. Arrange the angles in order of size from the largest to the smallest. Put in 1, 2, and 3 to show.

a.

____3   ____1   ____2

b.

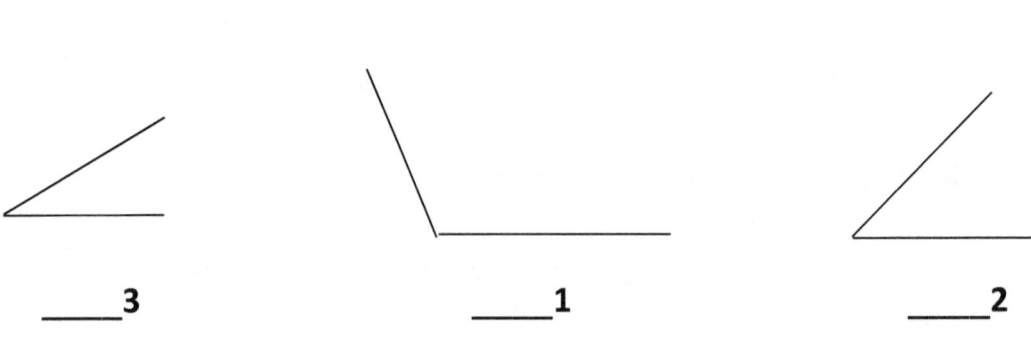

____3   ____1   ____2

2. Find the number of right angles of the following figures.

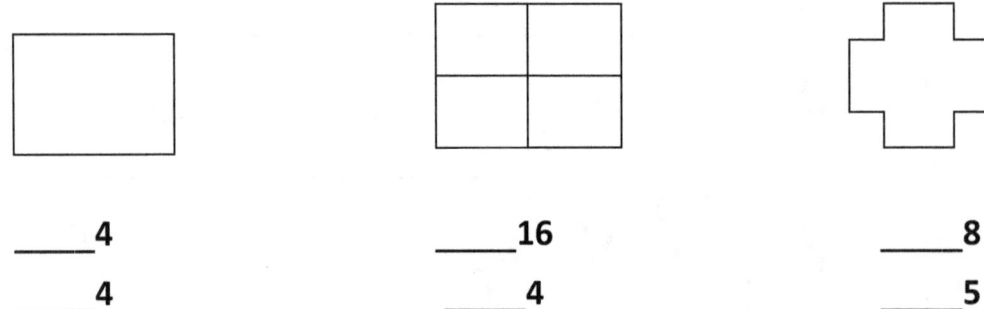

____4   ____16   ____8
____4   ____4   ____5

Shapes

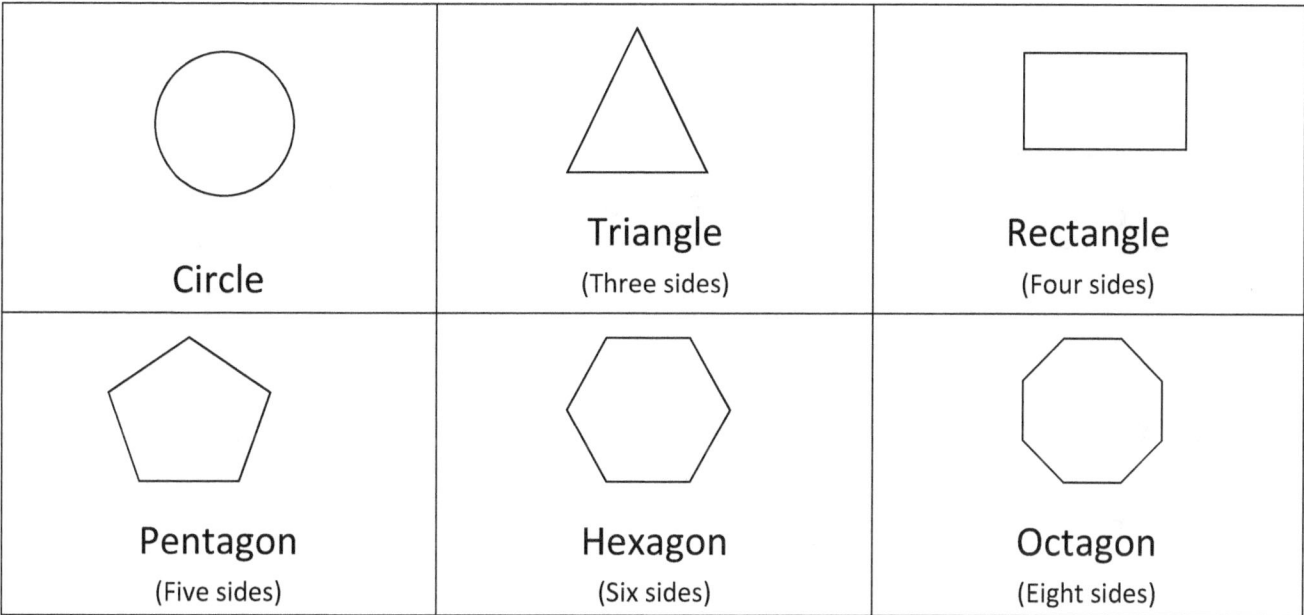

What types of shapes can you find in each figure?

	Circle	Triangle	Rectangle	Pentagon	Hexagon	Octagon
(triangle sign)	✓	✓	✓			
(stop sign)			✓			✓
(school bus)	✓		✓			

545

Quadrilaterals

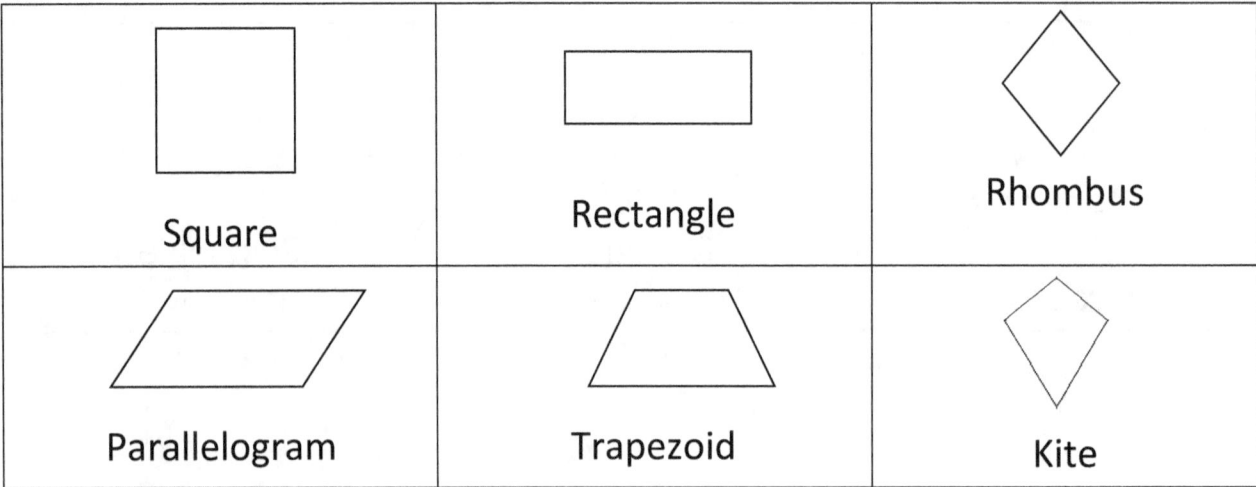

1. How many rectangles can you find?

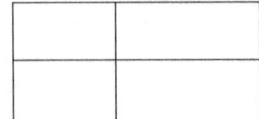

9

2. How many triangles can you find?

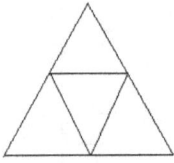

5

3. How many squares can you find?

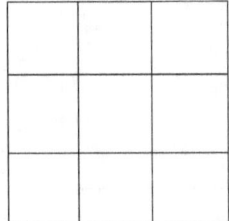

14

4. Matching.

Kite

Square

Trapezoid

Rectangle

Rhombus

Parallelogram

Shapes

The figure below is Chinese Tangram. It is made up seven pieces. Observe carefully and fill in the blanks below.

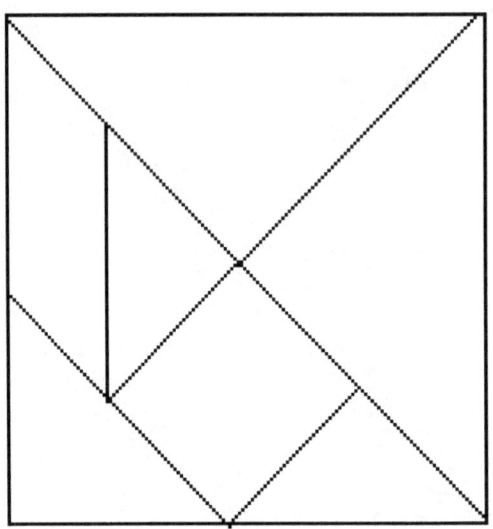

7
There are _____ triangles in the graph.

2
There are _____ squares in the graph.

1
There are _____ parallelograms in the graph.

5
There are _____ trapezoids in the graph.

Ho Math Chess 何数棋谜　妈！我会棋谜式加法啦！
Mom! I Learn Addition Using Math-Chess-Puzzles Connection
Contents include both traditional and Math-Chess-Puzzles combined methods. Extra strength
©2008 − 2018 Frank Ho, Amanda Ho　　All rights reserved. www.homathchess.com

Student Name _____ Date _____

Symmetry line

Find the lines of symmetry for these drawings

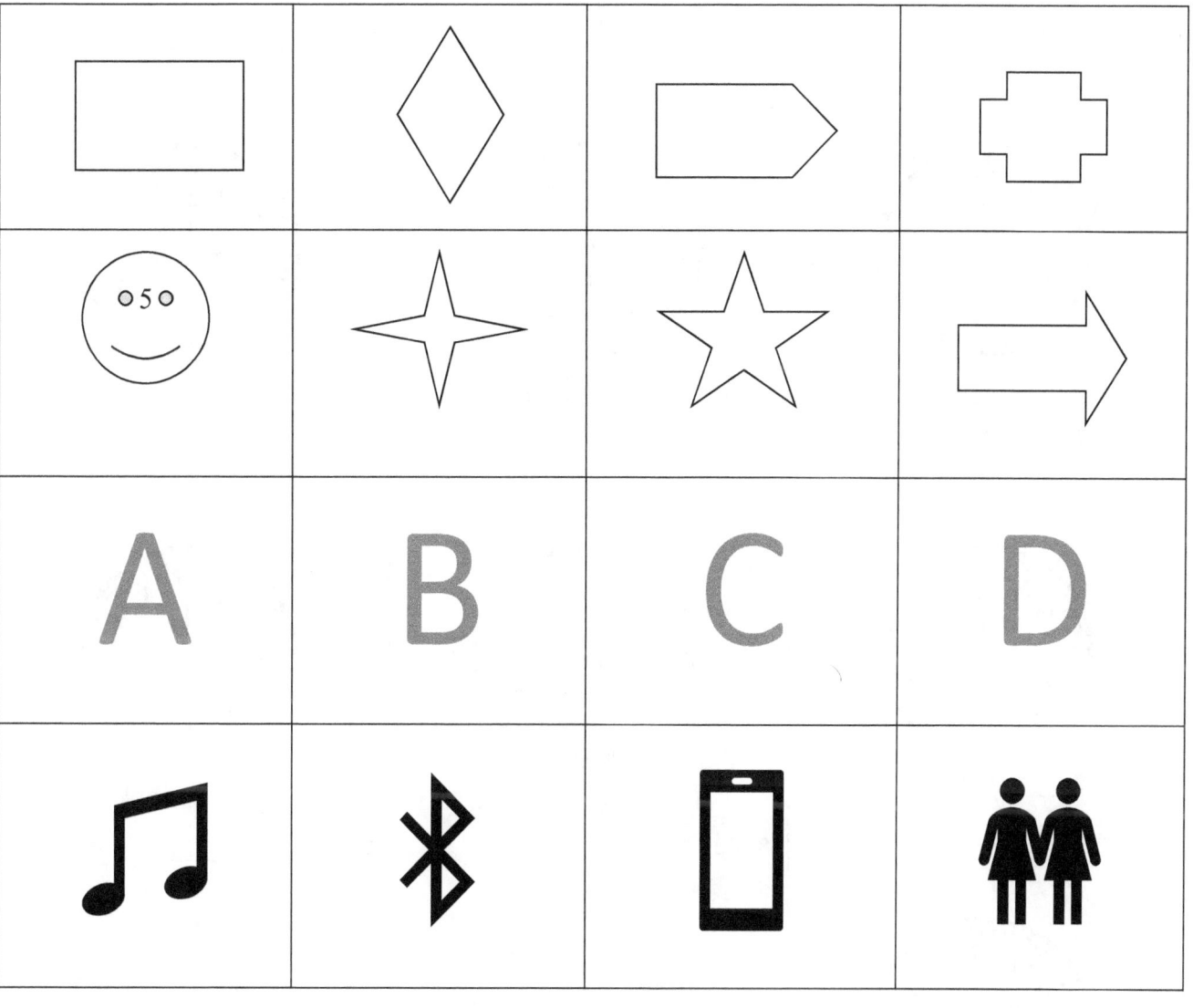

Part 8: Time and clock

Time

Telling time

1.

   3____o'clock
   3   00
   ____ : ____

2.

   6____o'clock
   6   00
   ____ : ____

3.

   9____o'clock
   9   00
   ____ : ____

4.

   2____o'clock
   2   00____ : ____

5.

   8____o'clock
   8   00____ : ____

6.

   5____o'clock
   5   00____ : ____

7.

   10____o'clock
   10   00____ : ____

8.

   4____o'clock
   4   00____ : ____

9.

   7____o'clock
   7   00____ : ____

550

Telling time

1.
half past ____
2   30
____ : ____

2.
half past ____
10   30
____ : ____

3.
half past ____
12   30
____ : ____

4.
Quarter past ____
1   15
____ : ____

5.
Quarter past ____
4   15
____ : ____

6.
Quarter past ____
11   15
____ : ____

7.
Quarter to ____
1   45
____ : ____

8.
Quarter to ____
1   45
____ : ____

9.
Quarter to ____
1   45
____ : ____

Telling time

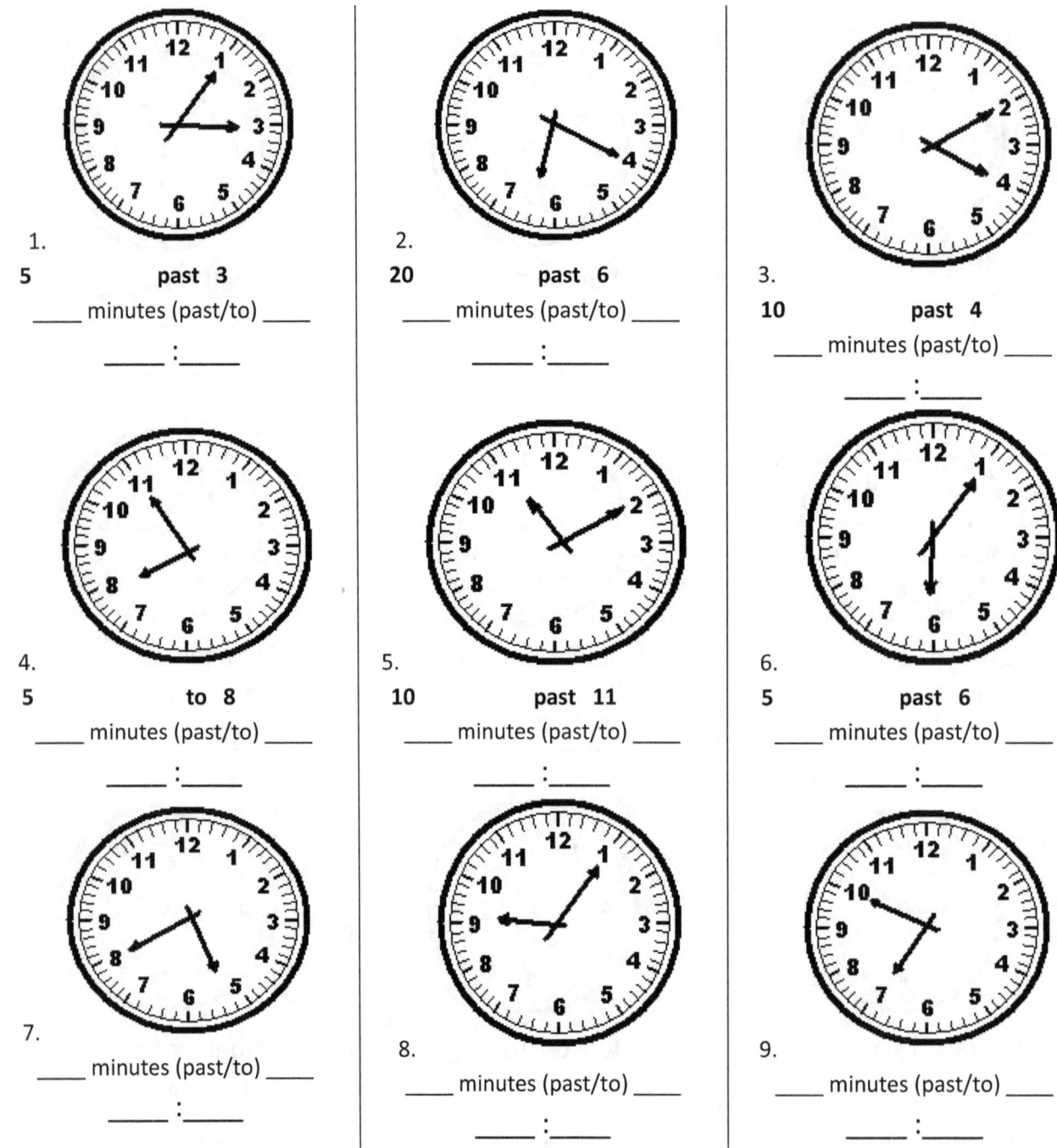

1.
5 _____ past 3
____ minutes (past/to) ____
____ : ____

2.
20 _____ past 6
____ minutes (past/to) ____
____ : ____

3.
10 _____ past 4
____ minutes (past/to) ____
____ : ____

4.
5 _____ to 8
____ minutes (past/to) ____
____ : ____

5.
10 _____ past 11
____ minutes (past/to) ____
____ : ____

6.
5 _____ past 6
____ minutes (past/to) ____
____ : ____

7.
____ minutes (past/to) ____
____ : ____

8.
____ minutes (past/to) ____
____ : ____

9.
____ minutes (past/to) ____
____ : ____

Ho Math Chess 何数棋谜 妈！我会棋谜式加法啦！
Mom! I Learn Addition Using Math-Chess-Puzzles Connection

Contents include both traditional and Math-Chess-Puzzles combined methods. Extra strength

©2008 – 2018 Frank Ho, Amanda Ho   All rights reserved. www.homathchess.com

Student Name _____ Date _____

Drawing the hands on the clock to show the time

1.

2:00

2.

4:00

3.

5:00

4.

8:00

5.

12:00

6.

3:00

7.

6:00

8.

9:00

9.

10:00

553

Drawing the hands on the clock to show the time.

1.

2:30

2.

half past 5

3.

9:30

4.

quarter past 5

5.

3:15

6.
10:15

7.

1:45

8.

quarter to 4

9.

quarter to 12

Ho Math Chess 何数棋谜   妈！我会棋谜式加法啦！
Mom! I Learn Addition Using Math-Chess-Puzzles Connection
Contents include both traditional and Math-Chess-Puzzles combined methods. Extra strength
©2008 — 2018 Frank Ho, Amanda Ho   All rights reserved. www.homathchess.com

Student Name_____ Date_____

Drawing the hands on the clock to show the time.

1.

2:05

2.

7:15

3.

12:10

4.

9:35

5.

1:40

6.

6:20

7.

5 minutes past 7

8.

10 minutes to 8

9.

10 minutes after 5

555

Digital Time

01 : 25 am
25 past 1
___ minutes past/to ___
in the morning/afternoon

09 : 55 am
5 to 10
___ minutes past/to ___
in the morning/afternoon

03 : 40 pm
20 to 4
___ minutes past/to ___
in the morning/afternoon

10 : 50 pm
10 to 11
___ minutes past/to ___
in the morning/afternoon

11 : 45 am
15 to 12
___ minutes past/to ___
in the morning/afternoon

04 : 25 pm
25 past 4
___ minutes past/to ___
in the morning/afternoon

07 : 55 pm
5 to 8
___ minutes past/to ___
in the morning/afternoon

06 : 10 am
10 past 6
___ minutes past/to ___
in the morning/afternoon

08 : 40 am
20 to 9
___ minutes past/to ___
in the morning/afternoon

Calendar

Fill in the following blank.

There are __12__ months in a year.

In a regular year, February has __28__ days. A regular year has __365__ days.

In a leap year, February has __29__ days. A leap year has __366__ days.

There are about __52__ weeks in a year.

January, March, May, July, August, October and December have __31__ days each.

April, June, September and November have __30__ days each.

There are __7__ days in a week. They are Monday, __Tuesday__, __Wednesday__, __Thursday__, __Friday__, __Saturday__ and __Sunday__. I like _____ the best.

Today is January 6th. Tomorrow will be __Jan 7th__. The day after tomorrow will be __Jan 8th__.

9. Today is March 10th. Yesterday was __March 9th__. The day before yesterday was __Mar 8th__.

Part 9: Fraction

Example: Divide 1 whole into equal parts.

  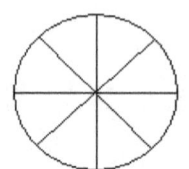

4 equal parts   9 equal parts   8 equal parts
4 fourth or 4 quarters   5 fifth   10 tenth

1. Circle the figures that show equal parts.

  √

 √   √

 √ √

 √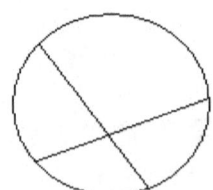

2. Name the equal parts of each whole.

2 halves

3 thirds

4 fourths

5 fifths

6 sixths

7 sevenths

8 eighths

9 ninths

10 tenths

3 thirds

2 halves

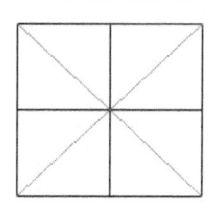

8 eighths

Equal Parts

1. Name the equal parts of each whole.

Example:

3 equal parts   3 thirds

1. 5 fifths
2. 2 halves
3. 4 fourths
4. 6 sixths
5. 7 sevenths
6. 8 eighths

2. Divide each square into 2 halves in five ways.

3. Divide each square into four fourths in five ways.

560

Ho Math Chess 何数棋谜　妈！我会棋谜式加法啦！
Mom! I Learn Addition Using Math-Chess-Puzzles Connection

Contents include both traditional and Math-Chess-Puzzles combined methods. Extra strength
©2008－2018 Frank Ho, Amanda Ho　All rights reserved. www.homathchess.com

Student Name _____　Date _____

Representing Fractions

$$\frac{\text{numerator}}{\text{denominator}} = \frac{\text{the number of favorite parts}}{\text{the number of equal parts in a whole}}$$

1. What fraction of each figure is shaded?

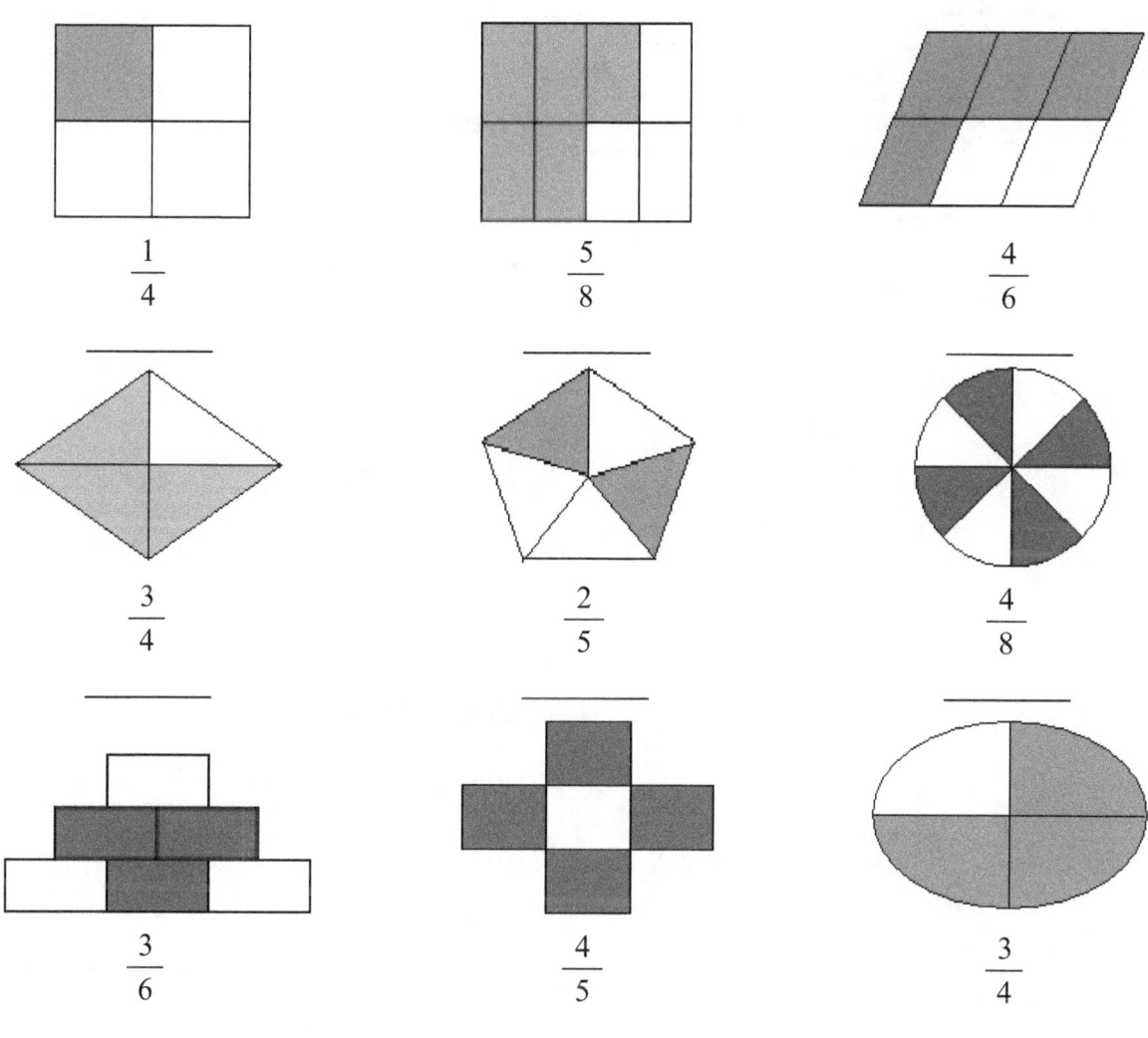

561

2. What fraction of each figure is shaded? Express in words and in fraction form.

1.   two thirds $\dfrac{2}{3}$

2.   **three fifths** $\dfrac{3}{5}$

3.   **five ninths** $\dfrac{5}{9}$

4.   **three-eighths** $\dfrac{3}{8}$

5.   **four fourths** $\dfrac{4}{4}$

6.   **three-eighths** $\dfrac{3}{8}$

7.   **three fifths** $\dfrac{3}{5}$

Ho Math Chess 何数棋谜   妈！我会棋谜式加法啦！
Mom! I Learn Addition Using Math-Chess-Puzzles Connection
Contents include both traditional and Math-Chess-Puzzles combined methods. Extra strength
©2008 — 2018 Frank Ho, Amanda Ho   All rights reserved. www.homathchess.com

Student Name _____ Date _____

Part 10: Word problem solving

Each hand has five fingers. I count 20 fingers. How many hands are there?

4

Each car can carry 5 people. There are 3 cars. How many people can the car take?

15

One marble is 5¢. Ashley bought 4 marbles. How much did she spend on the marbles?

20

Write from 1 to 40. How many times do you write digit 5?

4

One nickel is 5¢. Linda has 20¢ in nickels. How many nickels does she have?

4

Ben has 25 chocolates. He divides them into a group of 5. How many groups can he make?

5 groups

A rabbit eats 5 carrots each day. There are 30 carrots. Is it enough to feed the rabbit for 6 days?

Yes

Part 11: Tests

Test 1

If ◯ = 3 then ◯ + 1 = _____. 4	
If ◯ = 4 then ◯ + 1 = _____. 5	
If ● = 1 then ◯ = ●●● then ◯ + ● = _____. 4	
If ◯ = ●● then ◯ + ●● = _____ 5	
If ◯ = ●● and ● = 1 then ◯ − ● = _____ 2	

Match left to right by drawing lines.

```
  ●        4            ●        4
 ●●                    ●●
 ●●       2            ●●       2
 ●●       3            ●●       3
 ●●●      1           ●●●       1
                              answer
```

Out of 1, 2, and 3, what numbers added together will give answers of 3?

1, 2

A number is drawn as follows to show 1 + 1 = 2.

Look at the following diagram and come up with your equation.

3 - 1 = 2

If the following diagram shows 3 - 1 = 2,

Then what is the value of the following diagram?

 1

What is the value of the following diagram?

 0

Ho Math Chess 何数棋谜　妈！我会棋谜式加法啦！
Mom! I Learn Addition Using Math-Chess-Puzzles Connection

Contents include both traditional and Math-Chess-Puzzles combined methods. Extra strength
©2008 – 2018 Frank Ho, Amanda Ho   All rights reserved. www.homathchess.com

Student Name _____ Date _____

37 − 9 = 28
73 − 4 = 69
45 − 8 = 38
89 + 7 = 82
36 − 17 = 29
What is the largest one-digit number? 9
What is the largest two-digit number? 99
What is the smallest two-digit number? 10
Circle the following even numbers. 177, 46, 8, 91, 32, 21 46, 8, 32

Insert >, <, or = in ☐.

(67 − 19) ☐ 47 >	(31 − 18) ☐ 12 >
(77 − 28) ☐ (37 + 12) >	(34 − 26) ☐ (17 − 8) <

If ● + ● + ● + ● + 6 = 8 + ● + ● + ●, what is the value of ●? 2

| Ho Math Chess 何数棋谜　妈! 我会棋谜式加法啦!
Mom! I Learn Addition Using Math-Chess-Puzzles Connection
Contents include both traditional and Math-Chess-Puzzles combined methods. Extra strength
©2008 – 2018 Frank Ho, Amanda Ho   All rights reserved. www.homathchess.com
Student Name _____ Date _____

Test 1

Problem-solving

Harry Hiker saw 24 caterpillars when he was walking in the forest. When he stopped for lunch, he saw 5 more. How many caterpillars did he see altogether?

29

Wally Walker spotted and picked up 45 leaves on the grass. He then found 23 more in the park. What is the total number of leaves that Wally found?

68

Wendy Watcher saw 12 birds in her backyard today. Yesterday she saw 38 birds at the park. How many birds has she seen in all?

50

Tiffany owns 23 books. She received 2 more today for her birthday. What is the total number of books that Tiffany owns?

25

Tyler has 32 baseball cards. His friends give him 20 more. How many baseball cards does he have in all?

52

Christina counted 24 tulips in her garden. Her sister counted 15 roses. What is the sum number of flowers that they counted?

39

Anthony was on a Scouts trip and found 23 small twigs, 34 medium-sized twigs, and 25 large twigs for a campfire. How many twigs did he find altogether?

82

Tom, who was also on the camping trip, saw 10 squirrels climbing trees. He also saw 7 chipmunks scurrying to their homes. How many animals did Tom see in all?

17

Jennifer went fishing with her dad and she caught 3 fish. Her dad caught one. How many fish did Jennifer and her dad catch?

4

Larry, Curly, and Moe were playing a game of darts. Larry scored 10 points, Curly scored 12 points and Moe scored 8 points. What is the total number of points that they scored?

30

Yumiko helped her mother make four rolls of sushi. Her younger sister helped make two. How many rolls of sushi did Yumiko and her sister help make?

6

Dave found 17 chocolates in the community Easter egg hunt. He also helped his little sister find seven. How many chocolates did he find altogether?

24

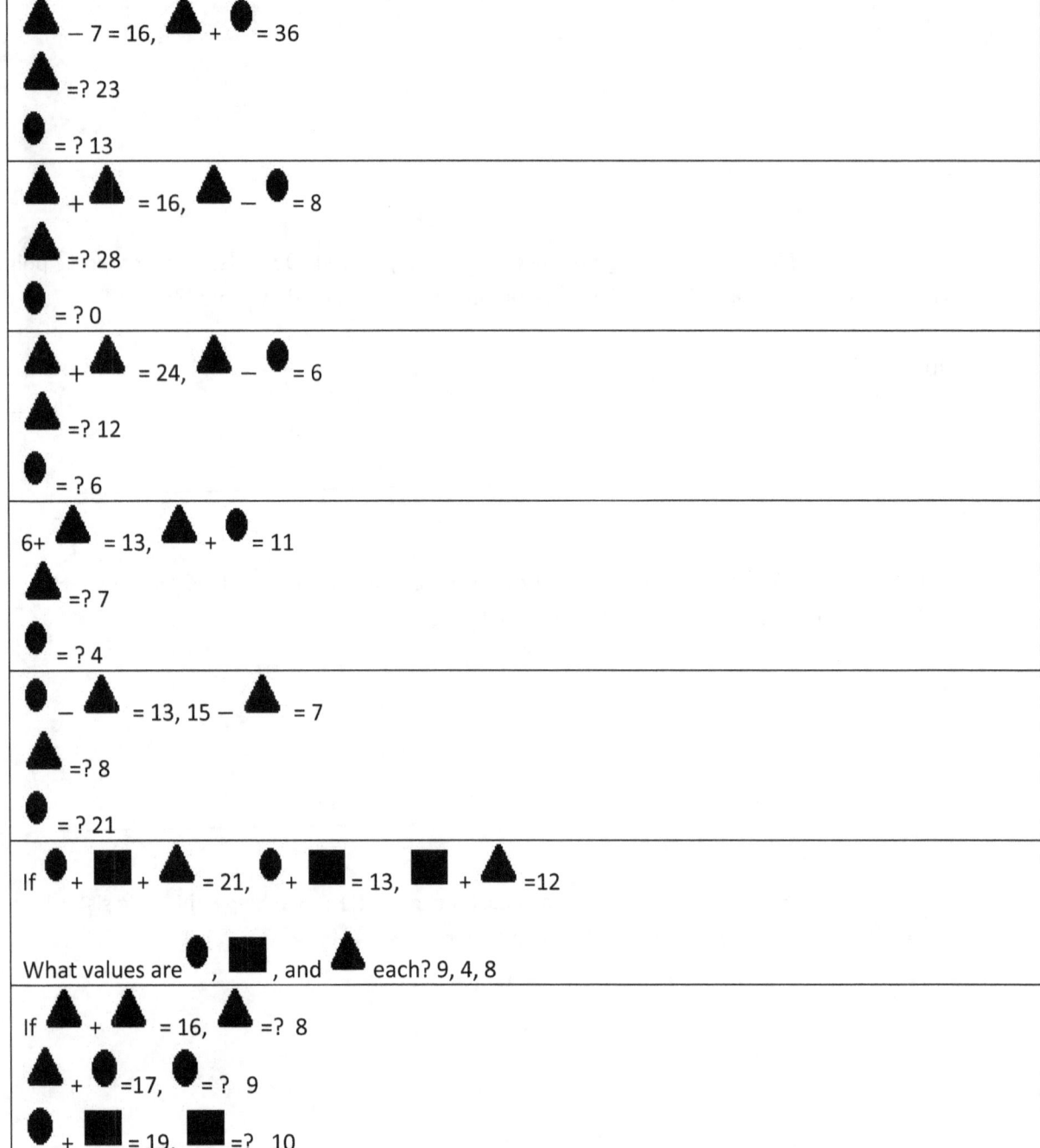

Test 2

■ + ▲ = 12, ■ + ▲ + ▲ + ▲ + ▲ = 36

▲ = ? 8

■ = ? 4

● − ▲ = 2

● + ● − ▲ = 10

● = ? 8

▲ = ? 6

▲ + ▲ + ▲ = 33

● + ● + ● = 21

▲ − ● = ? 4

▲ − ● = 10

▲ + ● = 22

▲ = ? 16

● = ? 6

What are the even numbers which are less than 20?

From 2 to 18

571

Test 2

How many squares are in the following figure?
[figure of 2x2 grid] 5
What are the odd numbers which are less than 20?   From 1 to 19
Ethan is in a lineup for a concert. There are 3 people in front of Ethan and 5 people behind Ethan. How many people are in the lineup?   9
Adam is 2 years older than Bob now. How old will Adam be more than Bob next year?   2
Meghan had 17 apples and she gave two of her friends every 4 apples then how many apples does she have now?   9
Andrew has $13 and he wishes to buy an applications software which costs $29. How much more he has to save in order to buy the software he wants to?   29-13 = 16

Test 3

Complete the pattern. 5, 51, 511, 5111, _____ 51111
Compute 17 + 15 + 13 + 11 + 9 + 7 - 16 - 14 - 12 - 10 - 8 - 6 = 6
Complete the pattern. 10, 101, 1001, 10001, _____ 100001
Circle the number which does not belong to the group. 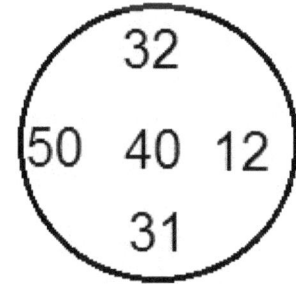 The odd number 31.
Find the next item.   answer
Complete the next number. 88, 96, 11, _____ 69

Ho Math Chess 何数棋谜   妈！我会棋谜式加法啦！
Mom! I Learn Addition Using Math-Chess-Puzzles Connection
Contents include both traditional and Math-Chess-Puzzles combined methods. Extra strength
©2008 – 2018 Frank Ho, Amanda Ho   All rights reserved. www.homathchess.com

Student Name _____ Date _____

Test 3

Questions	Solutions
Tammy has 15 books. Tina has 36 books. How many books do they have in all?	51
Mike has 37 stickers. Aaron has 19 stickers. How many fewer stickers does Aaron have than Mike?	18
Sandy Liang has 29 e-mail messages. Jeffrey Lai has 13 e-mail messages less than Sandy. How many messages they have in all?	45
Jun has 27 cookies and Veronica has 16 cookies more than Jun. How many cookies they have in all?	70
Shirley has finished reading 119 pages of a book and there are 293 pages left. What is the total number of pages of the book in all?	412
14 books were sold, and 29 books are left. How many books were there before selling?	43

51, 18, 45, 70, 412, 43

Ho Math Chess 何数棋谜　妈！我会棋谜式加法啦！
Mom! I Learn Addition Using Math-Chess-Puzzles Connection

Contents include both traditional and Math-Chess-Puzzles combined methods. Extra strength
©2008 – 2018 Frank Ho, Amanda Ho　All rights reserved. www.homathchess.com

Student Name _____ Date _____

Test 3

Questions	Solutions
Amanda has 139 oil paintings and 2878 watercolour paintings, how many paintings does she have in all?	
The plumber has 219 copper pipes. There are 139 more copper pipes than steel pipes. How many steel pipes are there?	
Altogether, Michelle and Rosalind have 219 marbles. Michelle has 19 marbles more than Rosalind. How many marbles does Rosalind have?	
Rosalind has 321 marbles. Michelle has 51 marbles less than Rosalind. How many marbles does Michelle have?	
23 people got on the bus, 15 people got off the bus, 10 people got on the bus and 5 people got off and there were 23 people left on the bus. How many people were on the bus originally?	

3017; **80**; Michelle has119 marbles, **Rosalind has 100 marbles**; Rosalind has 186 marbles, **Michelle has 270 marbles**, 23+5-10+15-23=**10 people**

Test 3

10 people got on the bus, 8 people got off the bus, 5 people got off and the bus was empty. How many people were on the bus originally (excluding the driver)?	
Jerry would have received as many e-mail messages as Julie if he had 16 more messages. Julie had 37 messages. How many messages did Jerry receive?	

3, 21

Test 4

One chick has 2 legs.
Three chicks, how many legs?   6

Circle the odd one.

Which figure should be in place of the "?"?   =

What number should replace ? to balance the weights on the scales?

(8 + 3)   (4 + ?)

7

I had 2 pairs of gloves, I lost one glove. How many gloves do I have now?
3

I am 8 years old and my brother is 5 years older than me. How old will my brother be next year?
14

How many rectangles can you see in the following figure?   9

Test 4

After taking six plums from a basket, six were left in the basket. How many plums were in the basket at the beginning? 12
Bob has 8 oranges more than Adam and Adam has six oranges. How many oranges do they have altogether? 22
Amy has nine more apples than Bryan. Bryan has 3 fewer apples than Cathy. Cathy has 13 apples. How many apples does each one of them have? A=19, B = 10, C = 13
Bill has four more apples than Bob and Bob has 8 more apples than Coco. Together, all three have 41 apples. How many apples does each one of them have? Coco = 7, Bob = 15, Bill = 19
Some birds were on the tree. Seventeen more birds flew back and eight flew away. Now there are 24 birds on the tree. How many birds were on the tree originally? 15
Fill in each the same box with the same number. 6 + ☐ = 13 ☐ + △ = 12 7, 7, 5

Test 5

If □ = 6 and ○ = 3, then □ + ○ = ?.   9
5 = 7 − □   2
□ + 3 − 1 = □ + ?   2
6 + 4 + □ = 13   3
7 + 7 + □ = 15   1
8 + 2 + 5 + 5 + □ = 21   1
A half dozen is 6. How many is one dozen?   12
Mom gave me one-half of $4. How much was it? $2
□□□ − □□□□ + □□□ = ? □   4
If □ = 2 and ○ = 3, then □ + ○ = ?.   5
4 + □ = 6   2
One dog has 4 legs. If there are 12 legs, then how many dogs are there?   6
One boy has 2 legs. Two boys have ____ legs.   4
One wagon has 4 wheels, ____ wagons have 8 wheels.   2
If □ = 7 and ○ = 5, then □ − ? = ○.   2

Test 5

If 13 + 7 + 6 + ? = 31, then what is ? ? 5

One boy has 2 legs and one dog has 4 legs. How many legs are there for 2 boys and 2 dogs?

12

The following large square is made of 16 small squares but some of them are missing. How many small squares are missing?

9

Each dot is two points. Find the value of one

●●● + ●●● = ○ 18

○ + □ = 31   18 + 13 = 31

What number should replace ? to balance the weights on the scale 3?

A   BB
  (1)

B   CC
  (2)

A   ?
  (3)

The first two scales are balanced. How many C's are needed to balance the scale 3?   4

580

Ho Math Chess 何数棋谜 妈！我会棋谜式加法啦！
Mom! I Learn Addition Using Math-Chess-Puzzles Connection
Contents include both traditional and Math-Chess-Puzzles combined methods. Extra strength
©2008 — 2018 Frank Ho, Amanda Ho   All rights reserved. www.homathchess.com

Student Name _____ Date _____

Test 5

Which of the following figure has the longest line in the following figures?

18, 11.5, 14, 20, The first one.

Last month, the cats Kiko and Snow together ate 12 cans of fish. This month they ate 5 more cans of fish than the last month. How many cans of fish did they eat in two months?

29

The cat Kiko likes to play pom pom balls but often she lost them and so far, she has lost 21 pom pom balls in two months. This month she lost twice as many as the last month. How many pom pom balls she lost last month?

7

The cat Snow likes to watch leaves falling from trees in the fall by turning his head up and down whenever he sees a leaf falling. This morning he saw 20 leaves falling and, in the afternoon, he saw one less than half of what he saw in the morning. How many leaves fell did Snow see in the morning and in the afternoon?

29

The cat Snow likes to jump up to catch his toy. In the morning, he jumped high 15 times which was one more than twice as many times as he jumped in the afternoon. How many times did he jump in the morning and in the afternoon altogether?
22

B

581

Find the values of the following figures.

○ + ○ + ○ + △ + △ + □ + □ + □ + □ = 423

○ = _____ 1

△ = _____ 10

□ = _____ 100

Fill in each square with a number from 1 to 9 such that the sum of each line is 9.

   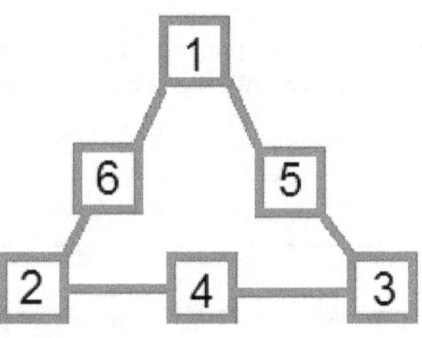

answer

Part 12: A free Paper Set of Ho Math Chess

Want to have a free (almost) Ho Math Chess set? It takes about 10 minutes to complete a DIY chess set.

You can use the following link to print the chess pieces and the chessboard.
https://1drv.ms/u/s!AnVFh_KY48OgtJsze8k6ipl9GeaCQ

Chess pieces

Print the following Ho Math Chess set (on a colour printer is preferred), then cut each piece in a squared shape and paste them on pieces of corrugated papers, foams, or bottle caps.

Ho Math Chess 何数棋谜 妈！我会棋谜式加法啦！
Mom! I Learn Addition Using Math-Chess-Puzzles Connection
Contents include both traditional and Math-Chess-Puzzles combined methods. Extra strength
©2008 — 2018 Frank Ho, Amanda Ho   All rights reserved. www.homathchess.com

Student Name _____ Date _____

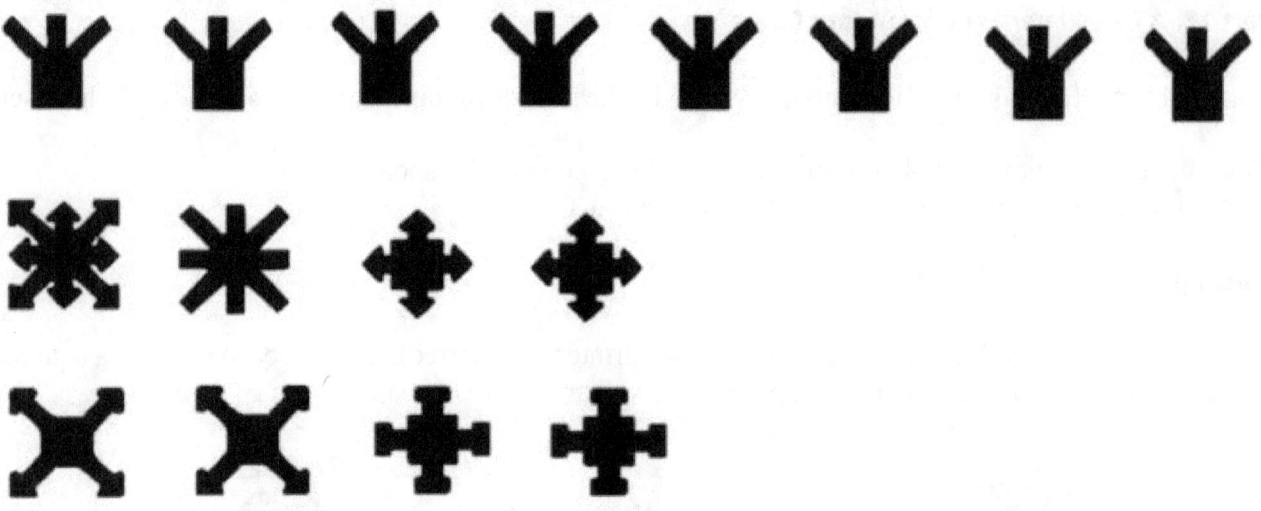

Chessboard

Ho Math Chess 何数棋谜　妈！我会棋谜式加法啦！
Mom! I Learn Addition Using Math-Chess-Puzzles Connection

Contents include both traditional and Math-Chess-Puzzles combined methods. Extra strength

©2008 – 2018 Frank Ho, Amanda Ho   All rights reserved. www.homathchess.com

Student Name _____ Date _____

Ho Math Chess 何数棋谜   妈！我会棋谜式加法啦！
Mom! I Learn Addition Using Math-Chess-Puzzles Connection
Contents include both traditional and Math-Chess-Puzzles combined methods. Extra strength
©2008 – 2018 Frank Ho, Amanda Ho   All rights reserved. www.homathchess.com

Student Name _____ Date _____

A picture of DIY Ho Math Chess teaching set is imaged as follows:

Ho Math Chess 何数棋谜　妈！我会棋谜式加法啦！
Mom! I Learn Addition Using Math-Chess-Puzzles Connection
Contents include both traditional and Math-Chess-Puzzles combined methods. Extra strength
©2008 – 2018 Frank Ho, Amanda Ho　　All rights reserved. www.homathchess.com
Student Name_____Date_____

介紹何数棋谜

何数棋谜 = 奧数棋谜 + 思唯腦力開發
英文教材，中英双语教学

什麼是何数棋谜?

上百篇科學論文巳發表國際象棋可以提高兒童問題解答能力. 並且訓練他們的專心及耐力. 所以我們已經知道下國際象棋對兒童有好處. 但是因為國際象棋與計算能力並無直接開係, 所以如何讓兒童能在一個歡樂的環境下也能利用下棋來提高數學的計算呢? 何老師首創並發明有版权的幾何棋藝符號並利用此符號發明了世界第一的独特結合數學与棋谜教材. 何数棋谜讓兒童能利用幾何棋藝符號進行邏輯推理及數字的運算. 棋藝與算術的綜合題含蓋了整數, 幾何, 集合, 抽象數, 對比異同, 函數, 座標, 多空間圖形資料, 及規則性數字分析. 並且把棋藝的趣味性和數學的知識性結合在一起.

何数棋谜如何幫助兒童腦力思唯的開發?

很簡單的一個道理就是讓學生自願地去用腦, 何数棋谜首創獨一無二的融合數學與棋谜的独特趣味寓教於樂教材, 利用國際象棋訓練右腦的座標, 空間分析及圖形處理, 並利用發明了整合棋子與數學的圖形語言, 讓兒童能利用符號圖形訓練左腦進行邏輯推理及數字的運算. 國際象棋與算術的綜合題含蓋了整數, 幾何, 集合, 抽象數, 對比異同, 函數, 多空間圖形資料. 所以枯燥無味的計算題變成了謎題, 學生需要通過更多的思考. 能讓腦去思考愈多則腦力也愈開發. 處里訊息, 分析資料才能發掘出題目. 做這些謎題式數學時可以訓練學生比較會專心及有耐心.

何数棋谜融合數學與國際象棋的教學理論巳在 BC 省數學教師刊物上發表. 科研報告已經證實何数棋谜教學法不但可以提高兒童數學解題及思維能力, 還可以開發兒童的腦力, 及分析問題的能力並且增加兒童學習的耐力, 學生的探索創造精神及求知欲. 判斷力, 及自信心等, 啓發思維訓練機警靈巧及加強手腦眼的靈活運用.

Ho Math Chess 何数棋谜   妈! 我会棋谜式加法啦!
Mom! I Learn Addition Using Math-Chess-Puzzles Connection
Contents include both traditional and Math-Chess-Puzzles combined methods. Extra strength
©2008 – 2018 Frank Ho, Amanda Ho   All rights reserved. www.homathchess.com
Student Name _____ Date _____

Introducing Ho Math Chess™

Ho Math Chess™ = math + puzzles + chess

Frank Ho, a Canadian math teacher, intrigued by the relationships between math and chess after teaching his son chess started **Ho Math Chess™** in 1995. His long-term devotion to research has led his son to become a FIDE chess master and Frank's publications of over 20 math workbooks. Today **Ho Math Chess™** is the world largest and the only franchised scholastic math, chess and puzzles specialty learning centre with worldwide locations. **Ho Math Chess™** is a leading research organization in the field of math, chess, and puzzles integrated teaching methodology.

There are hundreds of articles already published showing chess benefits children and that math puzzles are a very good way of improving brainpower. So, by integrating chess and mathematical chess puzzles together, the learning effect is more significant.

Parents send their children to **Ho Math Chess™** because they like **Ho Math Chess™** teaching philosophy – offering children problem-solving questions in a variety of formats. The questions could be pure chess, chess puzzles or mathematical chess puzzles in the nature of logic, pattern, tree structure, Venn diagram, probability and many more math concepts.

Ho Math Chess™ has developed a series of unique and high-quality math, chess, and puzzles integrated workbooks. **Ho Math Chess™** produced the world's first workbook **Learning Chess to Improve Math.** This workbook is not only for learning chess but also for enriching math ability. This sets **Ho Math Chess** apart from other math learning centres, chess club, or chess classes.

The teaching method at **Ho Math Chess™** is to use math, chess, and puzzles integrated workbooks to teach children fun math. The purposes of **Ho Math Chess™** teaching method and workbooks are to:

- Improve math marks.
- Develop problem-solving and critical thinking skills.
- Improve logic thinking ability.
- Boost brainpower.

Testimonials, sample worksheets, reports, and franchise information can be found at www.homathchess.com.

More information about **Ho Math Chess™** can also be found from the following publications:

1. Why Buy a **Ho Math Chess™** Learning Centre Franchise: A Unique Learning Centre?
2. **Ho Math Chess™** Sudoku Puzzles Sample Worksheets
3. Introduction to **Ho Math Chess™** and its Founder Frank Ho

The above publications can be purchased from www.amazon.com.

www.ingramcontent.com/pod-product-compliance
Lightning Source LLC
Chambersburg PA
CBHW081342230426
43667CB00017B/2699